Organic Solar Cells

Masahiro Hiramoto · Seiichiro Izawa
Editors

Organic Solar Cells

Energetic and Nanostructural Design

Springer

Editors
Masahiro Hiramoto
Department of Materials Molecular Science
Institute for Molecular Science
Okazaki, Aichi, Japan

Seiichiro Izawa
Institute for Molecular Science
Okazaki, Aichi, Japan

ISBN 978-981-15-9112-9   ISBN 978-981-15-9113-6 (eBook)
https://doi.org/10.1007/978-981-15-9113-6

This Springer imprint is published by the registered company Springer Nature Singapore Pte Ltd.
The registered company address is: 152 Beach Road, #21-01/04 Gateway East, Singapore 189721, Singapore

Preface

Scientists have a responsibility to address the world's energy problem as energy security in terms of energy resources from fossil fuels has become one of the main reasons for conflicts between nations. Organic solar cells present themselves as a potential candidate to solve the problem of energy access they have been intensively researched for more than half a century. The efficiency of organic solar cells has steadily been increasing by almost 1% per year since 2000 for 20 years and has not yet reached saturation. In 2020, the reported best efficiency and short-circuit photocurrent density reached 17.4% and 25 mA cm^{-2}, respectively.

I have been researching organic solar cells for 30 years since 1988, whereas up until 2000, few scientists around the world studied organic solar cells. The progress of organic solar cells has been far beyond my expectations. Although the science of organic semiconductors is approximately half a century behind compared to that of inorganic semiconductors, the research on organic solar cells has gradually been catching up to the research on inorganic ones. The current physics on organic solar cells is similar to the physics on inorganic solar cells around 1980 when I was still a student. As such, I am convinced that scientists, especially chemists, will be persistent with their attempts to develop and improve man-made solar energy conversion systems using organic compounds.

Recent driving forces for increases in efficiency have relied on the numerous repeats of the circulation around the syntheses of new organic semiconductors, blends and performance checks. However, a neglect of the basic physics, that underpins these complicated systems, has not permitted the future long-term development of organic solar cells.

I believe that original ideas are the most important in scientific research. As such, whilst writing this book, I was mindful of offering readers an insight into how essential ideas in organic solar cells have been created to date in efforts to provide a foundation for future progress in this field for readers such as students and amateur and experienced scientists.

The main topics this book covers include the fundamental principle of organic solar cells, invention of blended junction, lateral junction, final solution of blended cells, current status of polymer cells, carrier generation mechanism, cutting-edge theory on organic solar cells, photo voltage generation, doping and future development.

I was first approached regarding the writing of a book during a brief discussion with Claus Acheron, Ph.D., Executive Editor Physics in Springer. This occurred immediately before my presentation titled "Bandgap Science for Organic Solar Cell" at the 2015 European Material Research Society in Lille, France. In 2017, I was again requested to write a book by Shinichi Koizumi, Ph.D., a Senior Editor for Books Chemistry and Materials Science in Springer. It has taken approximately 2 years to complete this book. I am deeply appreciative of the efforts of my writers, Hideo Okita, Ph.D., Itaru Osaka, Ph.D., Takatoshi Fujita, Ph.D., Toshihiko Kaji, Ph.D. and Seiichiro Izawa, Ph.D. I am also very grateful to the Springer editors, Shinichi Koizumi, and Taeko Sato, for their assistance and encouragement.

It is my strong hope that young scientists of the next generation will greatly advance the research in this field.

平本昌宏

Okazaki, Aichi, Japan
June 2020

Masahiro Hiramoto

Contents

Chapter 1
Basic Principles of Modern Organic Solar Cells

Masahiro Hiramoto

1.1 Background

1.1.1 Current Status of Solar Cells: c-Si

Scientists have the responsibility to solve the energy problem since the security concerns about the energy resources from fossil fuels have been one of the main reasons for conflicts among nations. The solar cell, which has been studied intensively for more than half a century, is a candidate to solve this energy issue. Presently, almost all solar cell markets are dominated by the single-crystal silicon (c-Si). Figure 1.1a shows the global evolution of solar cell installations [1], which rapidly increased from around 2010 and reached 500 GW in 2018. Assuming that 1 GW (10^6 kW) corresponded to one nuclear power plant, the electricity equivalent of 50 power plants was supplied globally by solar cells by considering the net working rate of solar cells to be approximately 10% due to fickle weather conditions, as well as the absence of sunlight at night. Figure 1.1b shows the ratio of renewable energies, including solar cells (PV) [1] that supplied 2% of the global electricity demand in 2018.

By considering the best mix concept of energy resources based on risk management, the ratio of solar cells (PV) exceeding 20% of the total energy supply is appropriate. Thus, a 10 times solar cell ratio increase is necessary. One of the hurdles remains the higher cost of electricity produced by solar cells than that using fossil fuels. Another obstacle entails the management of the intermittent features of solar energy.

M. Hiramoto (✉)
Institute for Molecular Science, National Institutes of Natural Science, 5-1 Higashiyama, MyodaijiAichi, Okazaki 4448787, Japan
e-mail: hiramoto@ims.ac.jp

M. Hiramoto and S. Izawa (eds.), *Organic Solar Cells*,
https://doi.org/10.1007/978-981-15-9113-6_1

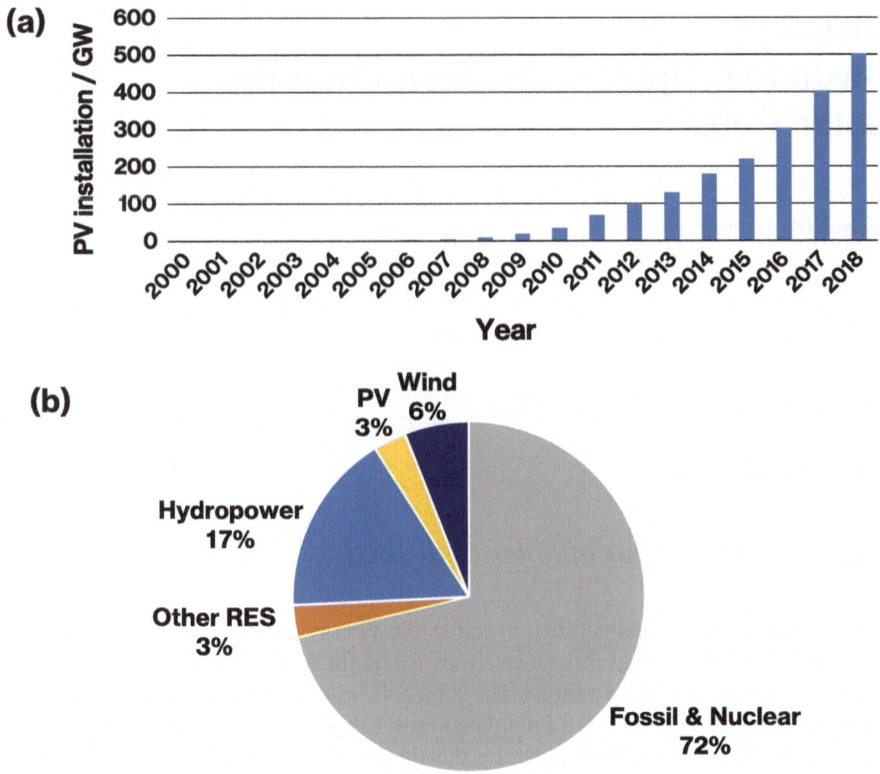

Fig. 1.1 **a** Global evolution of solar cell (PV) installation from 2000–2018. **b** Ratio of renewable energy, including solar cell (PV), in global demand of electricity in 2018

1.1.2 Chasers of c-Si

In 2020, the c-Si and single-crystal GaAs have the efficiencies of 26.7 and 27.8%, respectively. The main competitors of the c-Si are copper indium gallium selenide (CIGS) cells, consisting of Cu, In, Ga, Se (23.4%), amorphous Si:H (14.0%), dye sensitization (12.3%), perovskite (25.2%), organic (17.4%), and quantum dot (16.6%). The percentages shown in the parentheses above are the conversion efficiencies of each cell in 2020 obtained from the National Renewable Energy Laboratory (NREL) chart of "Best research-cell efficiencies" [2]. The efficiencies of CIGS, a-Si:H, dye-sensitization, and perovskite had reached their intrinsic limit and were almost saturated in 2020, while those of organic and quantum dot cells continue to increase in 2020. More than 20% is required for the next-generation solar cells. Most of the researchers investigating the c-Si are presumably convinced that it is not the only solar cell material and more suitable materials for solar cells are hidden in the nature. A compelling reason for this line of thought is that the c-Si is an intrinsically inadequate semiconductor for solar cells because of its indirect transition. This offers a strong motivation to develop the next-generation solar cells.

1.1.3 *Motivation for Organic Solar Cell*

The original motivation for organic solar cells was to obtain man-made solar energy conversion systems using organic compounds as an alternative to photosynthesis in plants. However, currently, the following motivations accelerate the development of organic solar cells.

(i) ***Construction of Novel Physics on Organic Semiconductors***
The novel physics on organic semiconductors emerged as a result of long-term research that lasted for more than a half century; researchers anticipated a vast and fruitful uncultivated field of organic semiconductors. The organic solar cell is one such spin-off.

(ii) ***Steady Efficiency Increase***
The efficiency of organic solar cells produced after 2000 is shown in Fig. 1.2 (red dots). A steady increase of approximately 1% per year was observed for 20 years, and there was no saturation. In 2020, it reached 17.3% [3], and by extrapolation, we can expect an increase of 20% by 2025. This steady increase in efficiency is a source of encouragement to researchers.

(iii) ***Unlimited Organic Compounds***
A unique feature of organic materials compared with other competitors is the numerous possible candidate compounds obtainable. Actually, the rapid increase observed after 2017 was as a result of newly synthesized organic

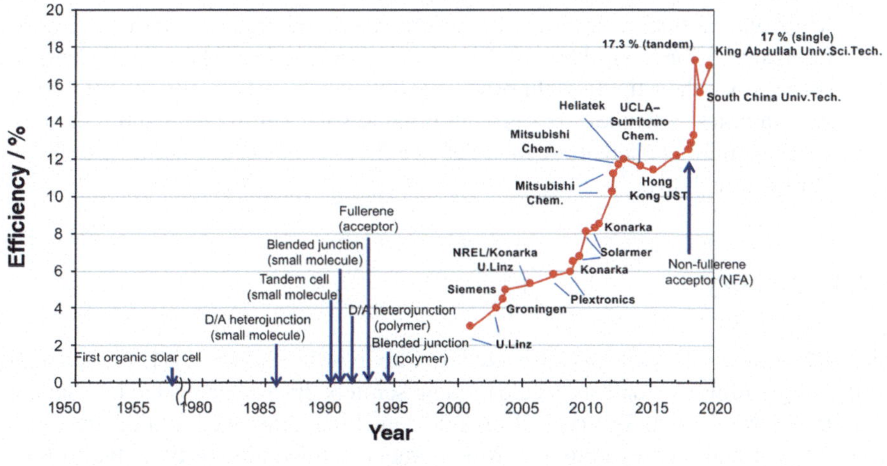

Fig. 1.2 Efficiency of organic solar cells after 2000 (red dots), according to the NREL chart. The main breakthroughs from 1957–2020 are indicated by blue arrows

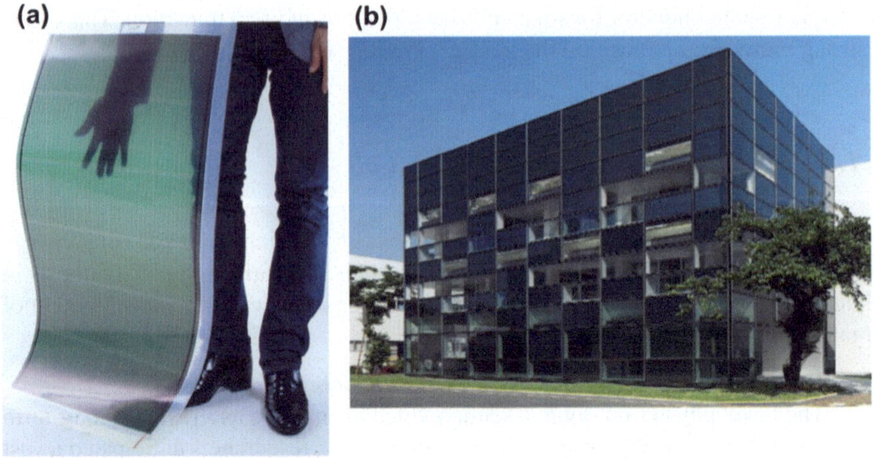

Fig. 1.3 **a** Commercial flexible see-through organic solar cell module. **b** A building using these modules on the exterior walls and windows

compounds. Undoubtedly, a vast number of unknown organic semiconducting compounds for organic solar cells are hidden in the nature.

(iv) ***Inherent Cost-Reducing Ability***
Organic solar cells are inherently cost-reducing because of the availability of the printing and the roll-to-roll processes, without the need to undergo high-temperature processing.

(v) ***Fascinating Advantages***
Organic solar cells possess fascinating advantages, such as being flexible, lightweight, colorful, transparent, and fashionable design. Printed organic solar cells can be attached to the roofs, windows, and walls of houses and buildings, while automobiles wrapped with colorfully printed organic solar cells can be fabricated. Moreover, they are suitable for constructing solar power plants in space, since their lightweight property allows them to be easily put into orbit. As examples, a flexible see-through organic solar cell module and a demonstration building using these modules as exterior walls and windows are shown in Fig. 1.3a, b [4].

1.1.4 History

The first organic solar cell was fabricated by Calvin in 1958 (Fig. 1.4) [5]. For a long while, single films of small-molecule organic semiconductors deposited by vacuum evaporation were used. The typical organic semiconductors were phthalocyanines (Fig. 1.5) [6] and merocyanine [7]. Most organic semiconductor films had *p*-type characteristics since oxygen molecules from the ambient air, acting as acceptors, inevitably doped the films. A photocurrent could be generated at the Schottky junction

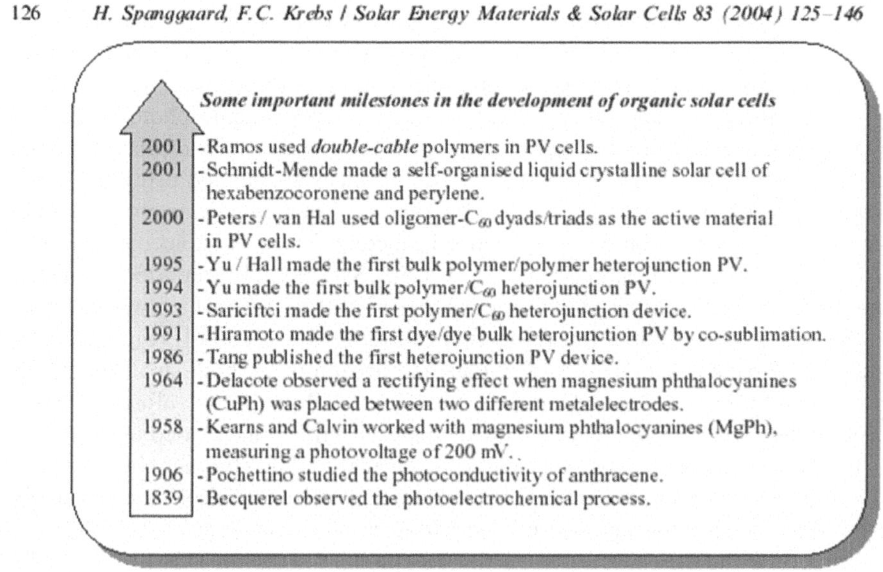
126 *H. Spanggaard, F.C. Krebs / Solar Energy Materials & Solar Cells 83 (2004) 125–146*

Some important milestones in the development of organic solar cells

2001 - Ramos used *double-cable* polymers in PV cells.
2001 - Schmidt-Mende made a self-organised liquid crystalline solar cell of hexabenzocoronene and perylene.
2000 - Peters / van Hal used oligomer-C_{60} dyads/triads as the active material in PV cells.
1995 - Yu / Hall made the first bulk polymer/polymer heterojunction PV.
1994 - Yu made the first bulk polymer/C_{60} heterojunction PV.
1993 - Sariciftci made the first polymer/C_{60} heterojunction device.
1991 - Hiramoto made the first dye/dye bulk heterojunction PV by co-sublimation.
1986 - Tang published the first heterojunction PV device.
1964 - Delacote observed a rectifying effect when magnesium phthalocyanines (CuPh) was placed between two different metalelectrodes.
1958 - Kearns and Calvin worked with magnesium phthalocyanines (MgPh), measuring a photovoltage of 200 mV..
1906 - Pochettino studied the photoconductivity of anthracene.
1839 - Becquerel observed the photoelectrochemical process.

Fig. 1.4 Some important milestones in the early stage development of organic solar cells appeared in the review by Spanggaard and Krebs in 2004. Reproduced with permission from H. Spanggaard et al., Solar Energy Materials and Solar Cells, Copyright 2004 Elsevier

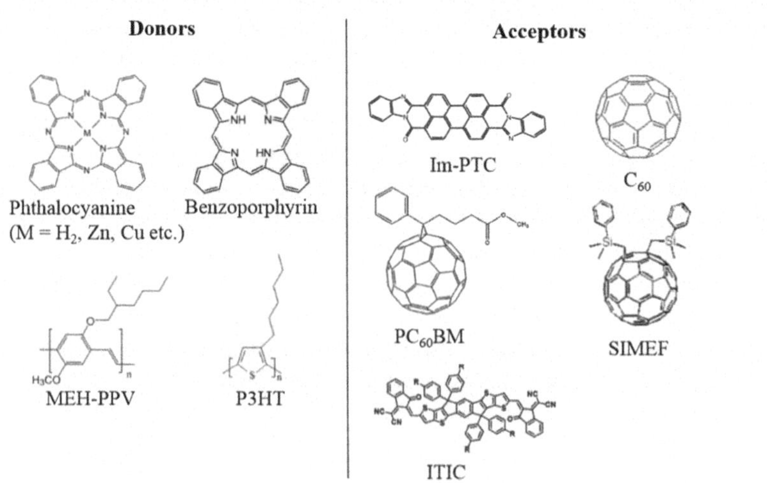

Fig. 1.5 Chemical structures of typical organic semiconductors acting as donors and acceptors

between the *p*-type organic film and a corrosive low work-function metal, such as Al. However, in the early stages of their development, organic solar cells had minimal photocurrent—typically less than several micro-amperes [8, 9].

A breakthrough occurred in 1986 when a two-layer organic photovoltaic cell was developed that had a large photocurrent density of the order of mA cm^{-2} and an efficiency of 1% (Figs. 1.2 and 1.4). This result was obtained using the donor (D)/acceptor (A) sensitization reported by Tang [10] that has massively influenced the field of organic solar cells. A blended junction for small-molecule cells was proposed in 1991 by Hiramoto [11, 12]. In 1992, Sariciftci [13] reported a polymer heterojunction cell composed of C_{60} and MEH-PPV (Fig. 1.5), which was followed in 1995 by a bulk polymer heterojunction cell reported by Yu [14] (Figs. 1.2, 1.4, and 1.5). Typical combinations of donors and acceptors for both small molecular systems and polymer systems are shown in Fig. 1.5. Based on the aforementioned basic studies, after 2000, the efficiency increased from 1% and started to escalate.

Furthermore, a tandem organic solar cell was reported by Hiramoto in 1990 [15]. The combination of the concepts of blended junction and tandem cell has consistently boosted efficiencies throughout the research on organic solar cells [3, 16, 17].

Fullerenes and their derivatives, acting as excellent acceptors [13, 14] (Fig. 1.5), have consistently been indispensable in the development of organic solar cells from 1993 to date. In 2016, a non-fullerene acceptor was developed [18, 19] and it ramped up the efficiency beyond 17% (Fig. 1.2). An example of a non-fullerene acceptor (ITIC) is shown in Fig. 1.5. Nowadays, the photocurrent density of organic solar cells produced by solar radiation exceeds 20 $mAcm^{-2}$ and the external quantum efficiency reached approximately 80%, which is comparable to the values produced by inorganic solar cells.

Today, researchers have started to focus on the open-circuit voltage (V_{oc}) since its increase is crucial to further enhance the efficiency. The main factors that determine the V_{oc} are the Highest-energy Occupied Molecular Orbital–Lowest-energy Unoccupied Molecular Orbital (HOMO–LUMO) gap (Sect. 1.2.2.1.) and the non-radiative carrier recombination loss (Sect. 1.2.2.2.).

In summary, the major factors for increasing the efficiency of organic solar cells thus far are as follows: (i) D/A sensitization; (ii) blended junction; (iii) percolation; (iv) tandem cells; (v) fullerenes; (vi) HOMO–LUMO gap tuning; (vii) non-fullerene acceptors; and (vii) non-radiative recombination suppression.

1.2 Principles

1.2.1 Photocurrent

1.2.1.1 Excitons

The dissociation of photogenerated electron–hole pairs (excitons) is a key factor for generating carriers in organic semiconductors. Exciton dissociation is affected by the relative permittivity of the solid (ε). The attractive force between an electron–hole pair is given by Coulomb's law:

$$F = (1/4\pi\varepsilon\varepsilon_0)(q_1q_2/r_2). \tag{1.1}$$

where q_1 and q_2 are the elementary charges on the electron and the hole, respectively, and ε_0 and r are the absolute permittivity and the distance between them, respectively. In a solid with a small value of ε, the positive and negative charges experience a strong attractive force, whereas, in a solid with a large value of ε, the positive and negative charges undergo a relatively weak attractive force. Inorganic semiconductors have large ε values. For example, the value for Si is 11.9; the exciton has a substantial diameter of 9.0 nm, and it is localized over approximately 10^4 Si atoms (Fig. 1.6a). The thermal energy at room temperature is sufficient for this Wannier-type exciton to

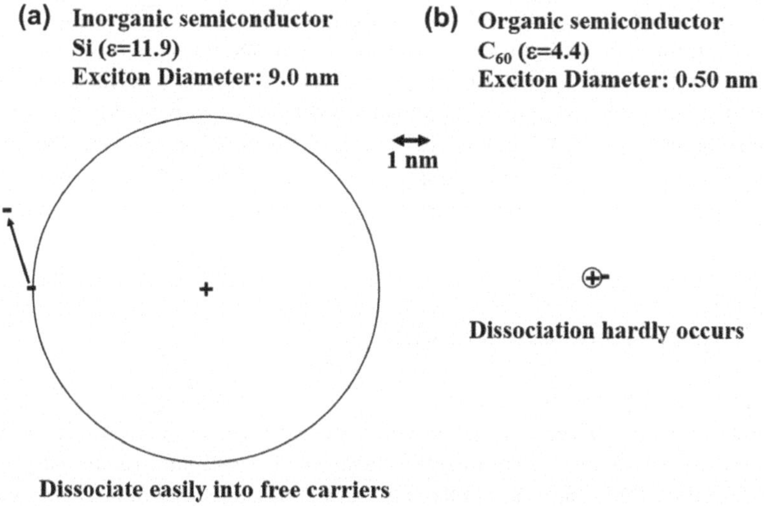

Fig. 1.6 Size of excitons for **a** an inorganic semiconductor (Si) and **b** an organic semiconductor (C_{60}). The former is a Wannier-type and easily dissociates into free carriers, while the latter is a Frenkel-type and dissociates into free carriers with difficulty

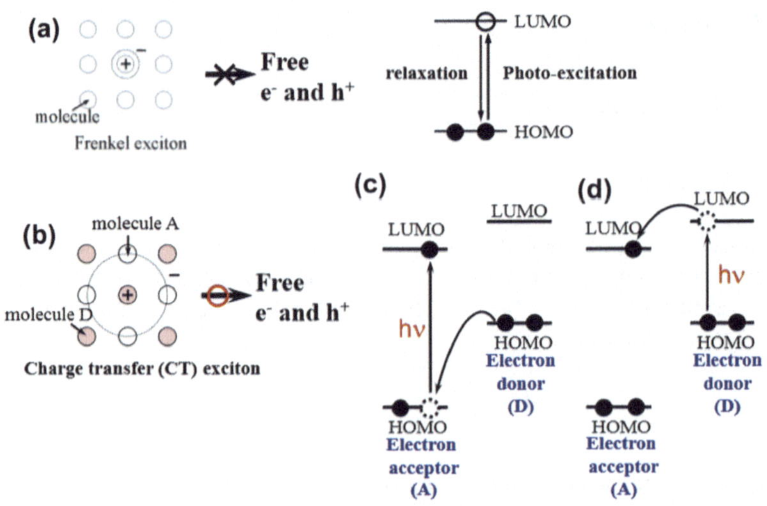

Fig. 1.7 Carrier generation in organic semiconductors. **a** Single molecule solids. **b** Donor (D)/acceptor (A) sensitization for carrier generation by mixing two kinds of organic semiconductor molecules. Efficient free carrier generation occurs from the charge transfer (CT) exciton. **c** Photoinduced electron transfer from the HOMO of the donor molecule (D) to the HOMO of the acceptor molecule (A). **d** Photoinduced electron transfer from the LUMO of the donor molecule (D) to the LUMO of the acceptor molecule (A)

dissociate immediately into a free electron and a hole, thereby generating a photocurrent. In contrast, organic semiconductors have small ε values. For example, the value for C_{60} is 4.4; the exciton diameter is merely 0.50 nm, and it is localized over a single C_{60} molecule (Fig. 1.6b). The thermal energy at room temperature is hardly sufficient for these Frenkel-type excitons to dissociate into free electrons and holes, and they can easily relax to the ground state (Fig. 1.7a). Therefore, organic semiconductors generate few photocarriers. The organic solar cells fabricated before the work of Tang [10] had extremely low photocurrents, with orders below 1 μA, because of this reason.

1.2.1.2 Donor–Acceptor Sensitization

The above-stated problem has been overcome in today's organic solar cells by combining two kinds of organic semiconductors. An electron-donating (D) molecule and an electron-accepting (A) molecule, whereof the energetic relationship of the HOMO and the LUMO shift parallelly (Fig. 1.7c, d). When the A molecule is excited, electron transfer from the HOMO of the D molecule to the HOMO of the A molecule occurs, and consequently, the A molecule charges negatively (A^-), whereas the D molecule charges positively (D^+), (Fig. 1.7c). Conversely, when the D molecule is excited, electron transfer from the LUMO of the D molecule to the LUMO of the A molecule occurs, and as a result, the A and D molecules charge negatively

(A^-) and positively (D^+), respectively (Fig. 1.7d). Irrespective of the excitation of the molecules (A) and (D), the charge-transferred states (D^+A^-) obtained are the same. Thus, a charge transfer (CT) exciton is formed, wherein the positive and negative charges are separated on neighboring D and A molecules due to photoinduced electron transfer (Fig. 1.7b). This CT exciton can dissociate to a free electron and a hole due to the thermal energy at room temperature. By utilizing this donor–acceptor (D/A) sensitization, organic semiconductors have become capable of generating photocurrents of significant magnitudes, of the order of milliamperes, by solar radiation. Note that the terminology, CT exciton and CT state that appeared in Sect. 1.2.1.2 and Sect. 1.2.2.2. are identical.

1.2.1.3 Exciton Diffusion

A two-layer organic solar cell (Fig. 1.8) [10] utilizes D/A sensitization at the heterojunction. The width of the photoactive region (shaded red) is, however, limited to approximately 10 nm in the vicinity of the heterojunction due to the extremely small exciton diffusion length of just several nm [16, 20]. So, when the thicknesses of the organic layers are increased, the so-called masking effect occurs, which entails the development of a dead region in front of the active region wherein the incident solar light is absorbed and no photocurrent is generated, and consequently, the magnitude of the photocurrent is severely suppressed. A 10 nm-thick organic film can only absorb a small part of the incident solar light. However, to increase the efficiency of organic solar cells, it is necessary for the entire incident solar light to be absorbed in the 10 nm-thick active layer.

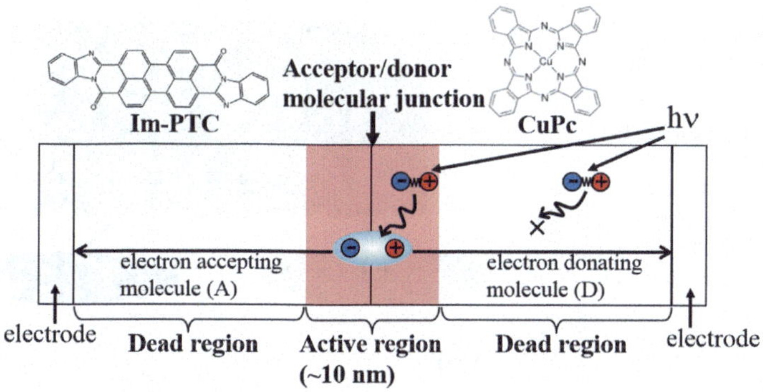

Fig. 1.8 Schematic illustration of a two-layer cell composed of perylene pigment (Im-PTC) acting as an acceptor molecule (A) and copper phthalocyanine (CuPc) acting as a donor molecule (D). Photocurrent is generated only in the active region (shaded red) close to the heterojunction and all other parts of the organic films act as dead regions

1.2.1.4 Blended Junctions

To overcome this problem, blended junctions in organic semiconductors, which contain both donors and acceptors, were fabricated using the co-deposition technique for small-molecule organic solar cells (Fig. 1.9a) [11, 12]. From the standpoint of photocarrier generation occurring at the molecular level, there are donor/acceptor molecular contacts acting as photocarrier generation sites due to donor/acceptor sensitization in the bulk of the co-deposited layer. Blended junctions, usually called "bulk heterojunctions," have become crucial in polymer organic solar cells [13], as well as small-molecule solar cells.

Blended junctions are physically similar to the porous TiO_2 in the dye-sensitized solar cells proposed by Gratzel in 1991 [21]. For both solid/solid and solid/electrolyte junctions, by transmitting the incident light through a vast number of heterointerfaces, the entire incident solar light can be absorbed into extremely thin active layers.

Figure 1.9b shows the energetic structure of the blended junction in a small-molecule cell. Photogenerated electrons in donor molecules are transferred from the LUMO of the donor molecule to that of the acceptor molecule, while photogenerated holes in acceptor molecules are transferred from the HOMO of the acceptor molecule to that of the donor molecule. This D/A sensitization approach promotes photocurrent generation. The original concept entails that the positive and negative charges from ionized donors and acceptors in n- and p-type organic semiconductors, respectively, are compensated by each other, and the resulting co-deposited interlayer behaves like an intrinsic (i) semiconductor. With regard to the formation of the built-in potential in a molecular solid, the built-in electric field is distributed across

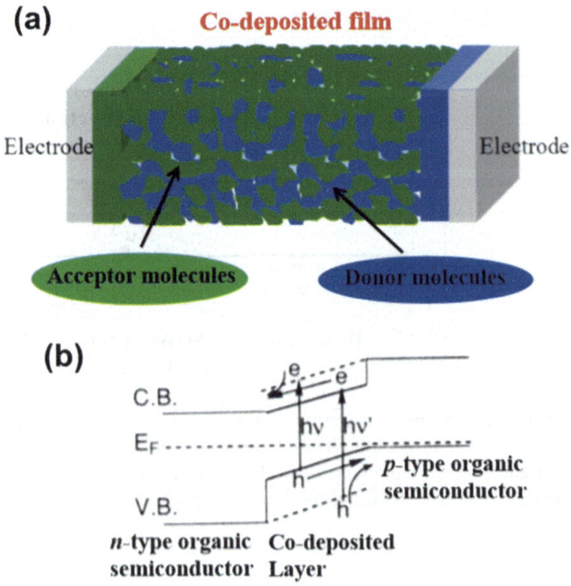

Fig. 1.9 **a** A small-molecule cell consisting of blended junctions in an organic semiconductor, containing both donor and acceptor molecules. The entire bulk of the blended layer acts as an active layer for photocarrier generation. **b** Energetic structure of a blended junction in a small-molecule cell

an i-interlayer sandwiched between n- and p-layers, similar to the case of amorphous silicon incorporating a p-i-n junction.

1.2.1.5 Route Formation

Irrespective of the occurrence of exciton dissociation, the nanostructure control of the co-deposited films, i.e., route formation for electrons and holes generated by exciton dissociation, is crucial to extract a significant portion of the photogenerated charge to the external circuit. After the two types of organic semiconductors acting as the donor and acceptor layers are blended, the extraction of photogenerated holes and electrons becomes an inevitable and difficult problem because they can only move through the A and D molecules.

(i) ***Molecular mixture***: When the two types of organic molecules are blended using the co-deposition technique at room temperature, the co-deposited films usually have an amorphous molecularly blended structure (Fig. 1.9a), and only a minimal number of carriers can be extracted because there are few routes for electron and hole transport.
(ii) ***Percolation***: The routes for the electrons and holes are formed by percolation, which is promoted by annealing, while maintaining D/A molecular contacts in the bulk of the organic film (Fig. 1.10a) (see Chap. 3, Sect. 3.2) [22–24].
(iii) ***Vertical superlattice***: An ideal nanostructure is the "vertical superlattice" structure (Fig. 1.10b) (see Chap. 3, Sect. 3.3). This structure, which can be made artificially by cutting out the cross section from the multilayer [20], enables the efficient dissociation of photogenerated excitons at the D/A interfaces within the exciton diffusion length (5–10 nm) and the transport of electrons and holes to the respective electrodes. However, it is a difficult task to construct such a structure by artificial design over a large area.
(iv) ***Lateral superlattice***: The photogenerated holes and electrons are laterally transported and extracted to the respective electrodes. The lateral superlattice structure (Fig. 1.10c) corresponds to the 90° rotated vertical superlattice structure (Fig. 1.10b). This structure eventually became possible by the emergence of high-mobility organic semiconductors. The lateral alternating multilayered junctions (Fig. 1.10c) can collect both excitons and carriers with an efficiency of approximately 100%, as well as fabricate cells that have sufficient areas. Therefore, the lateral junction is used as an alternative blended junction (see Chap. 3, Sect. 3.4.2) [25].

Here, two examples of the successful fabrication of ideal vertical superlattices (Fig. 1.10b) are introduced. When one succeeds in fabricating the pseudo-vertical superlattice, the photocurrent increases with the thickness of vertical superlattice, while the fill factor (FF) maintains a constant value. On the contrary, since the photocurrent and FF of the co-deposited film that has a molecular mixture steeply decrease for film thicknesses above several tens nm, thicker co-deposited films cannot be used for the solar cell.

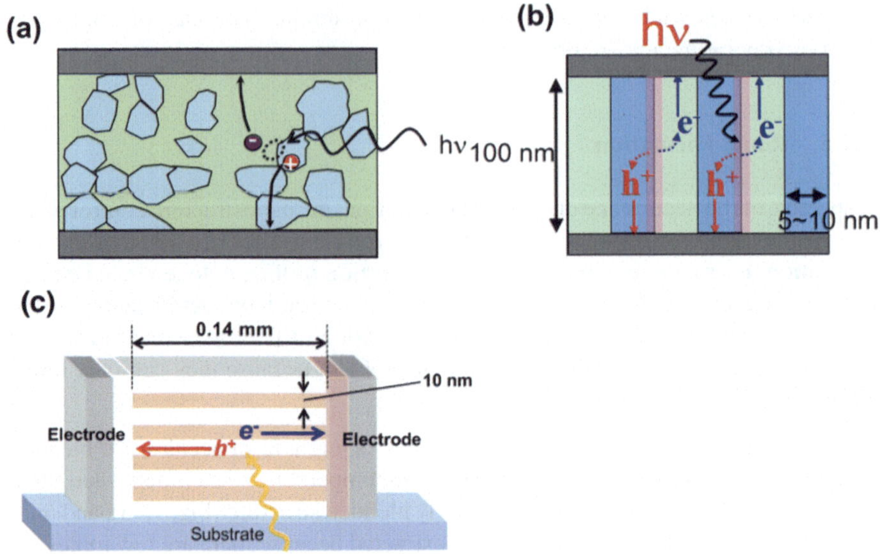

Fig. 1.10 **a** Phase-separated structure. **b** Vertical superlattice structure. **c** Lateral superlattice structure. **a** Reproduced with permission from K. Suemori et al., Appl. Phys. Lett., Copyright 2005 AIP Publishing. **b** Reproduced with permission from [20]. Copyright 2006 AIP Publishing. **c** Reproduced with permission from [25]. Copyright 2019 American Chemical Society

By introducing the co-evaporant molecule, acting as solvent in the vacuum chamber, a pseudo-vertical superlattice structure with the thickness of 10 μm was fabricated by the co-deposition of ZnPc and C_{60} (Fig. 1.11a) [26, 27]. Short-circuit photocurrent density (J_{sc}) increased with co-deposited layer thickness and reached approximately 20 mAcm^{-2}. Even at a thickness of 10 μm, J_{sc} and FF did not decrease. This result is discussed in Chap. 4. Figure 1.10b shows the photograph of H_2Pc:C_{60} cells with the thickness of 180 nm (left) and 1 μm (right) [28, 29]. The black color of the 1 μm-thick cell (right) means that the entire visible light is absorbed and utilized for photocurrent generation, whereas the transparent green color of the 180 nm-thick cell (left) means that a significant portion of visible light is transmitted and not utilized for photocurrent generation.

Another pseudo-vertical superlattice structure was fabricated by combining a new type of C_{60} derivative (SIMEF, Fig. 1.5) with benzoporphirine (BP, Fig. 1.5) [30]. BP is formed by thermal conversion from a precursor. Fabrication by wet processing was adopted, i.e., the blended structure (Fig. 1.9a) was produced via a spin-coating technique. After the removal of the SIMEF by washing in toluene, the desired interdigitated structure becomes evident (Fig. 1.11c). The round columns of crystalline BP with diameters of approximately 20 nm stand almost vertically. This structure allows excitons to dissociate within their diffusion length and ostensibly inhibits the recombination of photogenerated electrons and holes. In 2013, Mitsubishi Chemical Co. Ltd. succeeded in obtaining a power conversion efficiency of 12% with this

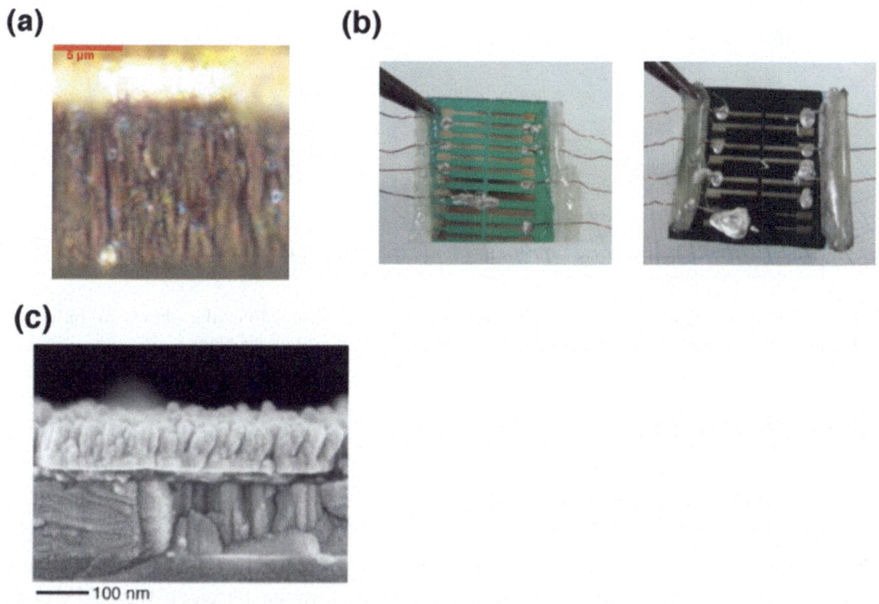

Fig. 1.11 **a** Cross-sectional optical microscope image of a pseudo-vertical superlattice structure with a thickness of 10 μm fabricated by the co-evaporant introduced co-deposition of ZnPc and C_{60}. **b** Photographs of H_2Pc:C_{60} cells with thicknesses of 180 nm (left) and 1 μm (right). **c** Interdigitated structure of the round columns of crystalline benzoporphyrine (BP) with diameters of approximately 20 nm, standing almost vertically, used in highly efficient cells. **a** Reprinted from [26] by M. Katayama(s) licensed under CC BY 4.0. **b** Reproduced with permission from [30]. Copyright 2009 American Chemical Society

system. A commercial version, using this system, is shown in Fig. 1.3a, b, which show a flexible see-through organic solar cell module and a demonstration building using these modules as exterior walls and windows [4].

1.2.1.6 $\pi-\pi$ Stacking Orientation

Holes are efficiently transported along the $\pi-\pi$ stacking orientation. The formation of the $\pi-\pi$ stacking of poly (3-hexylthiophene) (P3HT) in a PCBM:P3HT (Fig. 1.5) blend by means of a regular arrangement of long-chain substituents allowed for the fabrication of highly efficient polymer solar cells [31]. Since the organic films are sandwiched vertically between two electrodes on the substrate, face-on orientation (Fig. 1.12a) is more suitable for hole transport than edge-on orientation (Fig. 1.12b). Blended cells using polymers with face-on orientation enabled high photocurrents to be obtained with thicker active layers reaching 1 μm without reducing the FF, thereby resulting in an increase in efficiency [32]. The effects of the $\pi-\pi$ stacking are discussed in detail in Chap. 5.

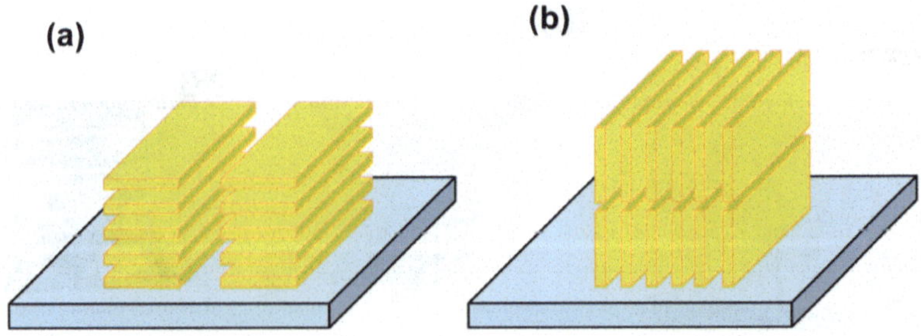

Fig. 1.12 **a** Face-on and **b** edge-on orientations of $\pi-\pi$ stacking. Molecular planes are indicated in the figure. $\pi-\pi$ stackings are formed perpendicular to the molecular planes

1.2.2 Photovoltage

1.2.2.1 HOMO–LUMO Gap Tuning

The maximum V_{oc} of organic solar cells consisting of a combination of D/A organic semiconductors is determined by the energetic difference between the LUMO of the A molecule and the HOMO of the D molecule ($V_{LUMO\text{-}HOMO}$) (Fig. 1.13a) [33]. In other words, to obtain high V_{oc}, a combination of D/A organic semiconductors having a larger $V_{LUMO\text{-}HOMO}$ difference is necessary (Fig. 1.13). By using the newly synthesized D molecules with deeper HOMO levels (Fig. 1.13b) [33] and the A molecules having shallower LUMO levels (Fig. 1.13c) [34], cell efficiency can be increased by the increase in V_{oc}. HOMO–LUMO gap tuning is discussed in detail in Chap. 5.

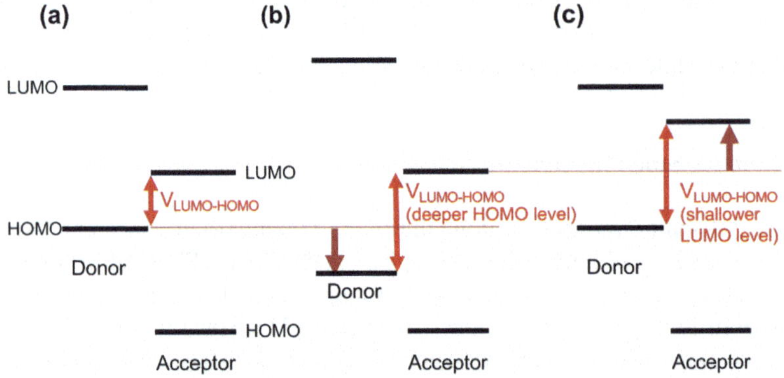

Fig. 1.13 **a** $V_{LUMO\text{-}HOMO}$ difference of donor and acceptor molecules. **b** $V_{LUMO\text{-}HOMO}$ difference using a donor molecule having a deeper HOMO level. **c** $V_{LUMO\text{-}HOMO}$ difference using an acceptor molecule having a shallower LUMO level

1.2.2.2 Non-radiative Recombination

Carrier recombination causing the V_{oc} loss is an essential process that determines the magnitude of the V_{oc}. Figure 1.14 shows the energetic diagram of the energy loss factors in D/A type organic solar cells (Fig. 1.7c, d). The V_{oc} loss is defined as the energy difference between the optical band gaps of the organic semiconductors (E_g) and eV_{oc} (Fig. 1.14, right vertical axis). CT states are formed by the dissociation of strongly bound (Frenkel) excitons in the single films of D and A cells (Figs. 1.6b, 1.7a) into electron–hole pairs (Fig. 1.7b) by the electron transfer from D to A at D/A interface.

CT states are the key to understanding the recombination process. When recombination occurs before the generation of free electrons and holes by dissociating the CT state, it is referred to as "geminate recombination" (Fig. 1.14). Recent reports have revealed that the energy difference between the optical band gaps and the CT states, i.e., electron transfer loss; ($E_g - E_{CT}$), which drives efficient free carrier generation, is smaller than 50 meV (Fig. 1.14, right vertical axis) [35, 36]. Thus, the main factor of the V_{oc} loss is the energy difference between CT states (E_{CT}) and eV_{oc} caused by the radiative recombination that dissipates the energy to emitting photons and the non-radiative recombination dissipating the energy to thermal energy, i.e., the molecular vibrations [37]. For the conventional organic semiconductor films, a large Stokes shift between the absorption and emission spectra was observed (Fig. 1.15a). Here, the vibrational coupling easily induces non-radiative recombination, which reduces the V_{oc}.

In addition, when the defects acting as carrier traps exist between the D/A interface, trap-assisted recombination occurs (Fig. 1.15b) [38–41]. Since this is an additional non-radiative recombination path, the V_{oc} decreases even more.

Figure 1.16 shows the dependence of V_{oc} on the CT state energy (E_{CT}) under 1 sun. The black solid line represents the thermodynamic theoretical limit of V_{oc}, which is calculated according to the Shockley–Queisser (SQ) assumption. On this ideal line, V_{oc} shows its intrinsically highest values, that is, the V_{oc} loss shows only those smallest and inevitable values that occur due to radiative recombination [42]. For all types of organic solar cells, the reported V_{oc} losses are always greater than approximately 0.6 eV (Fig. 1.16, crosses) [43]. For comparison, the V_{oc} of inorganic GaAs solar cells has a small V_{oc} loss of approximately 0.3 eV, which is achieved by suppressing the non-radiative surface recombination, is shown by the closed triangle [44]. For organic solar cells, it remains unclear how the non-radiative recombination lowers the V_{oc}. According to one study, the high crystallinity of the molecular layers in the vicinity of the D/A interface, which suppresses the molecular vibrations and lowers the defect concentration, has been reported to increase the V_{oc} close to the SQ limit [45]. The relationship between carrier recombination and V_{oc} is discussed in detail in Chap. 8.

Even after the free carrier generation, free electrons and holes are recombined by bimolecular recombination, the details of which are discussed in Chap. 6, Sect. 6.5.1.

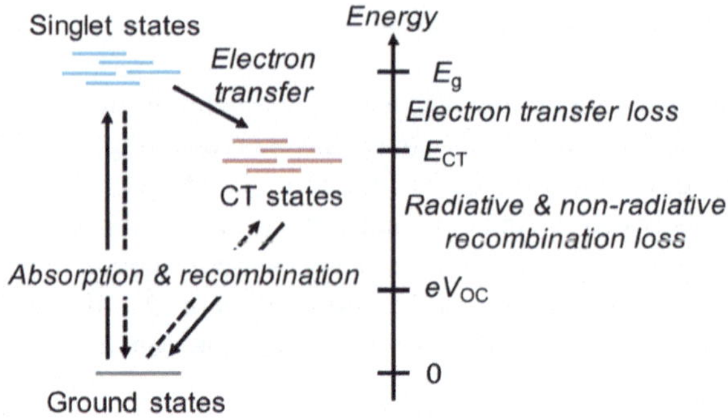

Fig. 1.14 Energetic diagram of the V_{oc} loss in D/A type organic solar cells. Reproduced with permission from [45]. Copyright 2019 AIP Publishing

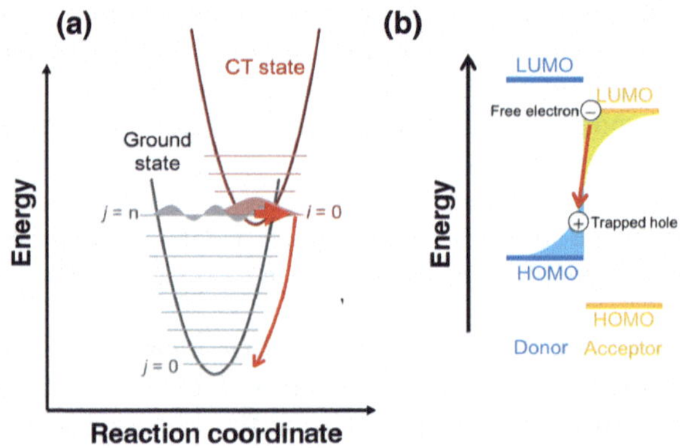

Fig. 1.15 **a** Schematic illustration of the mechanism of the non-radiative recombination from the CT to the ground states. **b** Schematic illustration of the mechanism of the trap-assisted non-radiative recombination. **b** Reproduced with permission from [41]. Copyright 2018 Elsevier

1.2.3 Cell Design

1.2.3.1 Tandem Cells

The first organic tandem cell wherein two unit cells were connected by metal nanoparticles was reported in 1990 (Fig. 1.17a) [15]. The V_{oc} can be doubled by connecting two unit cells. Due to the extremely small diffusion length of excitons, even joining two very thin donor/acceptor heterojunction cells with identical absorption spectra, which have the D and A films with less thickness than the exciton diffusion length (several nm), together can increase the total effective thickness of the cells (Fig. 1.17b) [46]. Since it has been difficult to increase the thickness of the photoactive blended layer, the tandem strategy has been effective to increase cell

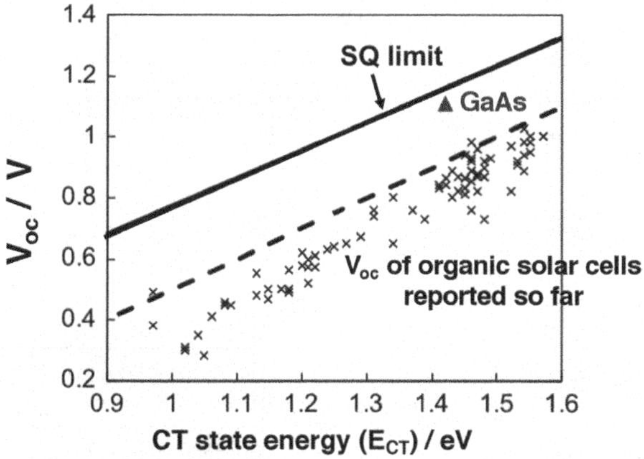

Fig. 1.16 Dependence of the open-circuit photovoltage (V_{oc}) on the CT state energy (E_{CT}). Black solid line is the theoretical (SQ) limit of the V_{oc}. The dashed black line shows $V_{oc} = E_{CT}/e - 0.5$ V. The crosses are the V_{oc} for all types of organic solar cells reported so far. A closed triangle represents the V_{oc} for GaAs

efficiency. Thus, many types of cells that connect two unit cells by various kinds of connecting materials have been fabricated [47].

Karl Leo et al. reported sophisticated small-molecule tandem solar cells (Fig. 1.17c) using transparent connecting layers of optimal thicknesses wherein the electric field of light inside the cell was maximized in the very thin photoactive layers of the co-deposited films with an efficiency of 12%, which was a world record in 2011 [17].

Originally, the objective of tandem cells is to utilize the entire solar spectrum by combining cells that absorb light from different parts of the spectrum. Polymer tandem cells having double junctions (Fig. 1.17d) [48] and triple junctions [49], utilizing the wide wavelength region from 300–900 nm, had a V_{oc} of 2.28 V and an efficiency of 11.55%. In 2018, a 17.3% efficiency was reported for a solution-processed tandem cell [3].

1.2.3.2 Junction Formation by Doping

Controlling the pn-type behavior of a semiconductor by adding an extremely small quantity of impurity (doping) is an indispensable technique for inorganic solar cells. The pn-control technique by doping has also resulted in remarkable progress in organic semiconductors (Chap. 9). All fundamental junctions, i.e., pn- and pin-homojunctions, as well as the ohmic junctions at organic/metal and organic/organic interfaces, can be intentionally fabricated only by doping in the single organic films and the D/A co-deposited organic films (Chap. 9). A tandem junction connecting two pin-junctions was designed and fabricated only by doping in the C_{60}:sexithiophene (6 T) blended film (Fig. 1.18) [50]. Karl Leo's group in Dresden has been intensively

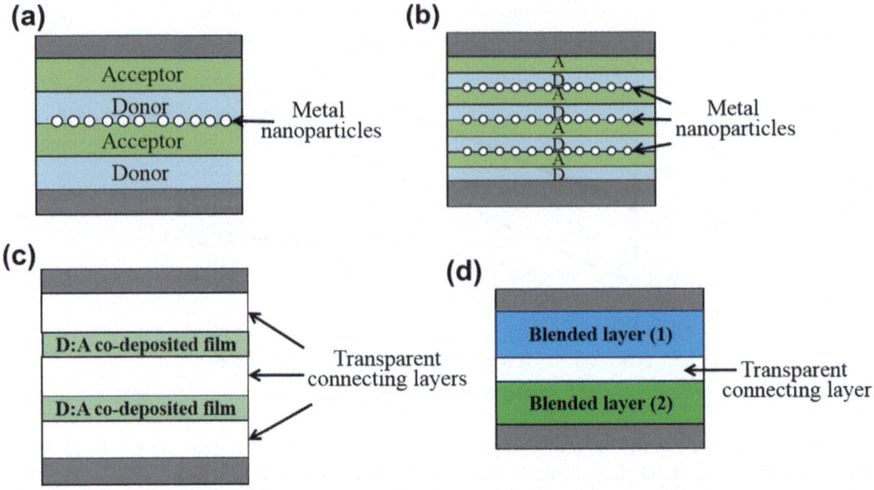

Fig. 1.17 Structures of organic tandem cells. **a** Two heterojunction cells connected together. **b** A series of many extremely thin heterojunction cells with identical absorption spectra. **c** Transparent connection layers with optimum thicknesses. **d** Two blended layers absorbing light from different parts of the spectrum

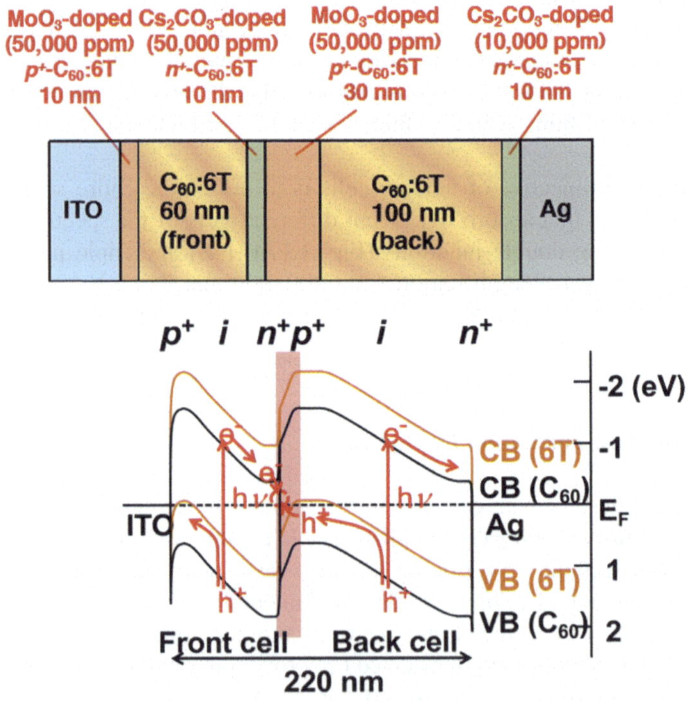

Fig. 1.18 Tandem junction, i.e., $p^{+}in^{+}p^{+}in^{+}$-homojunction fabricated in the C_{60}:6 T blended film by doping alone. Reproduced with permission from 3 [50]. Copyright 2013 Elsevier

investigating doped organic solar cells [51]. A tandem cell showing a 12% efficiency (Fig. 1.17c) [17] was also fabricated based on the doping technology. Nowadays, this technology is indispensable for the development of organic solar cells [28, 52, 53]. Junction formation by doping is discussed in Chap. 8 (Sects. 8.3.2. and 8.3.3) and Chap. 9.

1.3 Conclusions

The basic principles of modern organic solar cells are summarized. Namely, issues on exciton, D/A sensitization, exciton diffusion, blended junction, route formation, $\pi - \pi$ stacking orientation, HOMO–LUMO gap tuning, non-radiative recombination, tandem cells, and doping are discussed. Concepts of blended junctions and tandem cells have continuously boosted efficiencies, and the efficiency of organic solar cells reached 17.3% in 2020. Organic solar cells have sufficient potential to engender the next-generation solar cells.

References

1. Report of IEA (International Energy Agency) PVPS (Photovoltaic power Systems Programme), Trends in Photovoltaic Applications 2019. https://iea-pvps.org/trends_reports/2019-edition/
2. NREL (National Renewable Energy Lab).: Best Research-Cell Efficiency. https://www.nrel.gov/pv/cell-efficiency.html
3. Meng, L., Zhang, Y. Wan, X., Li, C., Zhang, X., Wang, Y., Ke, X., Xiao, Z., Ding, L., Xia, R., Yip, H.-L., Cao, Y., Chen, Y.: Organic and solution-processed tandem solar cells with 17.3% efficiency. Science **361**, 1094–1098 (2018)
4. https://www.taisei.co.jp/about_us/wn/2014/140324_3962.html
5. Kearns, D., Calvin, M.: Photovoltaic Effect and Photoconductivity in Laminated Organic Systems. J. Chem. Phys. **29**, 950–951 (1958)
6. Ghosh, A.K., Morel, D.L., Feng, T., Shaw, R.F., Rowe, C.A., Jr.: Photovoltaic and rectification properties of Al/Mg phthalocyanine/Ag Schottky-barrier cells. J. Appl. Phys. **45**, 230–236 (1974)
7. Ghosh, A.K., Feng, T.: Merocyanine organic solar cells. J. Appl. Phys. **49**, 5982–5989 (1978)
8. Chamberlain, G.A.: Organic solar cells: A review. Solar Cells **8**, 47–83 (1983) (and references therein)
9. Wohrle, D., Meissner, D.: Organic solar cells. Adv. Mater. **3**, 129–138 (1991) (and references therein)
10. Tang, C.W.: Two-layer organic photovoltaic cell. Appl. Phys. Lett. **48**, 183–185 (1986)
11. Hiramoto, M., Fujiwara, H., Yokoyama, M.: Three-layered organic solar cell with a photoactive interlayer of codeposited pigments. Appl. Phys. Lett. **58**, 1062–1064 (1991)
12. Hiramoto, M., Fujiwara, H., Yokoyama, M.: *p-i-n* like behavior in three-layered organic solar cells having c co-deposited interlayer of pigments. J. Appl. Phys. **72**, 3781–3787 (1992)
13. Sariciftci, N.S., Smilowitz, L., Heeger, A.J., Wudl, F.: Photoinduced electron transfer from a conducting polymer to buckminsterfullerene. Science **285**, 1474–1476 (1992)
14. Yu, G., Gao, J., Hummelen, J.C., Wudl, F., Heeger, A.J.: Polymer photovoltaic cells: Enhanced efficiencies via a network of internal donor-acceptor heterojunctions. Science **270**, 1789–1791 (1995)

15. Hiramoto, M., Suezaki, M., Yokoyama, M.: Effect of thin gold interstitial-layer on the photovoltaic properties of tandem organic solar cell. Chem. Lett. **19**, 327–330 (1990)
16. Peumans, P., Yakimov, A., Forrest, S.R.: Small molecular weight organic thin-film photodetectors and solar cells. J. Appl. Phys. **93**, 3693–3723 (2003)
17. (a) Riede, M., Uhrich, C., Widmer, J., Timmreck, R., Wynands, D., Schwartz, G., Gnehr, W.-M., Hildebrandt, D., Weiss, A., Hwang, J., Sundarraj, S., Erk, P., Pfeiffer, M., Leo, K.: Efficient organic tandem solar cells based on small molecules. Adv. Funct. Mater. **21**, 3019–3028 (2011). (b) Schueppel, R., Timmreck, R., Allinger, N., Mueller, T., Furno, M., Uhrich, C., Leo, K., Riede, M.: Controlled current matching in small molecule organic tandem solar cells using doped spacer layers. J. Appl. Phys. **107**, 044503 (6 pages) (2010). (c) Heliatek, H.P.: https://www.heliatek.com/en/press/press-releases/details/heliatek-consolidates-its-technology-leadership-by-establishing-a-new-world-record-for-organic-solar-technology-with-a-cell-effi
18. Hou, J., Inganäs, O., Friend, R., Gao, F.: Organic solar cells based on non-fullerene acceptors. Nat. Mater. **17**, 119–128 (2018)
19. Zhang, H., Yao, H., Hou, J., Zhu, J., Zhang, J., Li, W., Yu, R., Gao, B., Zhang, S., Hou, J.: Over 14% efficiency in organic solar cells enabled by chlorinated nonfullerene small-molecule acceptors. Adv. Mater. **30**, 1800613 (7 pages) (2018)
20. Hiramoto, M., Yamaga, T., Danno, M., Suemori, K., Matsumura, Y., Yokoyama M.: Design of nanostructure for photo-electric conversion by organic vertical superlattice.. Appl. Phys. Lett. **88**, 213105 (3 pages) (2006)
21. O'Regan, B., Grätzel, M.: A low-cost, high-efficiency solar cell based on dye-sensitized colloidal Ti02 films. Nature **353**, 737–740 (1991)
22. Padinger, F., Rittberger, F.R.S., Sariciftci, N.S.: Effects of postproduction treatment on plastic solar cells. Adv. Funct. Mater. **13**, 85–88 (2003)
23. Li, G., Shrotriya, V., Huang, J., Yao, Y., Moriarty, T., Emery, K., Yang, Y.: High-efficiency solution processable polymer photovoltaic cells by self-organization of polymer blends. Nat. Mater. **4**, 864–868 (2005)
24. Suemori, K., Miyata, T., Hiramoto, M., Yokoyama, M.: Enhanced photovoltaic performance in fullerene:phthalocyanine co-deposited films deposited on heated substrate. Jpn. J. Appl. Phys. **43**, L1014–L1016 (2004)
25. Kikuchi, M., Hirota, M., Kunawong, T., Shinmura, Y., Abe, M., Sadamitsu, Y., Moh, A.M., Izawa, S., Izaki, M., Naito, H., Hiramoto, M.: Lateral alternating donor/acceptor multilayered junction for organic solar cells. ACS Appl. Energy Mater. **2**, 2087–2093 (2019)
26. Katayama, M., Kaji, T., Nakao, S., Hiramoto, M.: Ultra-thick organic pigment layer up to 10 m activated by crystallization in organic photovoltaic cells. Frontiers Energy Res Sect Solar Energy **8**, 1–12 (2019)
27. Kaji, T., Zhang, M., Nakao, S., Iketaki, K., Yokoyama, K., Tang, C.W., Hiramoto, M.: Co-evaporant induced crystalline donor:acceptor blends in organic solar cells. Adv. Mater. **23**, 3320–3325 (2011)
28. Hiramoto, M., Kubo, M., Shinmura, Y., Ishiyama, N., Kaji, T., Sakai, K., Ohno, T., Izaki, M.: Bandgap science for organic solar cells. Electronics **3**, 351–380 (2014)
29. Sakai, K., Hiramoto, M.: Efficient organic p-i-n solar cells having very thick codeposited i-layer consisting of highly purified organic semiconductors. Mol. Cryst. Liq. Cryst. **491**, 284–289 (2008)
30. Matsuo, Y., Sato, Y., Niinomi, T., Soga, I., Tanaka, H., Nakamura, E.: Columnar structure in bulk heterojunction in solution-processable three-layered p-i-n organic photovoltaic devices using tetrabenzoporphyrin precursor and silylmethyl[60]fullerene. J. Am. Chem. Soc. **131**, 16048–16050 (2009)
31. Kim, Y., Cook, S., Tuladhar, S.M., Choulis, S.A., Nelson, J., Durrant, J.R., Bradley, D.D.C., Giles, M., McCulloch, I., Ha, C.S., Ree, M.: A strong regioregularity effect in self-organizing conjugated polymer films and high-efficiency polythiophene: fullerene solar cells. Nat. Mater. **5**, 197–203 (2006)

32. Osaka, I., Saito, M., Koganezawa, T., Takimiya, K.: Thiophene–thiazolothiazole copolymers: Significant impact of side chain composition on backbone orientation and solar cell performances. Adv. Mater. **26**, 331–338 (2014)
33. Scharber, M.C., Mühlbacher, D., Koppe, M., Denk, P., Waldauf, C., Heeger, A.J., Brabec, C.J.: Design rules for donors in bulk-heterojunction solar cells-Towards 10 % energy-conversion efficiency. Adv. Mater. **18**, 789–794 (2006)
34. Matsuo, Y.: Design concept for high-LUMO-level fullerene electron-acceptors for organic solar cells. Chem. Lett. **41**, 754–759 (2012)
35. Kawashima, K., Tamai, Y., Ohkita, H., Osaka, I., Takimiya, K.: High-efficiency polymer solar cells with small photon energy loss. Nat. Commun. **6**, 10085 (9 pages) (2015)
36. Ran, N.A., Love, J.A., Takacs, C.J., Sadhanala, A., Beavers, J.K., Collins, S.D., Huang, Y., Wang, M., Friend, R.H., Bazan, G.C., Nguyen, T.Q.: Harvesting the full potential of photons with organic solar cells. Adv. Mater. **28**, 1482–1488 (2016)
37. Benduhn, J., Tvingstedt, K., Piersimoni, F., Ullbrich, S., Fan, Y., Tropiano, M., McGarry, K.A., Zeika, O., Riede, M.K., Douglas, C.J., Barlow, S., Marder, S.R., Neher, D., Spoltore, D., Vandewal, K.: Intrinsic non-radiative voltage losses in fullerene-based organic solar cells. Nat. Energy **2**, 17053 (2017)
38. Nalwa, K. S., Kodali, H. K., Ganapathysubramanian, B., Chaudhary S.: Dependence of recombination mechanisms and strength on processing conditions in polymer solar cells. Appl. Phys. Lett. **99**, 263301 (4 pages) (2011).
39. Gorenflot, J., Heiber, M. C., Baumann, A., Lorrmann, J., Gunz, M., Kämpgen, A., Dyakonov, V., Deibel, C.: Nongeminate recombination in neat P3HT and P3HT: PCBM blend films. J. Appl. Phys. **115**, 144502 (9 pages) (2014)
40. Kirchartz, T., Pieters, B. E., Kirkpatrick, J., Rau, U., Nelson, J.: Recombination via tail states in polythiophene:fullerene solar cells. Phys. Rev. B 83, 115209 (13 pages) (2011)
41. Shintaku, N., Hiramoto, M., Izawa, S.: Effect of trap-assisted recombination on open-circuit voltage loss in phthalocyanine/fullerene solar cells. Org. Electron. **55**, 69–74 (2018)
42. Shockley, W., Queisser, H.J.: Detailed balance limit of efficiency of *p-n* junction solar cells. J. Appl. Phys. **32**, 510–519 (1961)
43. Li, W., Hendriks, K. H., Furlan, A., Wienk, M. M., Janssen, R.A.: High quantum efficiencies in polymer solar cells at energy losses below 0.6 eV. J. Am. Chem. Soc., 137, 2231–2234 (2015).
44. Queisser, H.J.: Detailed balance limit for solar cell efficiency. Mater. Sci. Eng. B **159–160**, 322–328 (2009)
45. Izawa, S., Shintaku, N., Kikuchi, M., Hiramoto, M.: Importance of interfacial crystallinity to reduce open-circuit voltage loss in organic solar cells. Appl. Phys Lett. **115**, 153301 (4 pages) (2019).
46. Yakimov, A., Forrest, S.R.: High photovoltage multiple-heterojunction organic solar cells incorporating interfacial metallic nanoclusters. Appl. Phys. Lett. **80**, 1667–1669 (2002)
47. Ameri, T., Dennler, G., Lungenschmied, C., Brabec, C.J.: Organic tandem solar cells: a review. Energy Environ. Sci. **2**, 347–363 (2009) (and references therein)
48. You, J., Dou, L., Yoshimura, K., Kato, T., Ohya, K., Moriarty, T., Emery, K., Chen, C., Gao, J., Li, G., Yang, Y.: A polymer tandem solar cell with 10.6% power conversion efficiency. Nat. Commun. 4, 1446 (10 pages), (2012).
49. Chen, C., Chang, W., Yoshimura, K., Ohya, K., You, J., Gao, J., Hong, Z., Yang, Y.: An efficient triple-junction polymer solar cell having a power conversion efficiency exceeding 11%. Adv. Mater. **26**, 5670–5677 (2014)
50. Ishiyama, N., Kubo, M., Kaji, T., Hiramoto, M.: Tandem organic solar cells formed in co-deposited films by doping. Org. Electron. **14**, 1793–1796 (2013)
51. https://forschungsinfo.tu-dresden.de/detail/professur/349
52. Lüssem, B., Riede, M., Leo, K.: Doping of organic semiconductors. Phys. Stat. Solidi a **210**, 9–43 (2013)
53. Hiramoto, M., Kikuchi, M., Izawa, S.: Parts-per-million-level doping effects in organic semiconductor films and organic single crystals. Adv. Mater. **30**, 1801236 (15 pages) (2018)

Chapter 2
A Path to the Blended Junction

Masahiro Hiramoto

2.1 Shock of Two-Layer Solar Cell

The author had been interested in the energy problem from the time of his undergraduate studies and was involved in the research on the photoelectrochemical conversion of solar light [1–5] and photocatalysis [6] during the early stage of his career from 1980 to 1988, focusing on inorganic semiconductors such as silicon [2–5] and cadmium sulfide [6]. The photoelectrochemical cells using such inorganic semiconductors typically showed the photocurrent density of approximately 20 $mAcm^{-2}$ under the irradiation of simulated solar light [1–5]. At that time, the author had the impression that compared to the high performance of inorganic solar cells, the performance characteristics of organic solar cells were poor. Organic solar cells of those days with the Schottky junction with corrosive metals such as Al showed extremely small photocurrents of the order of nano- and microamperes as summarized in review papers [7, 8].

In 1986, Ching W. Tang of Eastman Kodak Co. published a report in *Applied Physics Letters* titled "Two-layer organic solar cell" [9]. Surprisingly, he also reported his famous work on organic electroluminescence almost simultaneously in 1987 [10]. Following this report, in 1988 the author shifted his research field from inorganic semiconductors to organic semiconductors. At the time, the author was quite shocked when he first encountered the two-layer paper and was particularly surprised by the current–voltage characteristics shown in Fig. 2.1 of [9]. Tang reported that the short-

The original version of this chapter was revised: Figure 2.1 have been updated with new Figure. The correction to this chapter is available at https://doi.org/10.1007/978-981-15-9113-6_11

M. Hiramoto (✉)
National Institutes of Natural Sciences, Institute for Molecular Science, 5-1 Higashiyama Myodaiji, Aichi, Okazaki 444-8787, Japan
e-mail: hiramoto@ims.ac.jp

M. Hiramoto and S. Izawa (eds.), *Organic Solar Cells*,
https://doi.org/10.1007/978-981-15-9113-6_2

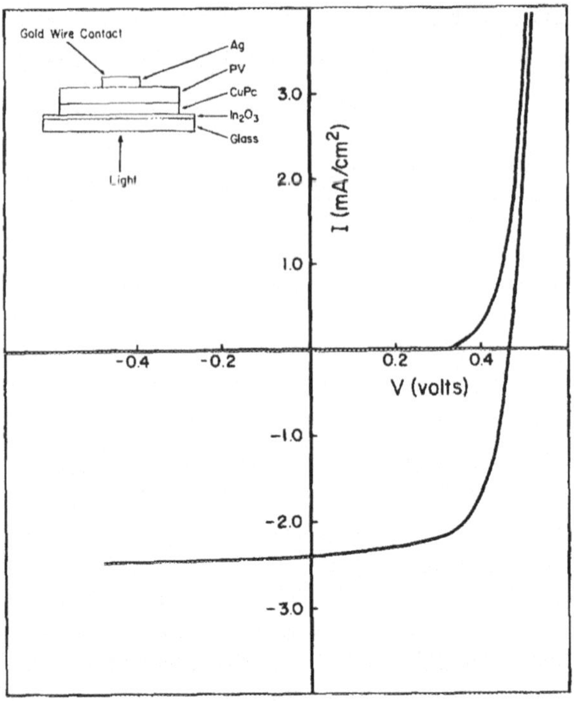

Fig. 2.1 Current–voltage characteristics reported in the paper, "Two-layer organic photovoltaic cell," by Ching. W. Tang [9]. Surprising performance, i.e., J_{sc}: 2.3 mAcm^{-2}, FF: 0.65, V_{oc}: 0.45 V, conversion efficiency: 1%, was reported. Inset shows the cell structure. The PV in this figure corresponds to Im-PTC in this chapter. Reproduced with permission from [9]. Copyright 1986 AIP Publishing

circuit photocurrent density (J_{sc}) reached 2.3 mAcm^{-2} under a simulated solar light intensity of 75 mWcm^{-2}, while the external quantum efficiency reached 15%, and the fill factor (FF) reached the record high value of 0.65. The author's first impression was that if only one order of magnitude increase in J_{sc} would be achieved, J_{sc} could reach more than 20 mAcm^{-2} that was a typical value for inorganic Si solar cells.

2.1.1 Motivation

The author considered that the intrinsic poor ability of photocurrent generation of organic semiconductors was a top priority, even though there were other issues that had to be addressed such as the unclear origin of the photovoltage and the high resistance of organic semiconductors that were called "insulating semiconductors." Since the Schottky junction-type cells using single layers of phthalocyanine and perylene pigments (Fig. 2.2) used in Tang's two-layer cell showed very small photocurrent, at first, it was difficult for the author to understand why such a large photocurrent over 2 mAcm^{-2} could be generated by the two-layer cell. Therefore, the author

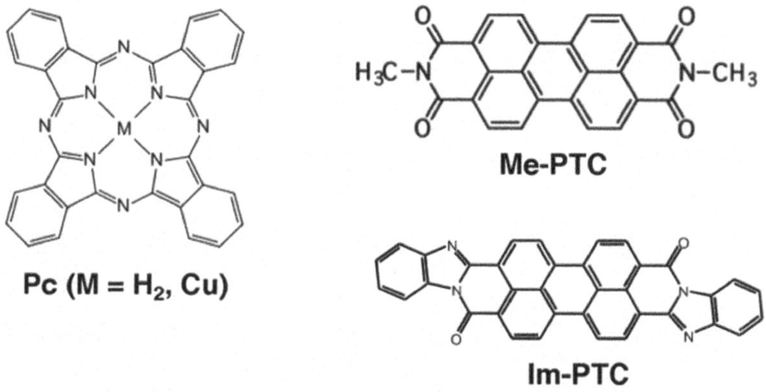

Fig. 2.2 Chemical structures of the phthalocyanine (H_2Pc, CuPc) and perylene derivatives (Me-PTC, Im-PTC). Reproduced with permission from [11]. Copyright 1992 AIP Publishing

initially focused on the elucidation of the processes at the interface between the phthalocyanine (Pc) (Fig. 2.2) and perylene (PTC) (Fig. 2.2) films.

2.2 Tracking During Interface Making

2.2.1 Expectation

We need to consider both the photovoltaic characteristics of the single layer in the Schottky junction and those of the two-layer cell. Apparently, these two situations were the two extreme cases because their photovoltaic characteristics were absolutely different. The author was convinced that transition between these two systems would provide the clues regarding the photocarrier generation process in the two-layer cell. It should be noted that the extremely thin phthalocyanine (Pc) layer with a thickness of 1 nm between the perylene pigment (PTC) film and metal electrode shall be regarded as interfacial doping. Based on the experience with vacuum deposition, the author considered that the H_2Pc molecules were doped at the interface with the minute island structure (Fig. 2.3a) and formed molecular-level contact with the Me-PTC molecules.

2.2.2 Tracking

We used a combination of perylene pigment (Me-PTC; Fig. 2.2) and metal-free phthalocyanine (H_2Pc; Fig. 2.2) [11]. The Me-PTC/Au interface forms a Schottky junction between the Me-PTC film and Au (Fig. 2.3a). The Au electrode exhibited

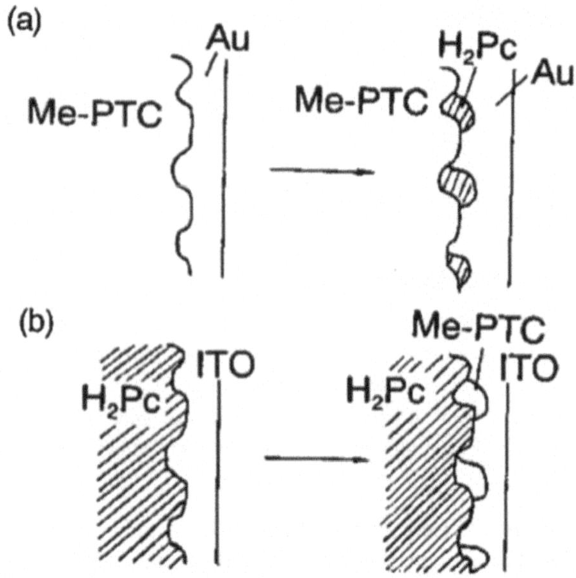

Fig. 2.3 Schematic illustrations of interfacial doping. **a** H_2Pc molecules are doped at the Me-PTC/Au interface. **b** Me-PTC molecules are doped at the H_2Pc/ITO interface

positive photovoltage with respect to the indium tin oxide (ITO) electrode (Fig. 2.4, curve a), implying that the Me-PTC film acts as an *n*-type semiconductor, and a small value of 0.2 $mAcm^{-2}$ was obtained for J_{sc}. By the interfacial doping of a monolayer-like H_2Pc layer with a thickness of 1.5 nm between Au and Me-PTC, J_{sc} was suddenly doubled to 0.4 $mAcm^{-2}$, increased steeply by applying the small reverse voltage, and reached close to 2 $mAcm^{-2}$ at −2 V (Fig. 2.4, curve b). At a reverse bias of only −0.4 V, the photocurrent quantum efficiency for 480-nm light reached a significantly high value of 40%. The author had a strong impression that only molecular H_2Pc doped at Me-PTC/Au interface has the dramatic ability for the photocurrent enhancement into the milliampere level. Figure 2.5 shows a photograph of a page of the author's laboratory notebook from that time, recording the result of interfacial doping. On the left middle side of the page, the author wrote a Japanese sentence meaning "Increase of carrier generation efficiency!".

Conversely, the interfacial doping of Me-PTC at H_2Pc/ITO interface also caused the drastic increase of carrier generation (Fig. 2.3b). It is clear that the molecular contact between the Me-PTC and H_2Pc molecules is a key for the photocarrier generation enhancement.

To quantitatively check this hypothesis, we measured the action spectra of the interfacial doped cells (Fig. 2.6b). Here, an observation window in the spectral region around 500 nm is present. Namely, in this range only Me-PTC absorbs light strongly while H_2Pc shows little light absorption (Fig. 2.7a). At approximately 500 nm, a drastic photocurrent enhancement was observed. The dramatic increase in J_{sc} at 480 nm reveals that a molecular amount of H_2Pc on the Me-PTC film apparently enhances the carrier generation efficiency of Me-PTC (Fig. 2.7b) until the thickness of 20 nm at which point the entire surface of Me-PTC is covered by H_2Pc

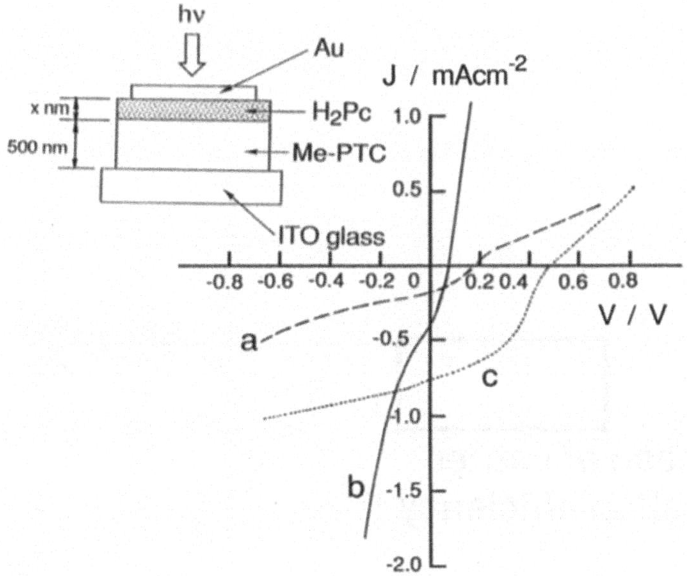

Fig. 2.4 Photocurrent–voltage (J–V) characteristics for the Me-PTC/Au cells with an H_2Pc layer with the thickness (x) of 0 (curve a), 1.5 (curve b), and 50 nm between Me-PTC and Au. The light intensity transmitted through the Au electrode was 50 $mWcm^{-2}$. Inset shows the cell structure with various thicknesses of H_2Pc from x = 0 to 50 nm at the interface between Me-PTC and Au. Reproduced with permission from [11]. Copyright 1992 AIP Publishing

(Fig. 2.7a). Thus, we could conclude that the enhancement of the photocarrier generation efficiency of Me-PTC was due to the formation of molecular contact with H_2Pc.

Alternatively, it is possible that the observed J_{sc} increase is caused by the increase in the built-in electric field at the Me-PTC/H_2Pc interface. V_{oc} representing the built-in potential difference between the Au electrode and Me-PTC layer showed consistently smaller values from x = 0 to 10 nm than the V_{oc} value of 0.2 V for the cell without H_2Pc (x = 0 nm) (Fig. 2.7b). Thus, we could exclude the possibility of photocarrier generation enhancement due to the built-in field increase.

2.2.3 *Key: Direct PTC/Pc Molecular Contact*

Doping of H_2Pc molecules at the Me-PTC/Au interface drastically enhances the photocarrier generation efficiency (Fig. 2.3a). Moreover, the doping of Me-PTC molecules at the H_2Pc/ITO interface also drastically enhances the photocarrier generation efficiency (Fig. 2.3b). Logically, it is obvious that a molecular contact between the Me-PTC and H_2Pc molecules with an excellent photocarrier generation ability is

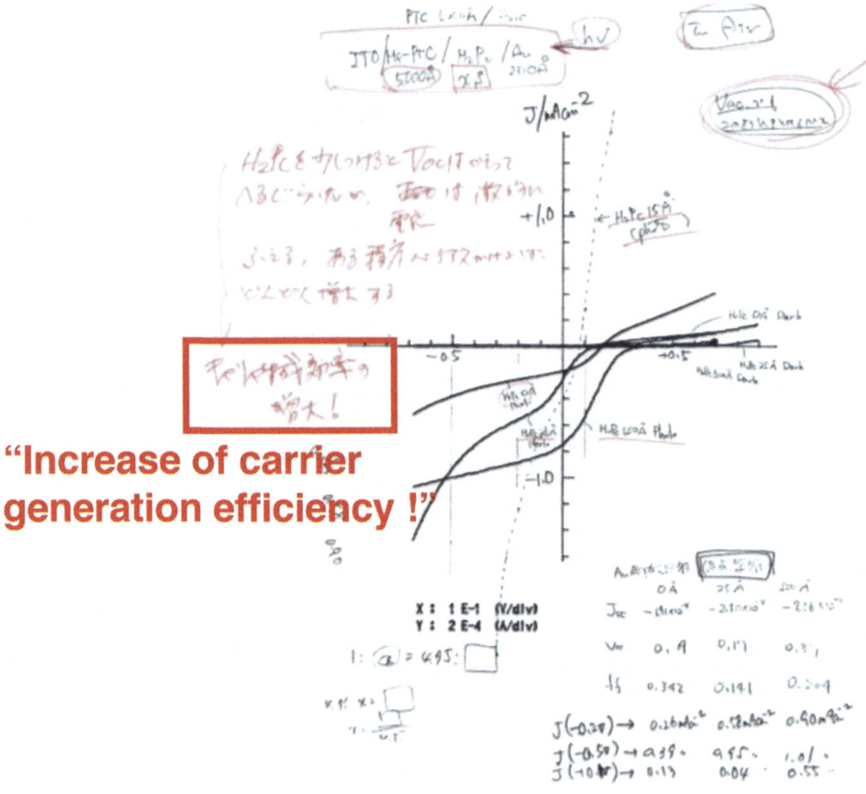

Fig. 2.5 A page from the laboratory notebook recording the interfacial doping result. A sentence in red stating "Increase of carrier generation efficiency!", is written on the left middle side in Japanese

a key. In the two-layer cell, the direct molecular contact at the interface between perylene (Me-PTC, Im-PTC) (Fig. 2.2) and the phthalocyanine pigments (H_2Pc, CuPc) (Fig. 2.2) plays a key role for the carrier photogeneration process.

This conclusion strongly suggests that there should be some kinds of interaction between PTC and Pc; this is discussed in detail in Sect. 2.3.7.

2.3 Blended Junction

The author proposed the first blended junction in 1991 [11, 21]. Figure 2.8 shows the abstract (in Japanese) for the master course presentation in 1991 of Fujiwara who was the author's student [11, 21] who experimentally examined the first blended junction.

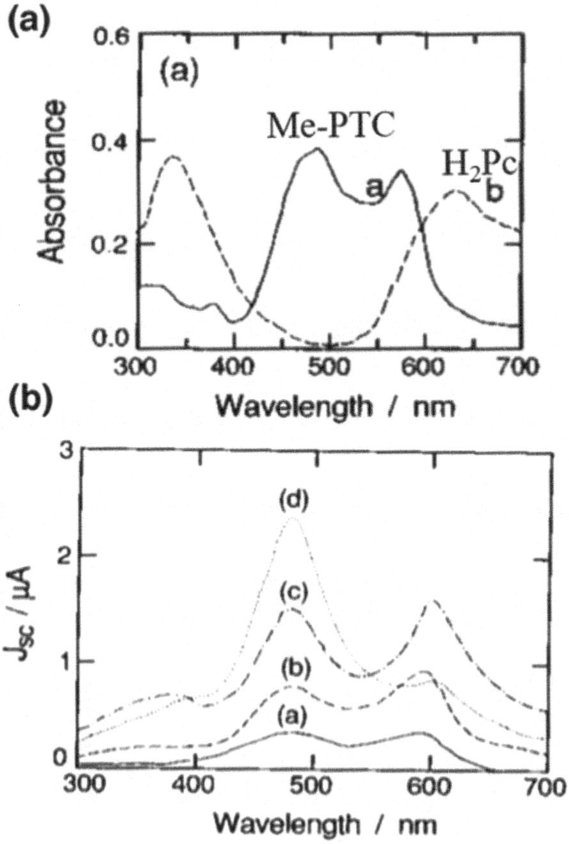

Fig. 2.6 **a** Absorption spectra of the films of Me-PTC (40 nm) (solid curve a) and H_2Pc (broken curve b). **b** Action spectra of the short-circuit photocurrent (J_{sc}) for the cells shown in Fig. 2.5. The thicknesses of the H_2Pc film (x) are 0 (curve (a)), 1.5 (curve (b)), 10 (curve (c)), and 50 nm (curve (d)). Monochromatic light was irradiated through the Au electrode. Reproduced with permission from [11]. Copyright 1992 AIP Publishing

2.3.1 *Expectation*

The author expected to observe the following two results.

(i) The fact that the direct molecular contact between Me-PTC and H_2Pc acts as a site for photocarrier generation suggests that a cell containing their blended layer that will inevitably contain a vast number of Me-PTC/H_2Pc sites will show higher photocarrier generation efficiency.

(ii) Since Me-PTC and H_2Pc films behave as *n*- and *p*-types, respectively, the blended film between Me-PTC and H_2Pc will show the intrinsic (*i*) character due to the compensation of the positively ionized donor and the negatively ionized acceptor. The author hypothesized that a *p-i-n* junction will be formed based on the knowledge regarding the *p-i-n* energetic structure that had been successfully used for amorphous silicon, [12] and was fascinated by the idea of fabricating a *p-i-n* junction using organic semiconductors.

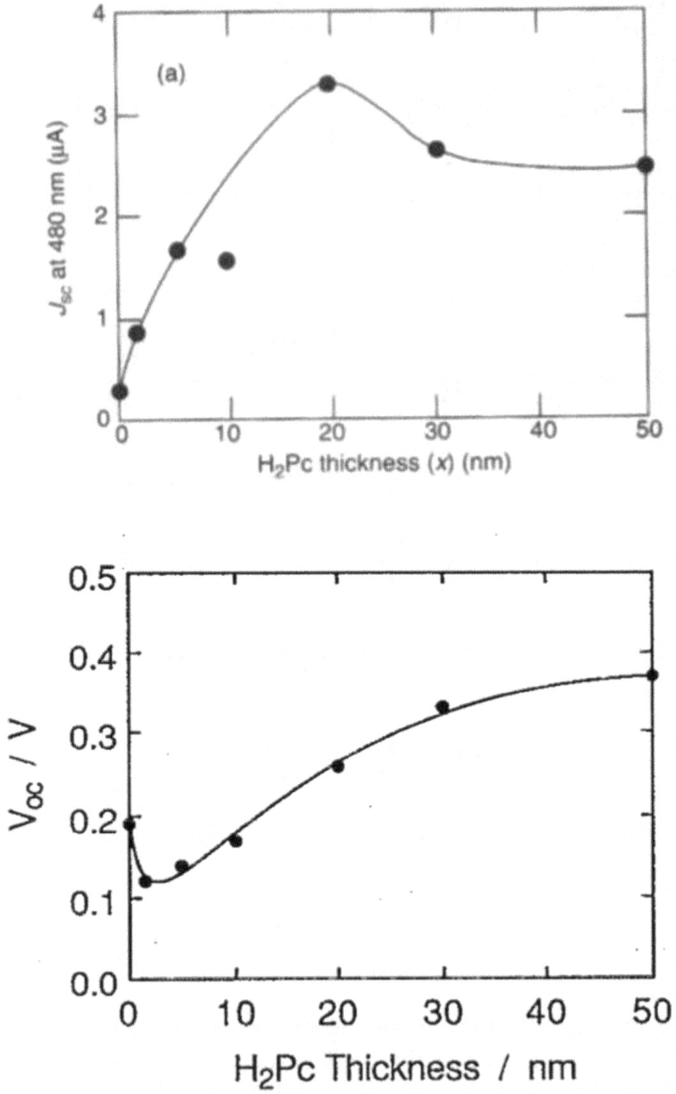

Fig. 2.7 **a** Dependence of J_{sc} on the H_2Pc film thickness (x). **b** Dependence of V_{oc} on the H_2Pc film thickness (x). **b** Reproduced with permission from [11]. Copyright 1992 AIP Publishing

Me-PTC and H_2Pc films behave as *n*- and *p*-types, respectively. For the *p*-type behavior of phthalocyanines (Pc), the oxygen molecules (O_2) from air were known to act as acceptors by extracting electrons from Pc through the formation of O_2^- [13, 14]. The acceptor impurity in Me-PTC was not identified. However, even though the impurities acting as dopants were unknown, the author could assume that the positively ionized donor impurity and the negatively ionized acceptor

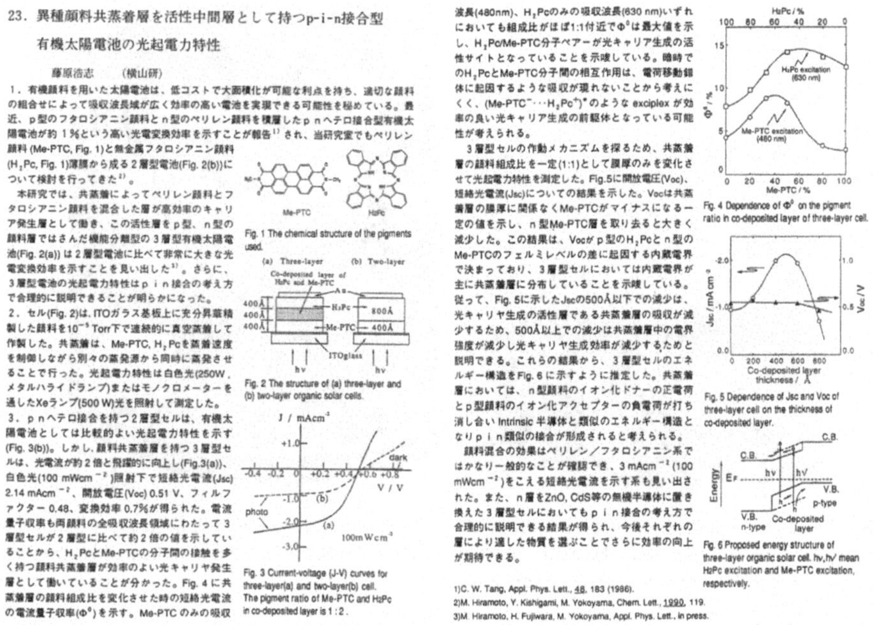

23．異種顔料共蒸着層を活性中間層として持つp-i-n接合型
有機太陽電池の光起電力特性

藤原浩志　　（横山研）

1．有機顔料を用いた太陽電池は、低コストで大面積化が可能な利点を持ち、適切な顔料の組合せによって吸収波長域が広く効率の高い電池を実現できる可能性を秘めている。最近、p型のフタロシアニン顔料とn型のペリレン顔料を積層したpnヘテロ接合型有機太陽電池が約1％という高い光電変換効率を示すことが報告[1]され、当研究室でもペリレン顔料(Me-PTC, Fig. 1)と無金属フタロシアニン顔料(H_2Pc, Fig. 1)薄膜から成る2層型電池(Fig. 2(b))について検討を行ってきた[2]。

本研究では、共蒸着によってペリレン顔料とフタロシアニン顔料を混合した層が高効率のキャリア発生層として働き、この活性層をp型、n型の顔料層ではさんだ機能分離型の3層型有機太陽電池(Fig. 2(a))は2層型電池に比べて非常に大きな光電変換効率を示すことを見い出した[3]。さらに、3層型電池の光起電力特性はpin接合の考え方で合理的に説明できることが明らかになった。

2．セル(Fig. 2)は、ITOガラス基板上に充分昇華精製した顔料を10^{-5} Torr下で連続的に真空蒸着して作製した。共蒸着は、Me-PTC, H_2Pcを蒸着速度を制御しながら別々の蒸発源から同時に蒸発させることで行った。光起電力特性は白色光(250W，メタルハライドランプ)またはモノクロメーターを通したXeランプ(500 W)光を照射して測定した。

3．pnヘテロ接合を持つ2層型セルは、有機太陽電池としては比較的よい光起電力特性を示す(Fig. 3(b))。しかし、顔料共蒸着層を持つ3層型セルは、光電流が約2倍と飛躍的に向上し(Fig.3(a))、白色光(100 mWcm^{-2})照射下で短絡光電流(Jsc) 2.14 mAcm^{-2}、開放電圧(Voc) 0.51 V、フィルファクター0.48、変換効率0.7％が得られた。電流量子収率も両顔料の全吸収波長領域にわたって3層型セルが2層型に比べて約2倍の値を示していることから、H_2PcとMe-PTCの分子間の接触を多く持つ顔料共蒸着層が効率のよい光キャリヤ発生層として働いていることが分かった。Fig. 4に共蒸着層の顔料組成比を変化させた時の短絡光電流の電流量子収率(Φ^0)を示す。Me-PTCのみの吸収波長(480nm)、H_2Pcのみの吸収波長(630 nm)いずれにおいても組成比がほぼ1:1付近でΦ^0は最大値を示し、H_2Pc/Me-PTC分子ペアーが光キャリア生成の活性サイトとなっていることを示唆している。暗時でのH_2PcとMe-PTC分子間の相互作用は、電荷移動錯体に起因するような吸収が現れないことから考えにくく、$(Me\text{-}PTC^- \cdots H_2Pc^+)^*$のようなexciplexが効率の良い光キャリア生成の前駆体となっている可能性が考えられる。

3層型セルの作動メカニズムを探るため、共蒸着層の顔料組成比を一定(1:1)として膜厚のみを変化させて光起電力特性を測定した。Fig.5に開放電圧(Voc)、短絡光電流(Jsc)についての結果を示した。Vocは共蒸着層の膜厚に関係なくMe-PTCがマイナスになる一定の値を示し、n型Me-PTC層を取り去ると大きく減少した。この結果は、Vocがp型のH_2Pcとn型のMe-PTCのフェルミレベルの差に起因する内蔵電界で決まっており、3層型セルにおいては内蔵電界が主に共蒸着層に分布していることを示唆している。従って、Fig. 5に示したJscの500Å以下での減少は、光キャリヤ生成の活性層である共蒸着層の吸収が減少するため、500Å以上での減少は共蒸着層中の電界強度が減少し光キャリヤ生成効率が減少するためと説明できる。これらの結果から、3層型セルのエネルギー構造をFig. 6に示すように推定した。共蒸着層においては、n型顔料のイオン化ドナーの正電荷とp型顔料のイオン化アクセプターの負電荷が打ち消し合いIntrinsic半導体と類似のエネルギー構造となりpin類似の接合が形成されると考えられる。

顔料混合の効果はペリレン／フタロシアニン系ではかなり一般的なことが確認でき、3 mAcm^{-2} (100 mWcm^{-2})をこえる短絡光電流を示す系も見い出された。また、n層をZnO, CdS等の無機半導体に置き換えた3層型セルにおいてもpin接合の考え方で合理的に説明できる結果が得られ、今後それぞれの層により適した物質を選ぶことでさらに効率の向上が期待できる。

Fig. 1 The chemical structure of the pigments used.

Fig. 2 The structure of (a) three-layer and (b) two-layer organic solar cells.

Fig. 3 Current-voltage (J-V) curves for three-layer(a) and two-layer(b) cell. The pigment ratio of Me-PTC and H2Pc in co-deposited layer is 1 : 2.

Fig. 4 Dependence of Φ^0 on the pigment ratio in co-deposited layer of three-layer cell.

Fig. 5 Dependence of Jsc and Voc of three-layer cell on the thickness of co-deposited layer.

Fig. 6 Proposed energy structure of three-layer organic solar cell. hv,hv' mean H2Pc excitation and Me-PTC excitation, respectively.

1)C. W. Tang, Appl. Phys. Lett., 48, 183 (1986).
2)M. Hiramoto, Y. Kishigami, M. Yokoyama, Chem. Lett., 1990, 119.
3)M. Hiramoto, H. Fujiwara, M. Yokoyama, Appl. Phys. Lett., in press.

Fig. 2.8 Abstract (in Japanese) for the master course presentation in 1991 of Fujiwara [11, 21] who was the author's student who obtained the first blended junction

impurity should compensate each other in the blended film of Me-PTC and H_2Pc that inevitably contains both donor and acceptor impurities.

2.3.2 *Overcoming Experimental Hurdles*

To fabricate blended films containing two kinds of organic semiconductors, the author adopted the co-deposition technique, namely the simultaneous evaporation technique. The author had been a fan of vacuum deposition since he was first strongly impressed by the fact that vacuum deposition can control the thickness with Å-level precision that is comparable to the size of molecules and had rich experience with technical methods such as gas-introduced evaporation at low pressure of 10^{-1} Pa of gases such as oxygen, hydrogen [15], vacuum deposition on the heated and cooled substrate, and other techniques.

Simultaneous evaporation, i.e., the co-evaporation of PTC and Pc from two separate controlled sources while controlling the deposition rate of each material, was the technique chosen for the fabrication of the blended layer. The author had a simple and cheap vacuum evaporator, attaching many ports for system extensibility (ULVAC KIKO, VPC-290) (Fig. 2.9). Two independently controlled evaporation sources and

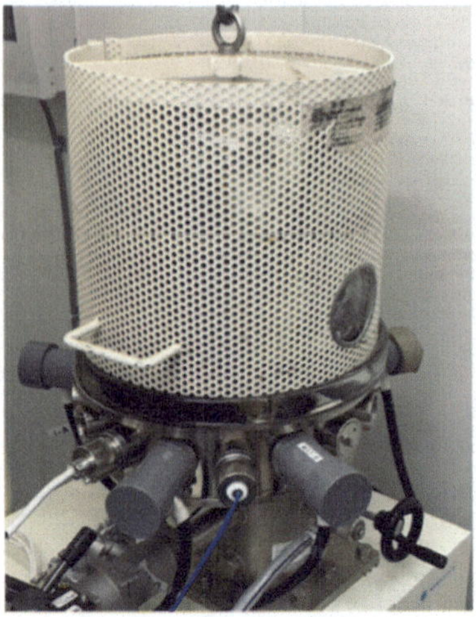

Fig. 2.9 (Left) Appearance of a vacuum evaporator used to fabricate the blended layer using the co-deposition technique. (Right) Inside the evaporation chamber. Electrodes (indicated by two red circles) for simultaneous evaporation can be seen

one quartz crystal microbalance (QCM) (ULVAC, CRTM-1000) were used in this equipment, and one of the voltage supplies was a simple variable Slidac.

Achieving simultaneous control of the evaporation rates of Pc and PTC using only one QCM was the main hurdle for achieving the fabrication of the blended layer and required the operator to master this technique. First, the operator evaporated PTC and waited to stabilize the evaporation rate to, e.g., 0.072 nms^{-1} and deposited the PTC layer (40 nm). Second, while maintaining the evaporation rate of PTC, H_2Pc was also evaporated, and the operator waited to stabilize the evaporation rate to, e.g., 0.185 nms^{-1}, corresponding to the sum of the H_2Pc and PTC rates, and co-deposited the blended layer of PTC and H_2Pc (40 nm). Third, PTC evaporation was stopped, and the operator waited to stabilize the evaporation rate at, e.g., 0.132 nms^{-1}, corresponding to the H_2Pc evaporation rate, and deposited the H_2Pc layer (40 nm). We could precisely estimate the compositional ratio of the co-deposited PTC:Pc film from the absorption spectra.

2.3.3 *Three-Layer Cell*

The structures of three- and two-layer organic solar cells are shown in Fig. 2.10. We adopted two kinds of combinations of Pc and PTC, namely H_2Pc:Me-PTC and CuPc:Im-PTC (Fig. 2.2). For both cells, the thickness of the PTC film was fixed to 40 nm for the precise comparison, avoiding the masking effect. The two-layer cell forms a heterojunction between the PTC and Pc films (Fig. 2.10b). Photocarrier

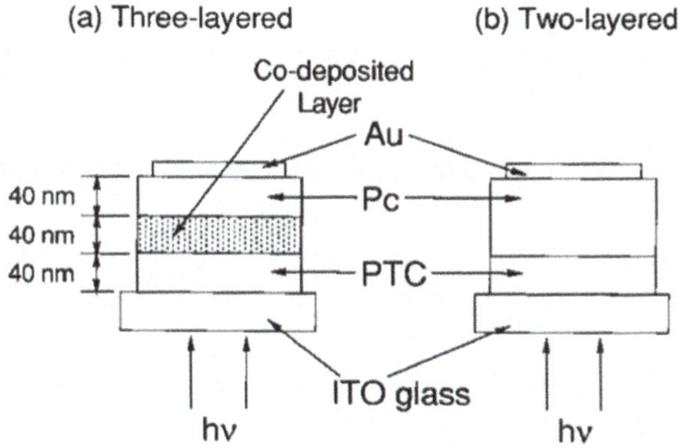

Fig. 2.10 Structures of the **a** three-layer and **b** two-layer organic solar cells. Reproduced with permission from [11]. Copyright 1992 AIP Publishing. Reproduced with permission from [21]. Copyright 1992 AIP Publishing

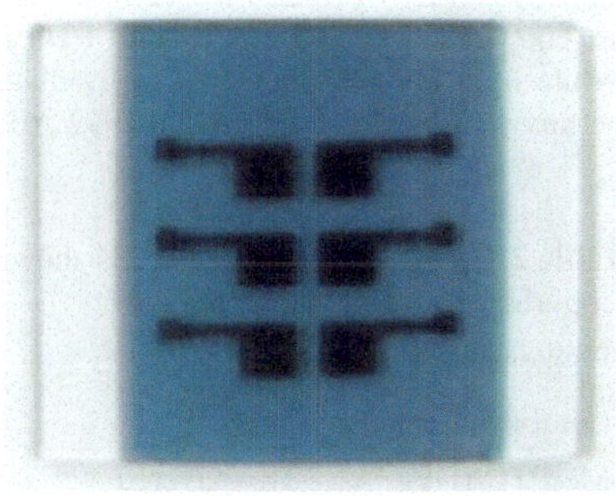

Fig. 2.11 A photograph of the first blended organic solar cell (reproduction)

generation occurs only close to the PTC/Pc interface [9, 15]. Three-layer cells have a blended interlayer of PTC and Pc pigments sandwiched between the PTC and Pc layers (Fig. 2.10a). Figure 2.11 shows a photograph of the three-layer cell of the Me-PTC/H_2Pc system (reproduction). This is the first blended organic solar cell.

Figure 2.12 shows the current–voltage (J–V) characteristics for the combinations of Me-PTC:H_2Pc (Fig. 2.12a) and Im-PTC:CuPc (Fig. 2.12b). Both combinations of three-layer cells with the co-deposited interlayer generated photocurrent densities (Fig. 2.11, **solid curves a)** that were twice as high as those of the heterojunction two-layer cells that showed relatively good performance (Fig. 2.11, **broken curves b).** Table 2.1 summarizes the performance of the cells. The photocurrent enhancement obtained by introducing the blended layer is universal as suggested by the clear effect observed for both combinations. The photovoltage remained almost unchanged for both three- and two-layer cells. Figure 2.13 shows a photograph of a page of the laboratory notebook recording the first result obtained for a blended cell at that time. The author was highly impressed by the clear indication of photocurrent doubling.

Figure 2.14a shows the spectral dependence of the internal quantum efficiency for the three- and two-layer cells of the Me-PTC/H_2Pc system. The internal quantum efficiency (Φ^0) is defined as the ratio of the number of carriers collected under the short-circuit condition to the number of photons absorbed by the organic layers.

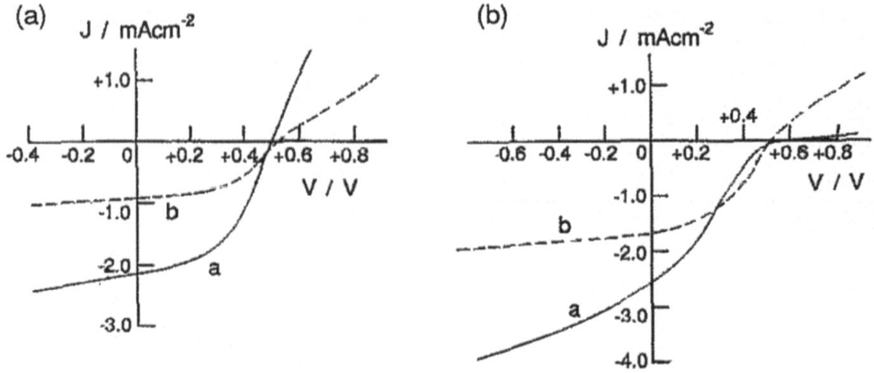

Fig. 2.12 Typical current–voltage (J–V) characteristics for a combination of H_2Pc/Me-PTC **a** and CuPc/Im-PTC (**b**). Three-layer cells (curves a) (Fig. 2.10a) and two-layer cells (curves b) (Fig. 2.10b) are compared. The volume ratios of PTC and Pc in the co-deposited layers are 1:2 (**a**) and 1:1 (**b**). White light (100 mWcm^{-2}) was irradiated through the ITO glass substrate. Reproduced with permission from [11]. Copyright 1992 AIP Publishing

Table 2.1 Performance of the cells shown in Fig. 2.10. Reproduced with permission from [11]. Copyright 1992 AIP Publishing

Pigment	Cell				
Combination	Type	J_{sc}/mA cm^{-2}	V_{oc}/V	ff	Φ^0/%
Me-PTC/H_2Pc	Two-layered	0.94	0.54	0.48	0.29
Me-PTC/H_2Pc	Three-layered	2.14	0.51	0.48	0.63
Im-PTC/CuPc	Two-layered	1.61	0.53	0.42	0.43
Im-PTC/CuPc	Three-layered	2.56	0.57	0.25	0.44

The quantum efficiency of the three-layer cell (curve b) was approximately twice as high as that of the two-layer cell (curve a) throughout the spectrum. Doubling of quantum efficiency was also confirmed for the Im-PTC/CuPc system. Since the masking effect due to the PTC film is the same for both three- (Fig. 2.10a) and two-layer cells (Fig. 2.10b), the author concluded that the photocurrent enhancement is attributed to an increase in the photocarrier generation efficiency obtained by the introduction of the co-deposited interlayer with a vast number of direct molecular contact sites between PTC and Pc.

2.3.4 p-i-n Junction

As mentioned in Sect. 2.3.1, the author already hypothesized that the blended film between the *n*-type Me-PTC and *p*-type H_2Pc will show intrinsic character due to the compensation of the positively ionized donor and the negatively ionized acceptor.

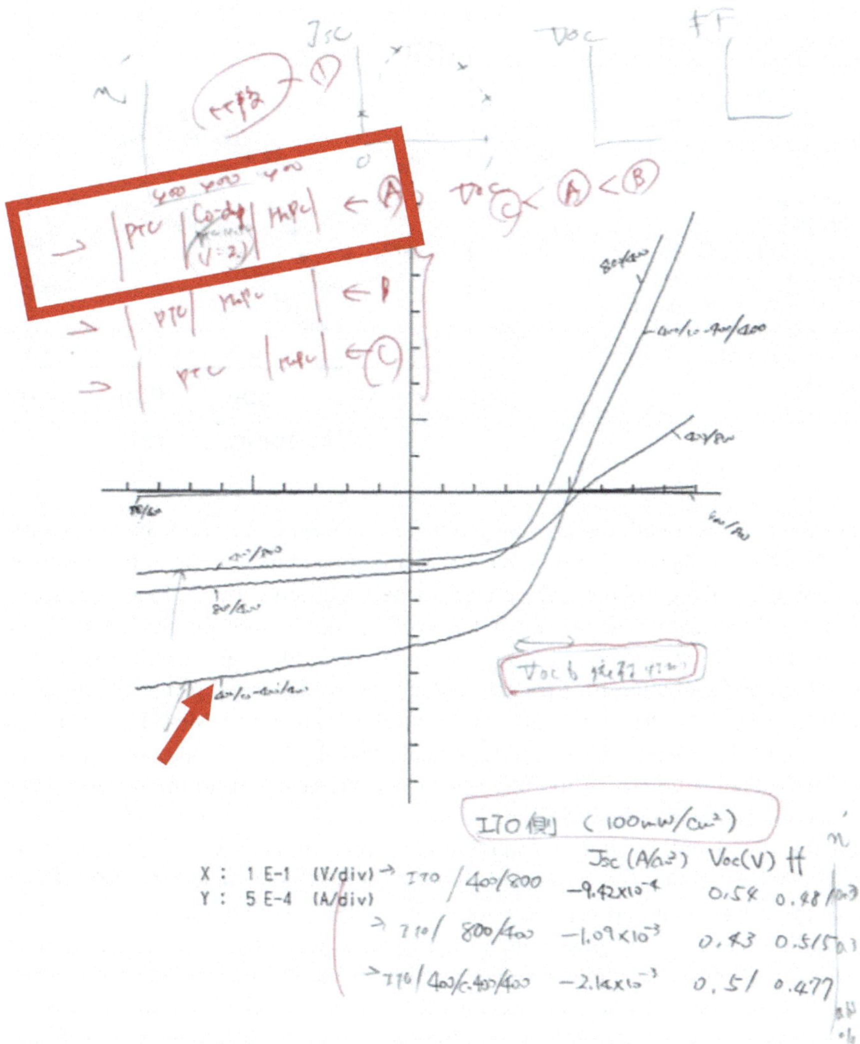

Fig. 2.13 A photograph of the page of the laboratory notebook recording the first blended cell result. The structure of the blended cell and the J–V curve is indicated by red rectangle and arrow

Therefore, the author immediately attempted to explain the photovoltaic properties of the three-layer cell based on the *p-i-n* junction. Figure 2.15a shows the energy diagrams of the H_2Pc and Me-PTC films prior to contact. The positions of the Fermi levels (E_F) indicated that H_2Pc and Me-PTC behave as *p*- and *n*-type semiconductors, respectively.

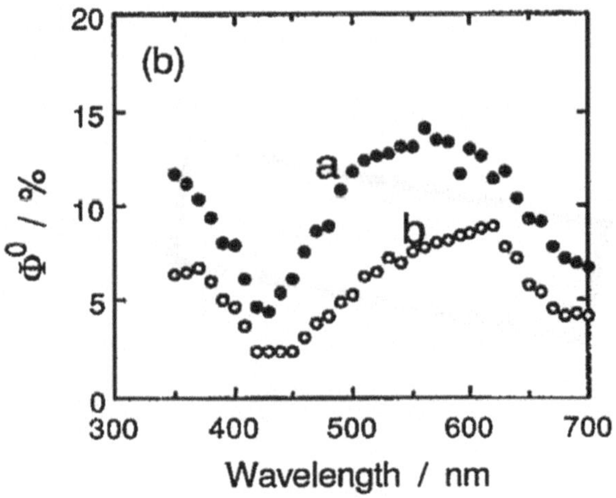

Fig. 2.14 Spectral dependence of the internal quantum efficiency (Φ^0) of J_{sc} for the three-layer cell (Fig. 2.10a) (curve a: solid circles) and the two-layer cell (Fig. 2.10b) (curve b: open circles) for the Me-PTC/H_2Pc system. Monochromatic light was irradiated through the ITO glass substrate. Reproduced with permission from [11]. Copyright 1992 AIP Publishing

For the two-layer cell, the organic *p-n* heterojunction is formed between *p*-H_2Pc and *n*-Me-PTC films (Fig. 2.15b, left) [15]. The sign and magnitude of the observed photovoltage in the two-layer cells are reasonably explained by the E_F difference.

For the three-layer cell, the dependence of J_{sc} and V_{oc} on the thickness of the co-deposited layer (x) (Fig. 2.16) provides essential information. A constant V_{oc} was maintained including for the case without the co-deposited layer, i.e., the two-layer cell (Fig. 2.16, triangles). By removing the front *n*-type Me-PTC layer, V_{oc} was strongly decreased. These results indicate that the built-in potential (V_{bi}) that directly generates the V_{oc} of the three-layer cell originates from the E_F difference between the H_2Pc and Me-PTC films.

On the other hand, J_{sc} showed a maximum at approximately $x = 50$ nm (Fig. 2.16, open dots). For a thinner co-deposited layer ($x < 50$ nm), the photocurrent decreased due to the weaker light absorption within the co-deposited interlayer with the Me-PTC/H_2Pc molecular contacts acting as active sites. For thicker co-deposited interlayers ($x > 50$ nm), the photocurrent decreased because of a smaller photocarrier generation efficiency due to the lowering of the electric field within the interlayer.

These explanations directly indicate that most of the built-in potential generated by the E_F difference in Me-PTC and H_2Pc is distributed across the interlayer. This picture is directly related to the hypothesis that the positive and negative charges of the donors and acceptors in the *n*- and *p*-type pigments are compensated by each other, and as a result, the co-deposited interlayer behaves as an intrinsic semiconductor. Thus, the energy structure of the three-layer cell that resembles the *p-i-n* junction for amorphous silicon cells can be depicted as shown in Fig. 2.15b, right. Here, a key point is that the built-in electric field across the interlayer efficiently drives the carrier photogeneration occurring at the M-PTC/H_2Pc molecular contacts.

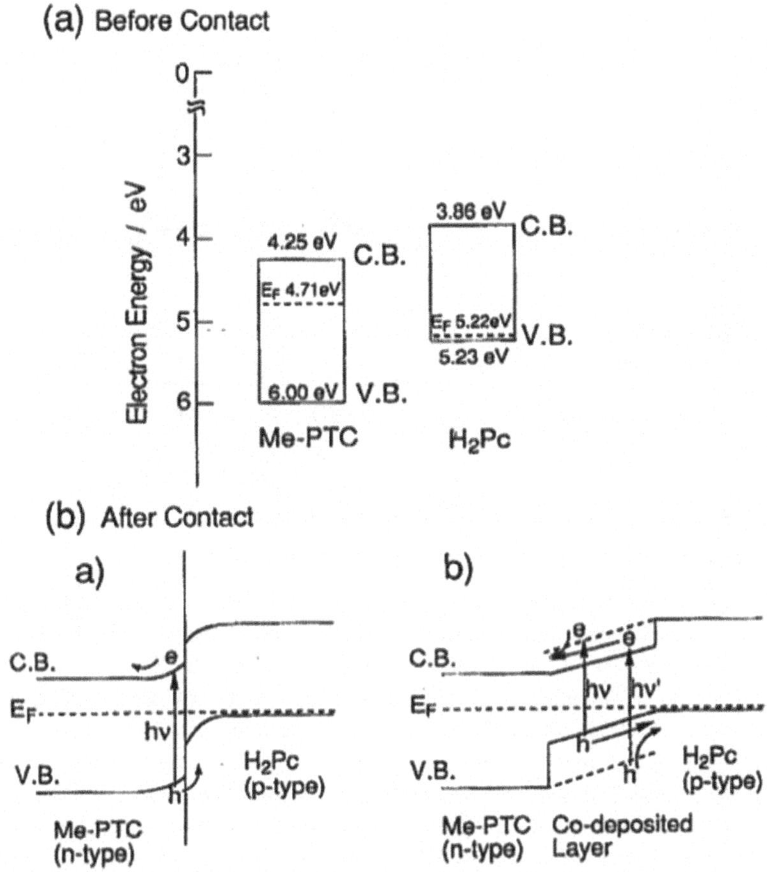

Fig. 2.15 **a** Energy band diagram of the H_2Pc and Me-PTC films prior to contact. CB, VB, and E_F denote the conduction band, valence band, and the Fermi level, respectively. E_F was evaluated by the Mott–Schottky plots for the Schottky junctions using Au ($\Phi_m = 5.1$ eV) for PTC and Al (4.3 eV) for the H_2Pc. **b** The energy structures of the two-layer cell (left) and the three-layer cell (right) after contact. Reproduced with permission from [11]. Copyright 1992 AIP Publishing

At the time of these experiments, the chemical species of the dopants were not identified and the precise control of the concentration of ionic charges in the co-deposited layer is difficult. The distribution of the built-in potential within the inter-layer may be easily distorted by the unstable balance between the ionic donors and acceptors that is determined by the pigment ratio as well as by the dopant concentrations in the respective pigments that are further affected during the vacuum evaporation. In fact, a distortion effect was observed by changing the pigment ratio. The precise control by the dopants must await the development of ppm-doping technology described in Chap. 9.

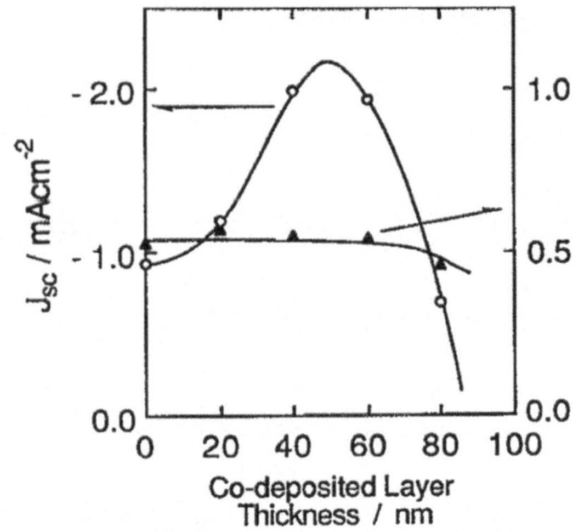

Fig. 2.16 Dependence of J_{sc} and V_{oc} of a three-layer cell on the thickness of the co-deposited layer with the Me-PTC:H_2Pc ratio of 1:1. The cell structure is the same as that shown in Fig. 2.10a except for the interlayer thickness. White light (100 mWcm^{-2}) was irradiated through the ITO electrode. A co-deposited layer thickness of zero corresponds to the two-layer cell. Reproduced with permission from [11]. Copyright 1992 AIP Publishing

2.3.5 Inorganic/Organic Hybrid p-i-n Junction

The *p-i-n* concept can be developed further. For example, the *n*-type organic layer can be replaced by transparent wide-bandgap inorganic semiconductors. Thus, the *n*-Me-PTC layer in the cells (Fig. 2.10a) was replaced by *n*-type ZnO and CdS films. The introduction of the H_2Pc:Me-PTC co-deposited interlayer gave rise to an enhancement of the photocurrent by a factor of approximately 20 compared to the ZnO/H_2Pc heterojunction cell (Fig. 2.17). Again, the co-deposited layer was confirmed to act as an efficient carrier generation layer, even if an inorganic semiconductor is used as the *n*-layer. The use of CdS for which the conduction band and Fermi-level energies are lower than those of ZnO [16] resulted in an increase in the V_{oc} from 0.47 to 0.61 V. This proved the suitability of the *p-i-n* junction concept for the development by using inorganic/organic hybrid cells, allowing the materials for each layer to be chosen independently. The *p-i-n* cell is a function separated device in which the co-deposited *i*-layer acts as the carrier generation layer, and the *n*- and *p*-layers act to provide the built-in potential.

The *p-i-n* junction model is based on the conventional band model of inorganic semiconductors. However, the organic semiconductors maintain their unique characteristics originating from their molecular-level structure and interactions, as mentioned in the subsequent sections.

2.3.6 Inevitably Conceived Issue: Percolation

The co-deposited layer has a vast number of photocarrier generation sites where the Pc and PTC molecules are in direct contact. However, if the two pigments are nested

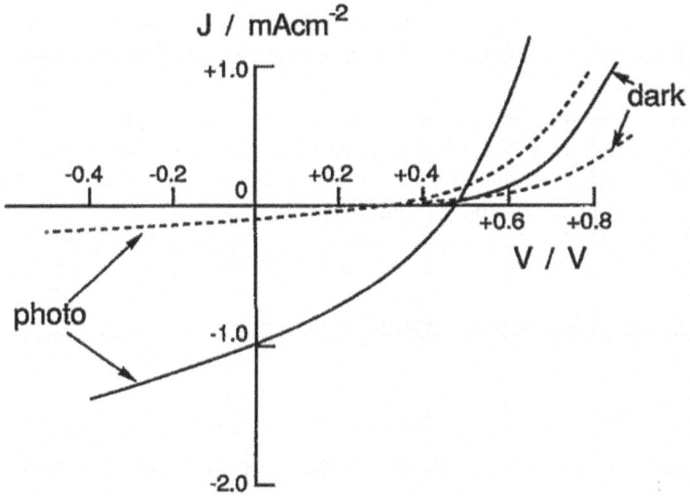

Fig. 2.17 J–V curves for a three-layer cell, ITO/ZnO (40 nm)/Me-PTC: H_2Pc co-deposited layer (40 nm)/H_2Pc (40 nm)/Au (solid curve), and a two-layer cell, ITO/ZnO (40 nm)/H_2Pc (80 nm)/Au (broken curve). The pigment ratio of Me-PTC and H_2Pc in co-deposited layer is 1:2. White light (100 $mWcm^{-2}$) was irradiated through the ITO electrode. Reproduced with permission from [11]. Copyright 1992 AIP Publishing

in the co-deposited interlayer, a recombination of electrons and holes was expected to occur easily. Therefore, we presumed that the blended layer shows an aggregation of small grains of both pigments (Fig. 2.18a). Once photogenerated, the electrons and holes diffuse into the different pigment grains; i.e., the electrons diffuse to the Me-PTC, and the holes diffuse to the H_2Pc. In addition, we presumed that the *p-n* junction barriers formed at the contact between the Me-PTC and H_2Pc possessing *n-* and *p*-type characters, respectively (Fig. 2.18b), may assist the effective transport of holes and electrons, respectively. At that time, we performed SEM observations for the Me-PTC:H_2Pc co-deposited film and observed the agglomeration of grains

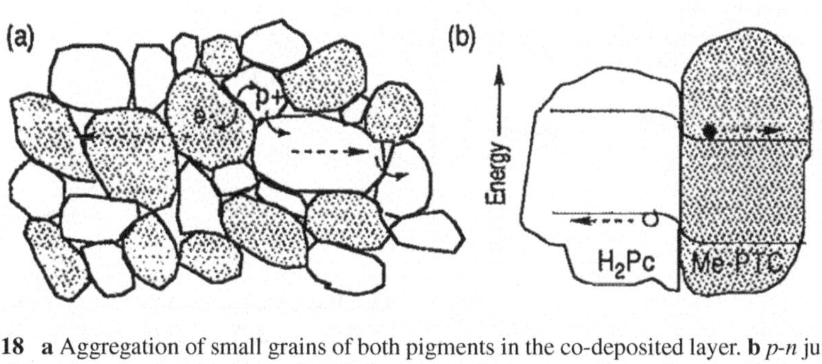

Fig. 2.18 **a** Aggregation of small grains of both pigments in the co-deposited layer. **b** *p-n* junction barrier formed at the grain contact of different types of pigments. Reproduced with permission from [11]. Copyright 1992 AIP Publishing

with a diameter of several tens of nm. However, the distribution of Me-PTC and H_2Pc could not be clarified by the elemental analysis based on the energy analysis of secondary electrons.

From the beginning, the concept of the co-deposited junction, i.e., the blended junction, inevitably raised the issue of percolation, i.e., nanoscale route formation, and the author was continuously forced to consider how to solve this issue. The possible solutions for this issue are summarized in Chap. 3.

2.3.7 *Carrier Generation Model*

2.3.7.1 Clues

The photocurrent enhancement by the blending of PTC and Pc means that there should be some type of interaction between these two organic semiconductors that accelerates the photocarrier generation. Several clues were available for elucidating the photocarrier generation process. The first was the internal quantum efficiency (Φ^0) data that showed maxima at the pigment ratio of approximately 1:1 irrespective of which pigment was excited (Fig. 2.19). This result strongly implies that a 1:1 molecular pair is the active site for charge separation. The second was the absorption spectra of the single H_2Pc and Me-PTC films, the H_2Pc/Me-PTC double-layered film,

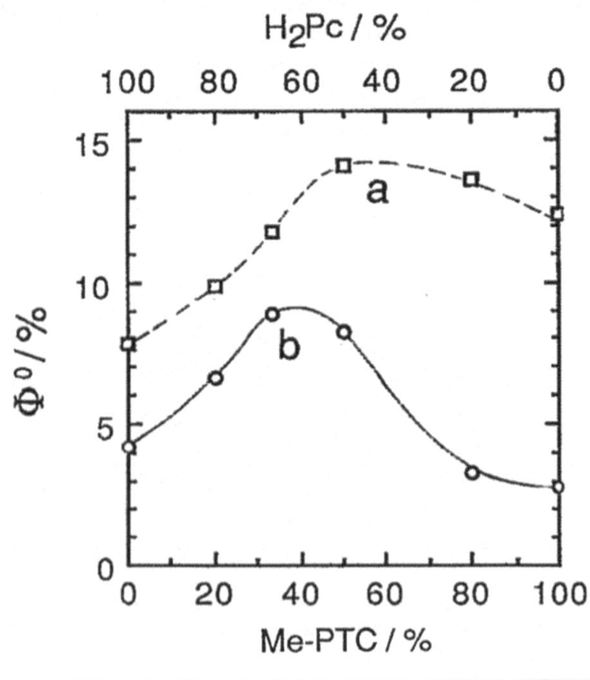

Fig. 2.19 Dependence of internal quantum efficiency (Φ^0) on the Me-PTC:H_2Pc ratio in the co-deposited layer of three-layer cell. Only H_2Pc was excited by a monochromatic light of 630 nm (curve a). Only Me-PTC was excited by a monochromatic light of 480 nm (curve b). (b) Reproduced with permission from [20]. Copyright 1992 AIP Publishing. Reproduced with permission from [21]. Copyright 1992 AIP Publishing

and H_2Pc:Me-PTC co-deposited film. No differences were observed between the calculated sum of the absorption spectra of the extremely thin (2.5 nm) single films of Me-PTC and H_2Pc, and the absorption spectrum of extreme thin two-layer film of Me-PTC (2.5 nm)/H_2Pc (2.5 nm). Moreover, the H_2Pc:Me-PTC co-deposited film showed no specific absorption that could be attributed to the charge–transfer interaction. This means that a strong interaction between H_2Pc and Me-PTC hardly occur under dark, i.e., in the ground state. Logically, we should consider the interaction between Me-PTC and H_2Pc in the excited states, and therefore, we discussed the exciplex model and the donor–acceptor model at that time [11].

2.3.7.2 Exciplex

The charge photogeneration via an exciplex of two pigments (Me-PTC$^-$-H_2Pc^+)* is suspected to be involved in the carrier photogeneration path (Fig. 2.20a). Such an exciplex appears to act as an efficient precursor for the generation of an ion pair that

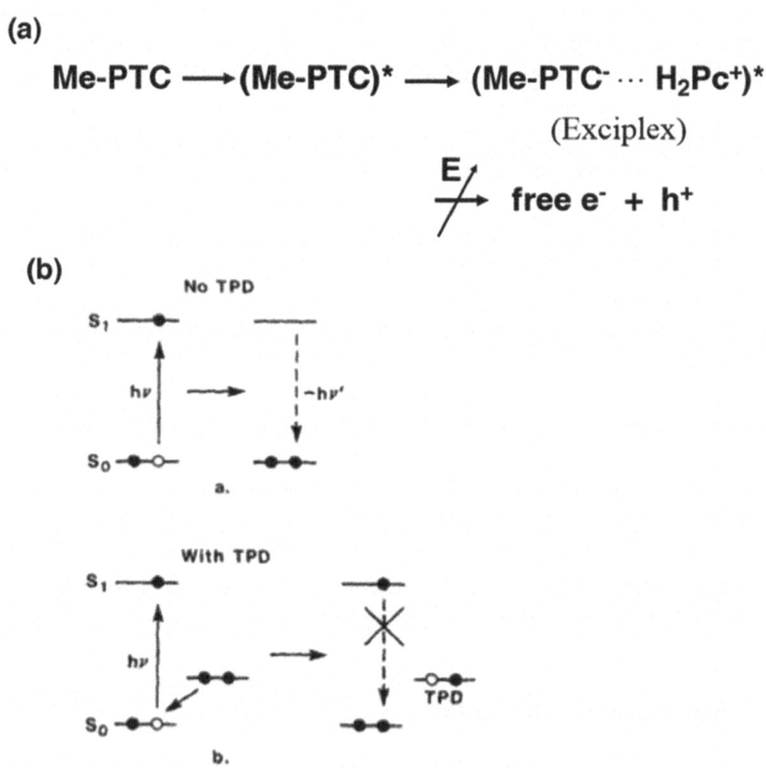

Fig. 2.20 **a** Charge photogeneration process through an exciplex of Me-PTC and H_2Pc. **b** Comparison of the first excited singlet state processes for the Im-PTC with (upper) and without TPD (lower). (b) Reproduced with permission from [20]. Copyright 1988 Elsevier

dissociates into free carriers, depending on the strength of the electric field [17, 18]. The author thought that the extremely high ability for the charge carrier generation through an exciplex is clearly manifested in Fig. 2.4 (curve b) since the photocurrent quantum efficiency increased drastically to 40% due to a small increase in the reverse bias of only −0.4 V.

2.3.7.3 Donor–Acceptor Pair

In the 1980s, few researchers in the field of electrophotography investigated the photocarrier generation mechanism in organic semiconductors since it was considered to be a very difficult and complicated issue. Popovic was the only researcher who energetically tackled this problem, for example, by using field-induced fluorescence quenching measurements [19]. He studied the carrier generation mechanism in electrophotographic device consisting of a photocarrier generation layer using Im-PTC and a hole transporting layer using tetraphenyldiamine (TPD) [20]. He found that the enhanced carrier generation efficiency in the Im-PTC/TPD system compared to the Im-PTC alone is due to a sensitization process. He then proposed a mechanism for this process as shown in Fig. 2.20b. His explanation was as follows. The generation mechanism involves exciton diffusion to the Im-PTC/TPD interface where the excitons dissociate by injecting an electron into the pigment and a hole into the TPD. In the absence of TPD (Fig. 2.20b, No TPD), the dominant mechanism for the decay of the first excited singlet state is radiative or radiationless relaxation to the ground state. A small fraction of excited states leads to photogenerated carriers. In the presence of TPD (Fig. 2.20b, With TPD), the excited state decay is dominated by a process in which charge exchange occurs between the excited Im-PTC molecule and a TPD at the interface. This leads to fluorescence quenching and generation of a geminate electron–hole pair that, aided by electric field, subsequently dissociates into free carriers.

This mechanism is the origin of the donor (D)–acceptor (A) sensitization (Chap. 1, Fig.1.7, and Chap. 9, Fig. 9.1b) which is a basic principle of the contemporary organic solar cells. TPD and Im-PTC can be regarded as donor (D) and acceptor (A), respectively. The photocurrent enhancement in the H_2Pc/Me-PTC and CuPc/Im-PTC systems can be attributed to the replacement of TPD with phthalocyanines (H_2Pc, CuPc). The offsets of the HOMO and LUMO levels formed between the donor and acceptor are essential for the exciton dissociation and thus the photocarrier generation.

2.3.7.4 Relationship to Recent Progress

Recent research on organic solar cells focused on the nature of the charge–transfer (CT) state between the donor (D) and acceptor (A). This CT state can be regarded as an exciplex, i.e., $(A^{-}–D^{+})^{*}$ shown in Fig. 2.20a. Energetically, the CT state can be illustrated as shown in Fig. 2.20b (with TPD). Recently, the CT states were revealed to dominate the magnitude of V_{oc}. These advances are described in detail in Chap. 8. Moreover, the carrier generation process via the CT states investigated by transient absorption spectroscopy is discussed in Chap. 6.

2.4 Conclusion

This chapter provides a detailed overview of the invention of the blended junction.

By recording the photovoltaic behaviors as a function of the fabrication of the interface, the molecular contact between the PTC and Pc molecules was revealed to be a key for the excellent photocarrier generation ability. This insight guided the author to develop the blended junctions (*p-i-n* junctions) with a vast number of PTC:Pc molecular contacts that doubled the photocarrier generation efficiency by overcoming the technical difficulties of co-deposition [11].

The three-layer organic solar cell was the first report of the blended junction [11, 21]. In those initial papers, the following predictions regarding the future organic solar cells already appeared: (1) *p-i-n* energetic structure and cells fabricated by doping, (2) percolation as an inevitable issue for the blended junction, (3) efficient carrier photogeneration via an exciplex $(PTC^{-}–PC^{+})^{*}$ and donor–acceptor sensitization that are related to the charge–transfer (CT) state.

Blended junctions have become indispensable for today's organic solar cells.

References

1. Matsumura, M., Hiramoto, M., Tsubomura, H.: Photoelectrolysis of water under visible light with doped $SrTiO_3$ electrodes. J. Electrochem. Soc. **130**, 326–330 (1983)
2. Nakato, Y., Yoshimura, M., Hiramoto, M., Tsumura, A., Murahashi, T., Tsubomura, H.: p-n junction silicon electrode coated with noble metal for efficient solar photoelectrolysis of hydrogen iodide. Bull. Chem. Soc. Jpn **57**, 355–360 (1984)
3. Nakato, Y., Hiramoto, M., Iwakabe, Y., Tsubomura, H.: ESCA and photoelectrochemical studies of p-n junction silicon electrodes protected by platinum deposition for use in solar energy conversion. J. Electrochem. Soc. **132**, 330–334 (1985)
4. Tsubomura, H., Nakato, Y., Hiramoto, M., Yano, H.: Metal oxide coated p-n junction silicon electrodes for photoelectrochemical solar energy conversion. Can. J. Chem. **63**, 1759–1762 (1985)
5. Hiramoto, M., Hashimoto, K., Sakata, T.: Visible-light induced water splitting on new semiconductor electrodes made by photolithography. Chem. Lett. **1986**, 899–902 (1986)
6. Matsumura, M., Hiramoto, M., Iehara, T., Tsubomura, H.: Photocatalytic and photoelectrochemical reactions of aqueous solutions of formic acid, formaldehyde, and methanol on platinized CdS powder and at a CdS electrode. J. Phys. Chem. **88**, 248–250 (1984)
7. Chamberlain, G.A.: Organic solar cells: A review **8**, 47–83 (1983). and references therein
8. Wohrle, D., Meissner, D.: Organic solar cells. Adv. Mater. **3**, 129–138 (1991). and references therein
9. Tang, C.W.: Two-layer organic photovoltaic cell. Appl. Phys. Lett. **48**, 183–185 (1986)
10. Tang, C.W., VanSlyke, S.A.: Organic electroluminescent diodes. Appl. Phys. Lett. **51**, 913–915 (1987)

11. Hiramoto, M., Fujiwara, H., Yokoyama, M.: *p-i-n* like behavior in three-layered organic solar cells having c co-deposited interlayer of pigments. J. Appl. Phys. **72**, 3781–3787 (1992)
12. Takahashi, K., Konagai, M.: Amorphous solar cells (book). Shoko-do. Chap. 5. p 127 (1983) (in Japanese)
13. Martin, M., Andre, J.-J., Simon, J.: Influence of dioxygen on the junction properties of metallophthalocyanine based devices. J. Appl. Phys. **54**, 2792–2794 (1983)
14. Tada, H., Touda, H., Takada, M., Matsushige, K.: Quasi-intrinsic semiconducting state of titanyl-phthalocyanine films obtained under ultrahigh vacuum conditions. Appl. Phys. Lett. **76**, 873–875 (2000)
15. Hiramoto, M., Kishigami, Y., Yokoyama, M.: Doping effect on the two-layer organic solar cell. Chem. Lett. **1990**, 119–122 (1990)
16. Vanden Berghe, R.A.L., Gomes, W.P.: Comparative study of electron injection into ZnO, CdS and CdSe single-crystal anodes. Ber. Bunsenges. Phys. Chem. **76**, 481–486 (1972)
17. Yokoyama, M., Endo, Y., Matsubara, A., Mikawa, H.: Mechanism of extrinsic carrier photogeneration in poly-N-vinylcarbazole. II. Quenching of exciplex fluorescence by electric field. J. Chem. Phys., 75, 3006–3011 (1981)
18. Yokoyama, M., Shimokihara, S., Matsubara, A., Mikawa, H.: Extrinsic carrier photogeneration in poly-N-vinylcarbazole. III. CT fluorescence quenching by an electric field. J. Chem. Phys., 76, 724–728 (1982)
19. Popovic, Z.D., Loutfy, R.O., Hor, A.-M.: Photoconductivity studies of perylene tetracarboxyldiimides. Can. J. Chem. **63**, 134–139 (1985)
20. Popovic, Z.D., Hor, A.-M., Loutfy, R.O.: A study of a carrier generation mechanism in benzimidazole perylene/tetraphenyldiamine thin film structures. Chem. Phys. **127**, 451–457 (1988)
21. Hiramoto, M., Fujiwara, H., Yokoyama, M.: Three-layered organic solar cell with a photoactive interlayer of codeposited pigments. Appl. Phys. Lett. **58**, 1062–1064 (1992)

Chapter 3
Percolation Toward Lateral Junctions

Masahiro Hiramoto

3.1 Motivation

Blended junctions are commonly used for modern organic solar cells [1–8]. Conventional organic solar cells are vertically built on indium tin oxide (ITO) glass substrates using a sandwich-type structure [9–11]. However, the nanostructure fabrication of the electron and hole transport routes in the blended donor (D) and acceptor (A) molecules developed by percolation is still a problem in spite of recent reports on blended systems where the thickness reached nearly a micron [6, 12–14]. To determine fundamental resolutions, we have demonstrated various new principles of donor/acceptor (D/A) junctions acting as alternative blended junctions.

This Chapter summarizes the junctions based on percolation (Sect. 3.2), vertical multilayers (Sect. 3.3), and lateral multilayers (Sect. 3.4).

3.2 Percolation

This section summarizes percolated junctions with crystalline–amorphous nanocomposite structures, which are formed by controlling the substrate temperature during the co-deposition of fullerene (C_{60}) and metal-free phthalocyanine (H_2Pc) [15, 16].

M. Hiramoto (✉)
National Institutes of Natural Sciences, Institute for Molecular Science, 5-1 Higashiyama
Myodaiji, Okazaki 444-8787 Aichi, Japan
e-mail: hiramoto@ims.ac.jp

M. Hiramoto and S. Izawa (eds.), *Organic Solar Cells*,
https://doi.org/10.1007/978-981-15-9113-6_3

3.2.1 Substrate Temperature Control

The effects of the substrate temperature on the photovoltaic properties of single-layered cells of C_{60}:H_2Pc co-deposited films sandwiched between ITO and Ag (Fig. 3.1a) were investigated. Figure 3.1b shows the current density–voltage (J-V) characteristics. Co-deposition was performed on the substrate under room temperature conditions of +20 °C (curve A) and a mildly heated condition of +80 °C (curve B). For +20 °C, the short-circuit photocurrent density (J_{sc}) was only 0.6 mA·cm^{-2} (curve A). For +80 °C, J_{sc} had increased 6.5 times and reached 3.9 mA·cm^{-2} (curve B). Figure 3.2 summarizes the dependence of photovoltaic parameters on the substrate temperature. The optimum temperature for J_{sc} (Fig. 3.2a), open-circuit photovoltage (V_{oc}) (Fig. 3.2b), fill factor (FF) (Fig. 3.2c), and photoelectric conversion efficiency (Fig. 3.2d) was concurrently found to be approximately +80 °C. Temperatures lower and higher than the optimum temperature caused a reduction in all parameters.

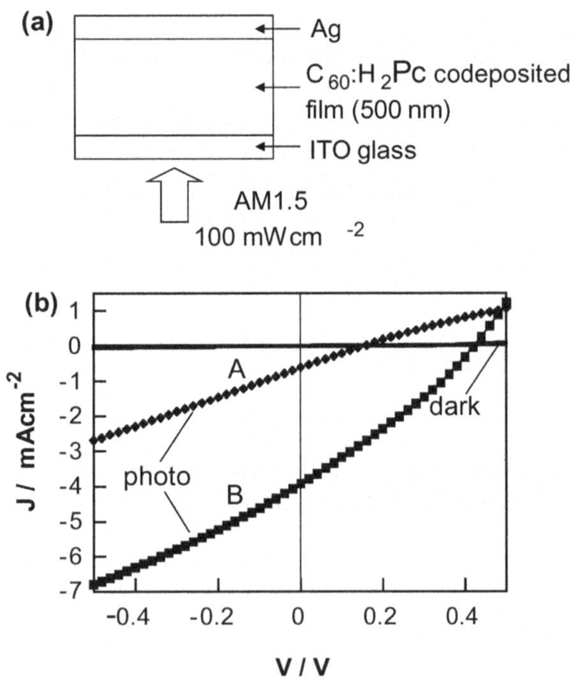

Fig. 3.1 **a** Structure of sandwich-type cell. **b** Current density-voltage (J-V) characteristics for C_{60}:H_2Pc (1:1) cells. Co-deposition was performed at +20 °C (curve A) and +80 °C (curve B). Simulated solar light (AM1.5, 100 mW·cm^{-2}) was irradiated to the ITO electrode. The direction of photovoltage is (+)ITO/C_{60}:H_2Pc/Ag(−)

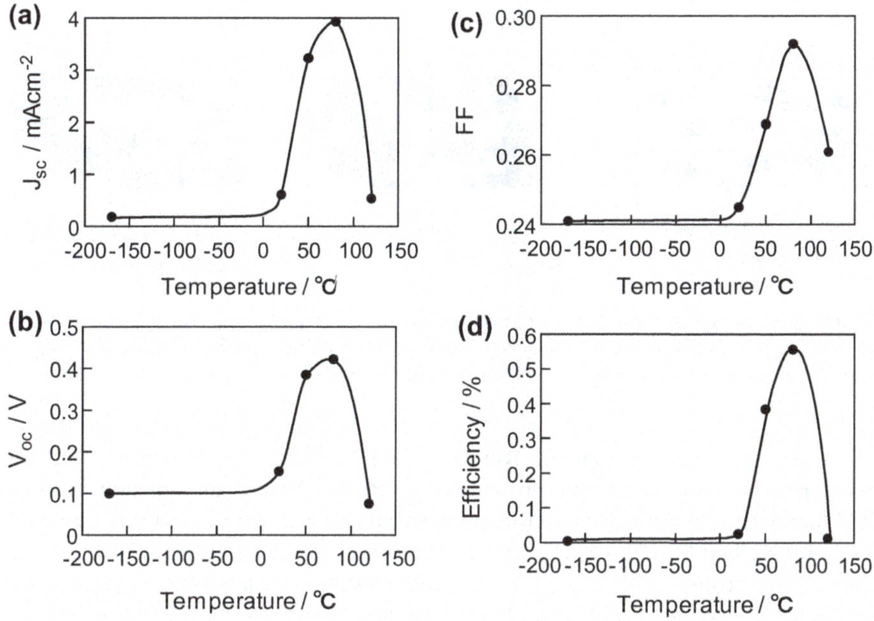

Fig. 3.2 Dependence of photovoltaic parameters on substrate temperature during co-deposition. **a** Short-circuit photocurrent density (J_{SC}). **b** Open-circuit photovoltage (V_{OC}). **c** Fill factor (FF). **d** Photoelectric conversion efficiency

3.2.2 Crystalline–Amorphous Nanocomposite

Figure 3.3 shows SEM images of a cross section of C_{60}:H_2Pc films co-deposited on the substrate at +20 °C (a), +80 °C (b), and +120 °C (c). Clear morphological differences were observed. At +20 °C, the entire cross section was smooth, and there was no recognizable structure. At +80 °C, many nanocrystals with a diameter of approximately 50 nm were surrounded by smooth sections. At +120 °C, the smooth sections disappeared, and the entire cross section consisted of nanocrystals. X-ray diffraction (XRD) measurements revealed that the nanocrystals at +80 °C

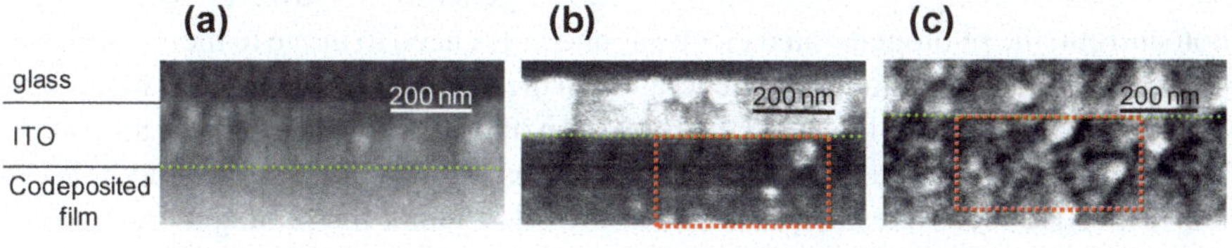

Fig. 3.3 SEM images of cross section of C_{60}:H_2Pc films co-deposited on the substrate at **a** +20 °C, **b** +80 °C, and **c** +120 °C. The cross sections of the co-deposited films are located below the green dotted lines

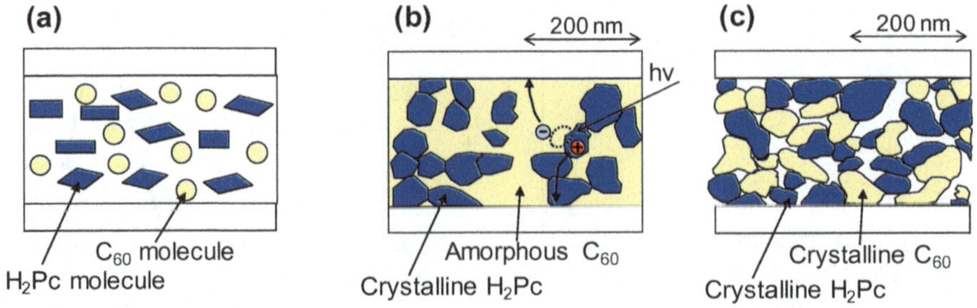

Fig. 3.4 Morphological illustrations of cross sections of C_{60}:H_2Pc co-deposited films. **a** Molecular mixture. **b** Crystalline–amorphous nanocomposite. **c** Crystalline–crystalline composite. In this Figure, b and c are drawn by tracing the images within the red dotted boxes of Fig. 3.3b and c

were H_2Pc, indicated by a clear XRD peak of H_2Pc ($2\theta = 7°$). Below +20 °C, the co-deposited films were amorphous with no observable XRD peak. Together with the SEM images, the morphological illustrations of the cross section of the co-deposited films sandwiched between ITO and Ag electrodes are depicted in Fig. 3.4. At low temperatures from −170 to +20 °C, C_{60} and H_2Pc were blended molecularly and formed an amorphous film [(a) molecular mixture]. At +80 °C, H_2Pc nanocrystals were surrounded by amorphous C_{60} [(b) crystalline–amorphous nanocomposite], and crystalline H_2Pc and crystalline C_{60} were blended [(c) crystalline–crystalline composite] at the high temperature of +120 °C. The substrate temperature of +80 °C, which induced the formation of the crystalline H_2Pc-amorphous C_{60} nanocomposite, coincides well with that of the maxima of the photovoltaic parameters (Fig. 3.2). Thus, a clear relationship between the nanostructure of the co-deposited films and their photovoltaic performance was confirmed. We investigated two other combinations of phthalocyanines and perylene pigments and revealed that the crystalline–amorphous nanocomposite films always generated the largest photocurrent irrespective of their formation temperature [8]. Therefore, we concluded that the formation of crystalline–amorphous nanocomposites is essential for photocurrent generation.

Because the direct heteromolecular contacts between C_{60} and H_2Pc offer photocarrier generation sites due to the dissociation of charge transfer (CT) excitons formed by the electron transfer from the donor (H_2Pc) molecules to the acceptor (C_{60}) molecules, co-deposited films containing a vast number of C_{60}/H_2Pc molecular contacts possess a high potential for photocarrier generation. However, to generate a photocurrent, the photogenerated electrons and holes need to move to their respective electrodes. Efficient photocurrent generation in the crystalline–amorphous nanocomposites was concluded to be due to the formation of routes for the electrons and holes. These routes allowed the spatially separated transport of the electrons and holes to their respective electrodes (Fig. 3.4b). On the other hand, the photogenerated electrons and holes of the molecular blend (Fig. 3.4a) easily encountered each other and recombined due to the absence of transport routes. For the crystalline–crystalline composites (Fig. 3.4c), the increase in the number of distinct grain boundaries formed between the nanocrystals seemed to obstruct carrier transport significantly.

3.2.3 p-i-n Cells

A C_{60}:H_2Pc layer deposited at +80 °C, which has a crystalline–amorphous nanostructure (Fig. 3.4b), was incorporated into three-layered cells (Fig. 3.5a). The percolated layer was sandwiched between *n*-type naphthalenetetracarboxylic anhydride (NTCDA) and *p*-type H_2Pc. A considerably thick NTCDA layer (600 nm), which was transparent throughout the visible region, was used to avoid short-circuiting the cells due to the migration of the vacuum-deposited Ag electrode [12, 17]. Figure 3.5b shows the current–voltage (J-V) characteristics. The best photovoltaic performance (J_{sc}: 9.9 mA·cm^{-2}, V_{oc}: 0.42 V, *FF*: 0.52, and photoelectric conversion efficiency: 2.5%) was observed when the thickness and the C_{60}:H_2Pc ratio of the co-deposited layer were optimized to 130 nm and 5:6, respectively. Figure 3.5c shows the spectral dependence of the internal quantum efficiency of J_{sc}(closed dots) and the absorption

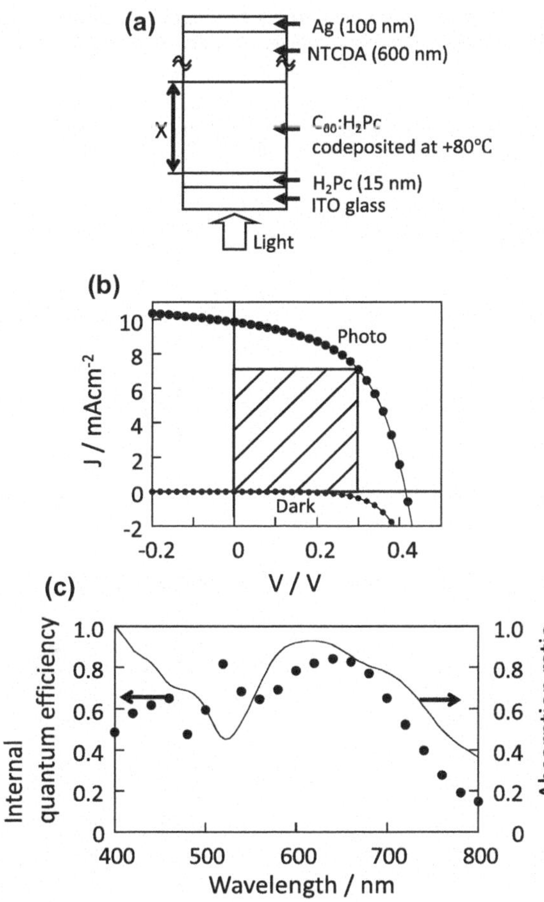

Fig. 3.5 **a** The structure of three-layered cell. **b** Current-voltage (J-V) characteristics of a three-layered cell incorporating a 130-nm-thick C_{60}:H_2Pc (5:6) interlayer co-deposited at +80 °C. The shaded rectangle indicates the maximum power output. Cell parameters; J_{sc}: 9.9 mA·cm^{-2}, V_{oc}: 0.42 V, FF: 0.52, and efficiency: 2.5%. **c** Spectral dependence of internal quantum efficiency of J_{sc} (closed dots) and absorption ratio of incident light by organic film (solid curve)

of the incident light absorbed by the organic film (solid curve). The quantum efficiency is calculated as the ratio of the number of carriers collected under short-circuit conditions to the number of photons absorbed by the organic layers. The maximum and averaged internal quantum efficiencies (400–800 nm) were 84% and 59.3%, respectively.

Figure 3.6 shows the dependence of J_{sc}, V_{oc}, and FF of the three-layered cells on the thickness (x) of the C_{60}:H_2Pc. J_{sc} increased with increasing x up to 130 nm and then started to decrease. V_{oc} showed a constant value of around 0.4 V, and FF decreased monotonically with increasing x. These characteristics can be reasonably explained based on the *p-i-n* energetic structure (Fig. 3.7). The V_{oc} value of 0.4 V corresponds well with the built-in potential (0.4 V) estimated by the Kelvin vibrating capacitor method [15, 16]. Even without the C_{60}:H_2Pc layer ($x = 0$ nm), i.e., Ag/NTCDA/H_2Pc/ITO cell, a similar value of V_{oc} was observed. Additionally, V_{oc} is not dependent on the type of electrode metal. Thus, we concluded that a built-in electric field, created by the difference in the Fermi levels (E_F) of NTCDA and H_2Pc, was distributed across the C_{60}:H_2Pc interlayer and drove the efficient charge carrier generation and charge transport in the co-deposited layer.

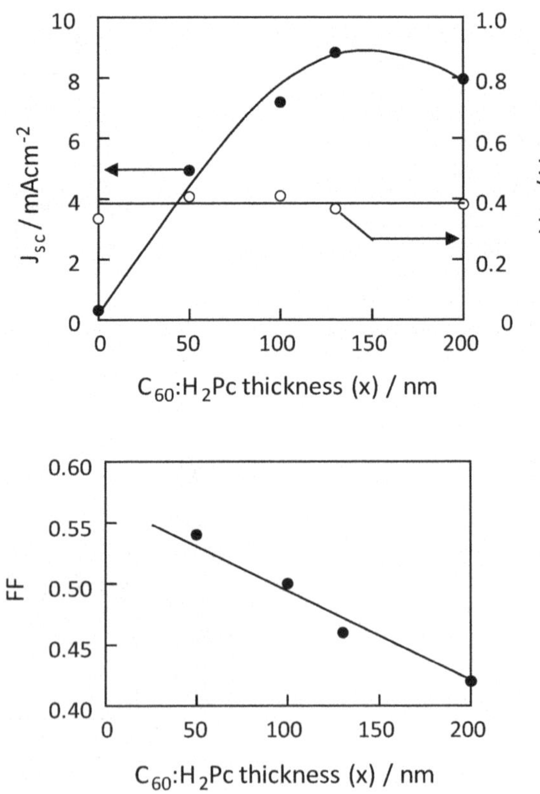

Fig. 3.6 Dependence of J_{sc}, V_{oc}, and FF of the three-layered cells on C_{60}:H_2Pc thickness (x). C_{60}:H_2Pc ratio was 1:1

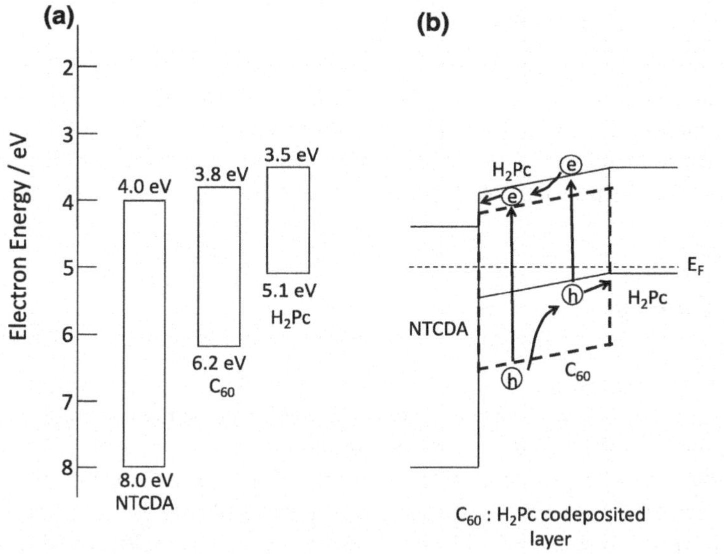

Fig. 3.7 The p-i-n energetic structure of the three-layered cell. **a** Before contact. **b** After contact. The energetic relationships are precisely illustrated with respect to the scale bar. The C_{60} in the co-deposited layer is depicted using a broken line. Numerical values are energetic positions of levels for respective H_2Pc, C_{60}, and NTCDA films before forming junctions, which were estimated from the ionization potentials measured by atmospheric photoelectron emission analysis and from the optical band gaps

The increase in J_{sc} between $x = 0$ and 30 nm was due to the increase in the absorption ratio of solar light induced by the C_{60}:H_2Pc interlayer. The decrease in J_{sc} above $x = 130$ nm was due to an increase in the resistance of the co-deposited layer with increasing x, which also caused a monotonic decrease in FF. Those observations support the *p-i-n* energetic structure. Additionally, a negative shift in E_F and simultaneous increase in the V_{oc} of the three-layered cells by doping with Na acting as donor in the NTCDA film and a positive shift in E_F and simultaneous increase in the V_{oc} by increasing the doping level of oxygen acting as acceptor in H_2Pc were observed. These observations suggest that the doping technique can increase the built-in field in the *p-i-n* cells.

Recent progress on the doping effects on organic solar cells is summarized in Chaps. 8 and 9. A more sophisticated percolation procedure using co-evaporant molecules acting as a solvent for vacuum co-deposition is summarized in Chap. 4.

3.3 Vertical Junctions

An ideal nanostructure is the "vertical superlattice," i.e., the vertical alternating multilayered junction (Fig. 3.8a) [18]. This structure enables the efficient dissociation of photogenerated excitons at the D/A interfaces within the exciton diffusion length (5–10 nm) and the transport of electrons and holes to the respective electrodes.

3.3.1 Concept

In the vacuum deposition of an alternating multilayered structure of two kinds of materials, each film layer must be deposited in sequence on the substrate (Fig. 3.8b). The thicknesses of the layers can be controlled only in the film thickness direction with angstrom-order precision by monitoring with a quartz crystal microbalance (QCM). Here, if we consider rotating the multilayered film by 90° so that it is vertical with respect to the substrate with the cross section of the multilayered film as the exposed surface (Fig. 3.8c), then nanometer-scale control of the structures parallel to the surface is possible by controlling the deposition film thickness.

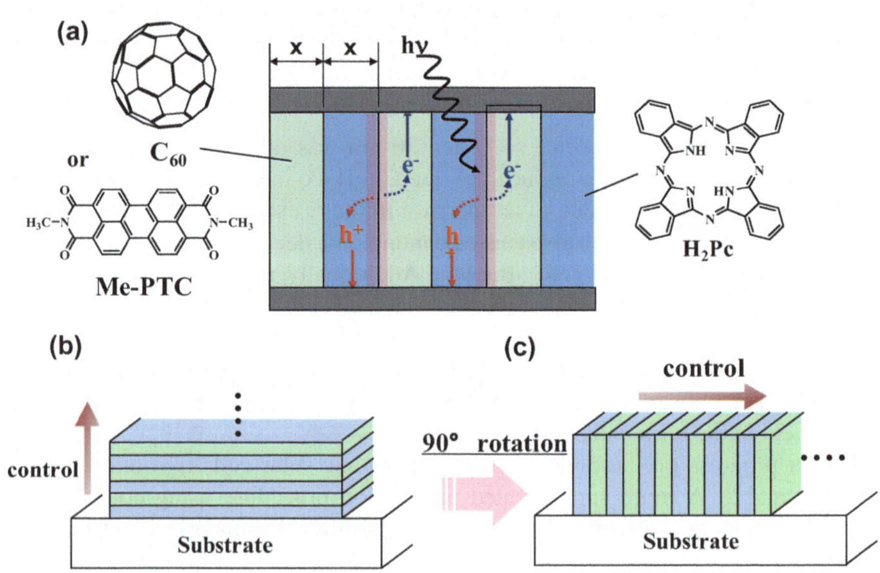

Fig. 3.8 **a** Vertical superlattice structure. **b** Multilayered film and **c** vertical multilayered film. Reprinted with permission from M. Hiramoto, *Electron. Comm. Jpn.* Part 2, **89**, 13–18 (2006). Copyright© 2006 Wiley Periodicals, Inc

3.3.2 Cell Fabrication

Regarding photoelectric conversion, when a vertical superlattice structure (Fig. 3.8a) consisting of alternating perylene pigment (C_{60}, Fig. 3.8a) layers acting as acceptor molecules and metal-free phthalocyanine (H_2Pc, Fig. 3.8a) layers acting as donor molecules is exposed to light, efficient charge separation occurs at the interface of the D/A organic semiconductors. Further, the photogenerated electrons and holes are transported such that they are spatially separated.

The fabrication method is shown in Fig. 3.9. A substrate with a flat surface was produced using an epoxy resin. C_{60} and H_2Pc were deposited alternately on the substrate by vacuum evaporation, and the multilayered film was embedded using the same epoxy resin as the substrate. By using a microtome (Leica, ULTRACUT UCT) equipped with a diamond knife (Diatome, Ltd.), the embedded film could be sliced perpendicular to the film surface. An epoxy film containing an embedded vertical superlattice with a height of 2 μm in the center was obtained. Finally, silver electrodes were deposited on both sides of this epoxy film. The completed cell had a structure consisting of a vertical multilayered film with height of 2 μm sandwiched by the Ag electrodes (Fig. 3.8a).

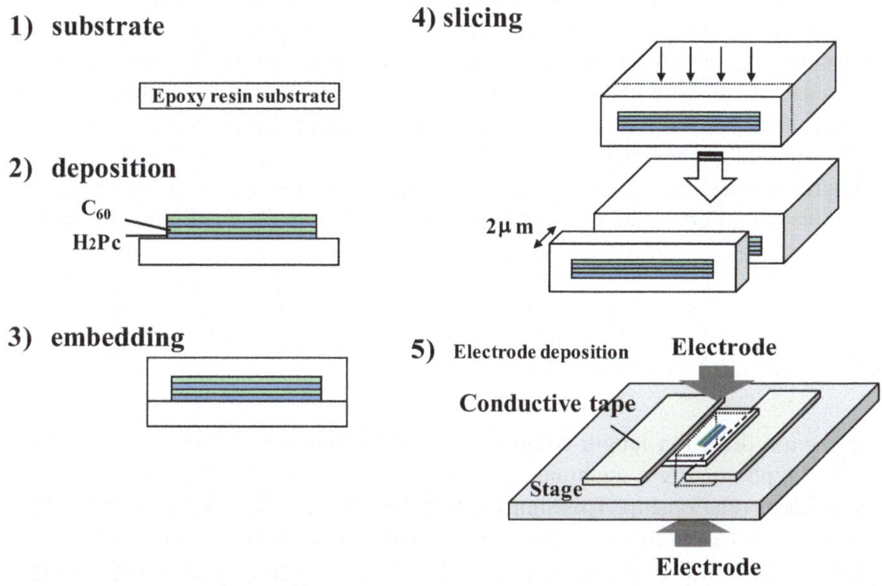

Fig. 3.9 Fabrication process. **1** Substrate. **2** Deposition. **3** Embedding. **4** Slicing. **5** Electrode deposition. Reprinted with permission from M. Hiramoto, *Electron. Comm. Jpn.* Part 2, **89**, 13–18 (2006). Copyright© 2006 Wiley Periodicals, Inc

3.3.3 Layer Width

When only the C_{60} or the H_2Pc were selectively excited using monochromatic lights of 500 and 600 nm, respectively, under a relatively small applied electric field, the photocurrent generated in the vertical superlattice was reproducibly observed. Figure 3.10a shows the dependence of the photocurrent generation quantum efficiency on the layer width (x; Fig. 3.8a) of the vertical superlattice. The layer width (x) was varied down to 2.5 nm. A decrease in x means an increase in the number of interfaces between C_{60} and H_2Pc. At $x = 2.5$ nm, the number of layers would be 4 million layers per 1 cm. The quantum efficiency rapidly increased as x decreased, and the maximum, 56% for C_{60}, was observed at 5–10 nm. However, it was also found that the efficiency decreased when the layer width was decreased to 2.5 nm. Thus, we could experimentally determine the optimal width (x) to be 5–10 nm, at which the performance of photoelectric conversion in the C_{60}/H_2Pc system was maximized.

The increase in quantum efficiency by decreasing the layer width (x) is related to the exciton diffusion length. A photocurrent is generated when the photogenerated excitons dissociate into free electrons and holes when they reach the C_{60}/H_2Pc interface (Fig. 3.10c). The excitons that could not reach the boundary surface are deactivated. The range where the excitons can reach the interface is the active area in which the photocurrent is generated (shaded area), and the remainder where the excitons are deactivated is the dead area (non-shaded area). The increase in the photocurrent quantum efficiency occurs due to an increase in the active area and a decrease in the dead area (Fig. 3.10b). In other words, when x is thick, i.e., the number of interfaces is small, the majority of the region will be the dead area, and only a small photocurrent is generated. When x becomes thin, meaning the number of interfaces increases, the photocurrent gradually increases due to the increase in the active area. Further, at $x = 5$–10 nm, the dead area in C_{60} is eliminated, and the maximum photocurrent is generated (Fig. 3.10b).

3.3.4 Exciton Diffusion Length

The exciton diffusion length could be obtained quantitatively. A semilogarithmic plot of the photocurrent quantum efficiency as a function of x (Fig. 3.10d) showed a clear linear relationship. This signifies that the concentration of the excitons could contribute to the photocurrent decreased exponentially by increasing the distance from the interface. The distance at which 90% of the excitons could reach the interface for C_{60} was precisely determined to be 1.7 nm, and similarly, that for H_2Pc was determined to be 4.6 nm by the selective excitation of H_2Pc. As given in Table 3.1, by adopting the C_{60}/H_2Pc system, the exciton diffusion lengths for C_{60} and H_2Pc were determined to be 1.7 and 4.6 nm, respectively. Furthermore, by adopting the Me-PTC/H_2Pc system, the exciton diffusion lengths for Me-PTC and H_2Pc were determined to be 4.4 and 5.0 nm, respectively. Similar values of the exciton diffusion

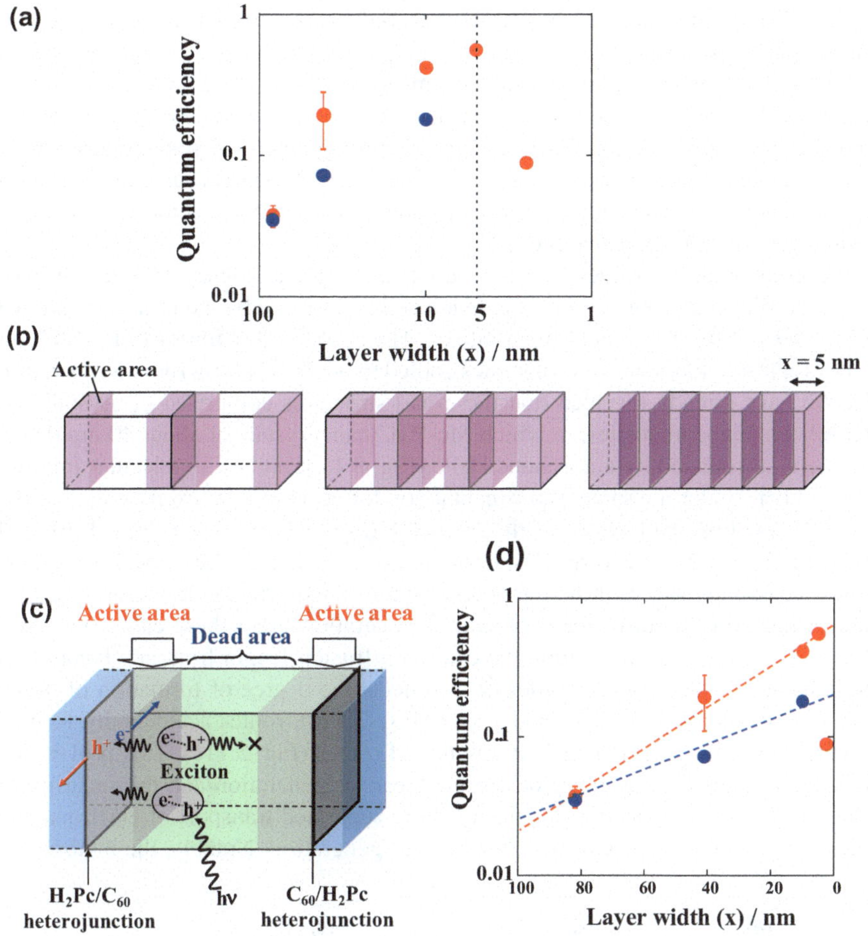

Fig. 3.10 **a** Dependence of photocurrent quantum efficiency on layer width (x). Only the C_{60} (red dots) and the H_2Pc (blue dots) were selectively excited using monochromatic lights of 500 and 600 nm, respectively, in an electric field of 2×10^4 V·cm^{-1}. **b** Schematic illustration of the active area (shaded) in the vicinity of the C_{60}/H_2Pc junction. The dead area (not shaded) in each layer decreases as the number of heterojunctions increases and is eliminated at x = 10 nm. **c** Schematic illustration of active area (shaded) and dead area (not shaded) formed between H_2Pc/C_{60} interfaces. **d** Semi-logarithm plot of the data in this figure a. Reprinted with permission from M. Hiramoto, *Electron. Comm. Jpn.* Part 2, **89**, 13–18 (2006). Copyright© 2006 Wiley Periodicals, Inc

Table 3.1 Distance for which 90% excitons can reach a D/A heterojunction

Vertical superlattice	C_{60}	H_2Pc	Me-PTC
C_{60}/H_2Pc	1.7 nm	4.6 nm	–
Me-PTC/H_2Pc	–	5.0 nm	4.4 nm

length of H_2Pc (4.6 and 5.0 nm) were observed irrespective of the C_{60}/H_2Pc and Me-PTC/H_2Pc systems.

In Fig. 3.10a, when the layer width became as narrow as $x = 2.5$ nm, the quantum efficiency decreased. When the film was less than 2.5 nm thick, each layer had not completely formed as a single sheet, and the probability of growth remaining as island regions became higher. Such spatial incompleteness of the extraction route decreased the efficiency of transporting the electrons and holes without recombination over a distance of 2 μm to each electrode.

We compared the values of the internal quantum efficiency (Φ) for the co-deposited films incorporated into sandwich cells fabricated by the procedure shown in Fig. 3.9 and for the vertical superlattices. The co-deposited films of Me-PTC and H_2Pc (1:1) were fabricated on substrates cooled to −170 °C and at room temperature [8]. We have already revealed that these had structures comprised of crystalline–amorphous nanocomposites, in which Me-PTC nanocrystals of about 20 nm diameter were surrounded by amorphous H_2Pc (Fig. 3.11b) and of a molecular mixture (Fig. 3.11c). Under the same experimental conditions (Me-PTC excitation, 2×10^4 V·cm^{-1}), the observed values of Φ were 41% (Fig. 3.11a), 17% (Fig. 3.11b), and 4% (Fig. 3.11c) for the vertical superlattice of $x = 5$ nm, the crystalline–amorphous nanocomposite, and the molecular mixture, respectively. Because the entire bulk of the films is active for photocarrier generation in all these cases (i.e., Me-PTC/H_2Pc interfaces exist within the exciton diffusion length from any location in the bulk films), the observed order of Φ reflects the degree of formation of transport routes: intentionally designed routes (Fig. 3.11a), routes accidentally formed by percolation (Fig. 3.11b), and an absence of routes (Fig. 3.11c). This proves that both the absence of a dead region for photocarrier generation and the securing of routes with no disconnection for the spatially separated transport of electrons and holes are crucially important for photocurrent generation. Notably, the nanoscopic

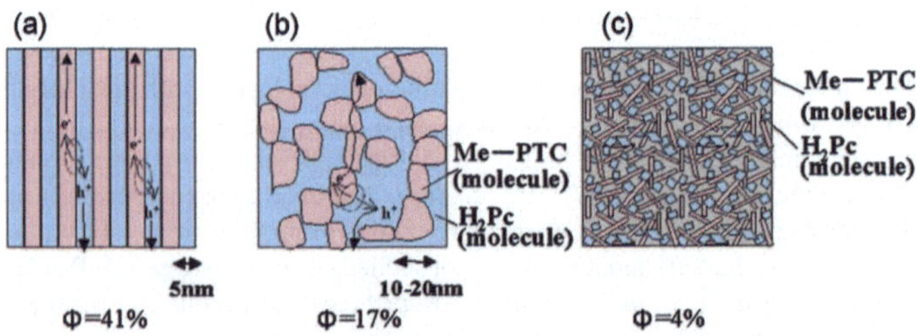

Fig. 3.11 Schematic illustration of nanoscopic structures. **a** Me-PTC/H_2Pc vertical superlattice of x = 5 nm. **b** Amorphous-crystalline nanocomposite formed by the co-deposition of Me-PTC and H_2Pc on a substrate cooled at −170 °C. **c** Molecular mixture formed by the co-deposition of Me-PTC and H_2Pc on a substrate at room temperature. All of these structures were incorporated in sandwich cells fabricated using the procedure shown in Fig. 3.9. The observed values of the internal quantum efficiencies (Φ) under the same conditions [Me-PTC excitation (480 nm); electric field: 2×10^4 V·cm^{-1}] are also shown

routes formed by percolation, which depends on accidental processes, are significantly inferior to the intentionally designed nanostructures with respect to the value of Φ [compare Fig. 3.11a and b]. This is a far from ideal structure.

The results of this study show that eliminating any dead areas, which do not generate photocurrents, as well as ensuring the formation of a continuous route to transport the spatially separated electrons and holes is critically important to obtaining high quantum efficiency in photoelectric conversion using organic semiconductors.

The areas of the present vertical junctions utilizing the cross section of multilayers are very small. From the standpoint of the practical fabrication of large area cells, the required aspect ratio reaches 100 because the required distance between adjacent D/A interfaces is less than 10 nm owing to the short exciton diffusion length, and the film thickness required to absorb all incident solar light is 1,000 nm. Therefore, the designated fabrication of vertical alternating multilayered structures when using current lithography techniques remains very difficult.

3.4 Lateral Junctions

If carrier transport over a long distance parallel to the substrate is possible, the formation of a vertical route is unnecessary. Based on this idea, we proposed and developed lateral junctions. First, we investigated the possibility of lateral electron transport in organic single crystals (Sect. 3.4.1) and then applied it to the lateral junctions (Sect. 3.4.2).

3.4.1 Lateral Electron Transport in Organic Single Crystals

In this section, the operation of single-crystal organic solar cells using lateral electron transport is demonstrated [19]. The spacing between the collection electrodes, which is determined by the lateral range of the electrons, was estimated to be 30 μm, and the possibility of millimeter-order electron ranges was indicated.

3.4.1.1 Concept

The concept of the lateral type cell is shown in Fig. 3.12b (right). Electrons move laterally through the single-crystal substrate. A pair of electrodes separated by a distance L is deposited in parallel on the crystal surface. A layer of the organic semiconductor acting as a donor is inserted between the organic single-crystal substrate, which acts as an acceptor and the electrode for collecting holes. At the D/A interface, the excitons dissociate into electrons and holes under light irradiation. Because the photogenerated holes move in the vertical direction, the distance to the electrode is

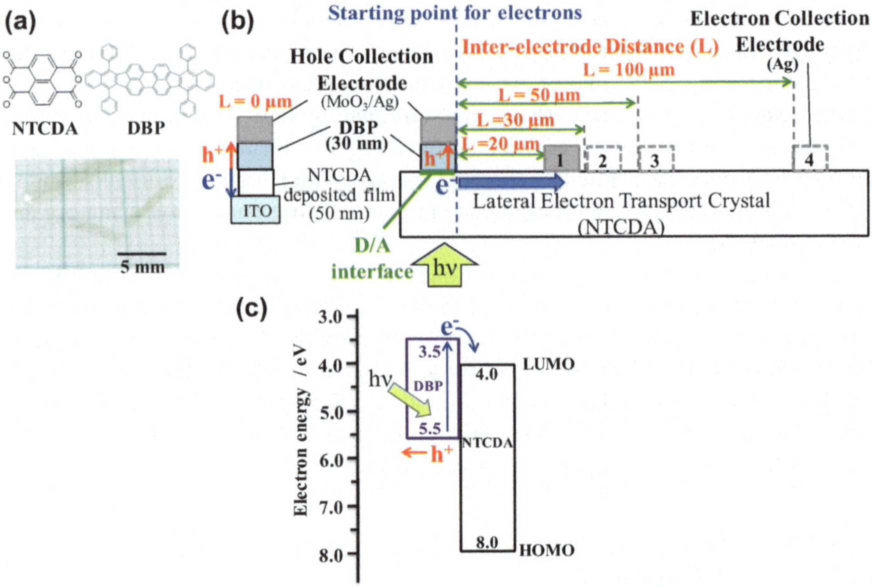

Fig. 3.12 **a** Chemical structures of NTCDA and DBP. A photograph of NTCDA single crystal is also shown. **b** Configurations of the lateral cell (right) and the vertical cell (left). Thickness of NTCDA single crystal in the lateral cell and that of vacuum deposited NTCDA film in the vertical cell are 50 μm and 50 nm, respectively. **c** Energy diagrams for DBP and NTCDA. Reproduced with permission from [23]. Copyright 2016, Elsevier B.V

equal to the film thickness of 30 nm. On the other hand, because the photogenerated electrons move laterally, the inter-electrode distance (L) is at least 30 μm. Thus, the lateral distance is 1,000 times the vertical distance, which means that the cell characteristics are dominated by electron transport. The right-hand edge of the hole collection electrode can be regarded as the starting point for the electrons (broken blue line). The range of the electrons (L_e) is expressed by Eq. (3.1).

$$L_e = \mu_e \tau_e E \tag{3.1}$$

Here, μ_e, τ_e, and E are the electron mobility, the electron lifetime, and the electric field, respectively. As shown in Fig. 3.12b (right), the inter-electrode distance L can be varied. When L is within the range L_e ($L_e > L$), the electrons can be collected at the electrode, but cannot when $L_e < L$. So, the photocurrent should decrease when L is approximately L_e. Conversely, L_e can be determined by observing the sudden drop in photocurrent.

3.4.1.2 NTCDA Single Crystal

We adopted single crystal of naphthalenetetracarboxylic dianhydride (NTCDA, Fig. 3.12a) for the electron-transporting material for two reasons. (i) NTCDA is a well-known electron-transporting organic semiconductor, and its use as an extremely thick (2 μm) electron transport layer in organic solar cells has been reported [17]. (ii) A NTCDA single crystal is transparent in the visible region, allowing direct irradiation of the D/A interface. The charge separation energy relationship is shown in Fig. 3.11c, and electrons are photogenerated only beneath the hole collection electrode. Hence, the starting point for electrons is aligned with the right-hand edge of the hole collection electrode.

Plate-shaped transparent NTCDA single crystals with the sizes of around 2 mm × 5 mm and thicknesses of around 50 μm were obtained (Fig. 3.11a) by physical vapor transport [20] in N_2 (1 atm). The NTCDA molecules were stacked in a herringbone structure in a single crystal with a monoclinic lattice [21].

3.4.1.3 Cell Fabrication

Figure 3.12b (right) shows the configuration of the NTCDA single-crystal cell employing lateral electron transport and collection. Tetraphenyl dibenzoperiflanthene (DBP) acting as donor, MoO_3 and Ag were deposited at 10^{-6} Pa in an oil-free vacuum evaporator (EpiTech, VTS-350 M/ERH) built in a glove box purged with N_2 gas. Metal masks with pairs of square apertures separated by distances of 20, 30, 50, and 100 μm were used to form the electrode pairs. A 30 nm-thick layer of DBP and a MoO_3 (10 nm)/Ag (100 nm) hole collection electrode were deposited on the NTCDA single crystal through one of the apertures. Then, an electron collection electrode (Ag (100 nm)) was deposited through the other aperture. For reference, a vertical cell (Fig. 3.12b (left)) was fabricated on an ITO substrate, which could be regarded as a standard cell with $L = 0$.

3.4.1.4 Operation of Lateral Cell

Figure 3.13a shows the J-V characteristics of a lateral type solar cell with $L = 30$ μm at solar light intensities of 1, 8, and 10 suns. The photocurrent is shown in the first quadrant. Open-circuit voltage (V_{OC}) values of 0.42, 0.62, and 0.61 V were observed for intensities of 1 (blue curve A), 8 (orange curve B), and 10 suns (green curve C). For the vertical-type cell (Fig. 3.12b (left)), V_{OC} values of 0.61, 0.66, and 0.67 V were observed under the same light intensities. The magnitudes of V_{OC} for the lateral cell ($L = 30$ μm) were slightly smaller compared with the vertical cell ($L = 0$ μm). This is a clear demonstration of the operation of the lateral cell with an extremely long electron transport distance of 30 μm.

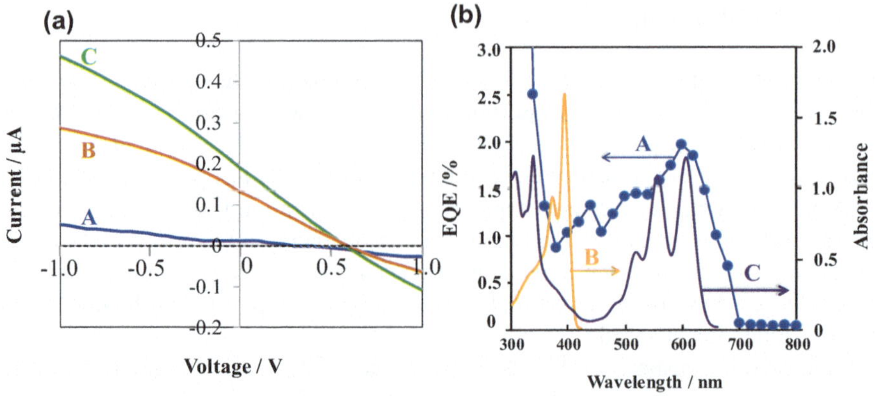

Fig. 3.13 **a** Current-voltage (J-V) characteristics at irradiation intensities of 1 sun (blue curve A), 8 suns (orange curve B), and 10 suns (green curve C). The broken black curve is the dark current. **b** Action spectrum of the external quantum efficiency (EQE) of J_{SC} for the lateral cell (blue dots A). The solid curves are the absorption spectra of the NTCDA film (50 nm)(yellow curve B) and the DBP film (30 nm)(purple curve C). Reproduced with permission from [23]. Copyright 2016, Elsevier B.V

Figure 3.13b shows the action spectrum (blue dots) of the external quantum efficiency (EQE) of the short-circuit photocurrent (J_{SC}) for the lateral cell. The sensitivity of the action spectrum is aligned with the absorption spectrum of the DBP film (purple curve). When the DBP layer, which acts as a donor, was removed, no photocurrent was observed throughout the spectral range (not shown). Apparently, without the DBP/NTCDA (D/A) interface, no excitons were formed, and no carriers were generated. Thus, we concluded that excitons were generated only in the DBP layer dissociating at the D/A interface (Fig. 3.12c), i.e., electrons and holes were generated only at the D/A interface. Therefore, we can regard the right-hand edge of the D/A interface as the starting point for the electrons (Fig. 3.12b, broken blue line).

3.4.1.5 Electron Range

In order to determine the electron range (L_e), the inter-electrode distance (L) was varied as shown in Fig. 3.12b. Figure 3.14a shows the J-V characteristics of cells with $L = 20$, 30, 50, and 100 μm under irradiation of 10 suns. When L was increased from 20 (orange curve A) to 30 μm (red curve B), the photocurrent magnitudes in the saturated region of the reverse direction from −1 to −3 V coincided well. When L was increased from 30 (red curve B) to 50 μm (blue curve C), the photocurrent decreased significantly. A further increase in L from 50 (blue curve C) to 100 μm (green curve D) caused a further moderate decrease in the photocurrent. This result suggests that the effective electron transport occurred within 30 μm in the lateral direction, i.e., the electron range (L_e) was approximately 30 μm. Because the electric field (E) depends on L, the photocurrent should be plotted as a function of L at constant E to evaluate L_e

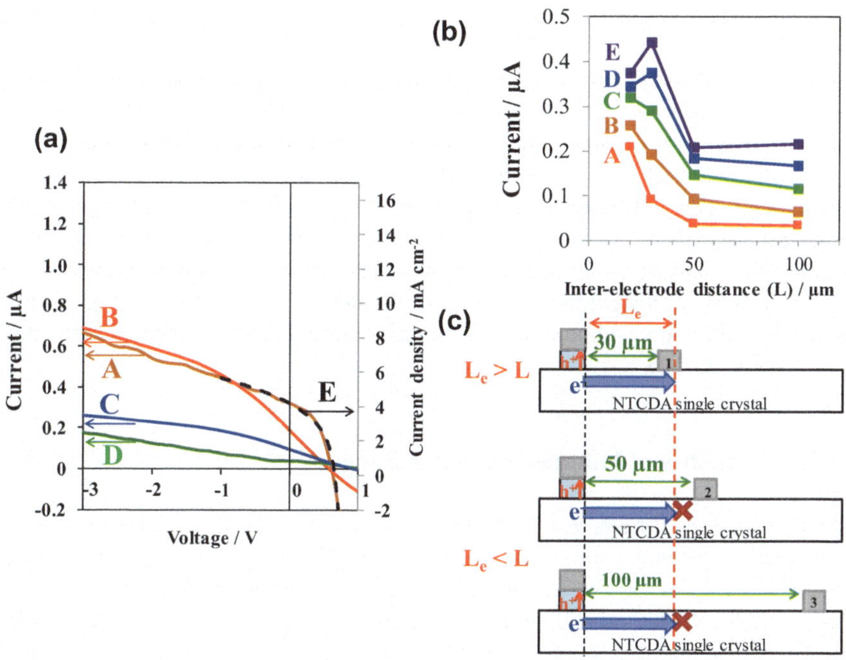

Fig. 3.14 **a** *J-V* characteristics of lateral cells with $L = 20$ μm (orange curve A), 30 μm (red curve B), 50 μm (blue curve C), and 100 μm (green curve D) and that of the vertical reference cell (broken black curve E) under irradiation of 10 suns. **b** Dependence of the photocurrent on L at constant values of E of 1×10^2 (red dots A), 2×10^2 (orange dots B), 3×10^2 (green dots C), 4×10^2 (blue dots D), and 5×10^2 V·cm^{-1} (purple dots E). **c** Schematic illustrations of the relationship between L and L_e in the lateral cells. Reproduced with permission from [23]. Copyright 2016, Elsevier B.V

precisely. E was determined from $E = (V_{oc} + V_{appl})/L$ where V_{appl} is the externally applied voltage. Figure 3.14b shows the photocurrent-L relationships at constant values of E of 1, 2, 3, 4, and 5 × 10^2 V·cm^{-1}, which are around the magnitude of the built-in field (2 × 10^2 V·cm^{-1}) estimated by V_{OC}(0.6 V)/L(30 μm). Sharp decreases in the photocurrent were again confirmed between $L = 30$ and 50 μm. The photocurrent was almost constant below $L = 30$ μm under sufficient electric fields (green dots C, blue dots D, purple dots E). Therefore, we concluded that the electron range (L_e) is approximately 30 μm.

Figure 3.14c shows schematic illustrations of lateral cells with $L = 30$, 50, and 100 μm and $L_e = 30$ μm. When $L_e < L$, a small number of electrons could reach the collection electrode, i.e., $L = 50$ (Fig. 3.14c, middle) and 100 μm (Fig. 3.14c, bottom). When $L_e > L$, most of the electrons can reach the collection electrode, i.e., $L = 30$ μm (Fig. 3.14c, top).

3.4.1.6 Active Area of Lateral Cell

Notably, in the lateral cell, the active area for generating photocurrent was only near the right-hand edge of the hole collection electrode (Fig. 3.12b). The maximum photocurrent density obtained for the lateral cell was comparable to that for the vertical reference cell (Fig. 3.12b, $L = 0$ μm). The shape of J-V curve for vertical reference cell (Fig. 3.14a, broken black curve E) coincided well with that for the lateral type cell (Fig. 3.14a, orange curve A). This strongly suggests that the photocurrent density of lateral-type cell ($L = 20$ μm) reached that of vertical reference cell ($L = 0$ μm). Based on this comparison, we could calculate the width of the active area as 8 μm for generating a photocurrent near the right-hand edge of the hole collection electrode (Fig. 3.12b).

3.4.1.7 Possibility of Millimeter Carrier Range

Based on Eq. (3.1), the electron lifetime τ_e was calculated to be 0.52 ms using $L_e =$ 30 μm. Here, the electron mobility μ_e of the NTCDA single crystal was determined to be 2.9×10^{-2} $cm^2{\cdot}V^{-1}{\cdot}s^{-1}$ from the field-effect transistor measurement, and the built-in field for the lateral cell with $L = 30$ μm was used for the value of E (2×10^2 $V{\cdot}cm^{-1}$). The value of τ_e obtained (0.52 ms) is of the order of milliseconds, which would enable L_e to be of the order of millimeters if single-crystal organic materials with the μ_e of 1 $cm^2{\cdot}V^{-1}{\cdot}s^{-1}$were used. There is a strong probability of achieving millimeter-order L_e because an electron mobility of 1.7 $cm^2{\cdot}V^{-1}{\cdot}s^{-1}$ has already been reported for a perylene derivative [22].

3.4.2 *Lateral Junctions*

In this section, a lateral alternating multilayered junction using a high-mobility crystalline organic semiconductor is proposed and demonstrated [23]. Essentially, the photogenerated holes and electrons are laterally transported and extracted to the respective electrodes. A total of 93% of the photogenerated electrons and holes were laterally collected over a millimeter-scale distance of 0.14 mm. The exciton-collection efficiency reached 75% in a lateral alternating multilayered junction with a layer thickness of 10 nm. Therefore, a lateral organic alternating multilayered junction that completely collects both excitons and carriers could be an alternative blended junction for organic solar cells.

3.4.2.1 Concept

Figure 3.15a shows the concept of a lateral alternating multilayered junction cell consisting of hole pathways (Fig. 3.15b) and electron pathways (Fig. 3.15c), which

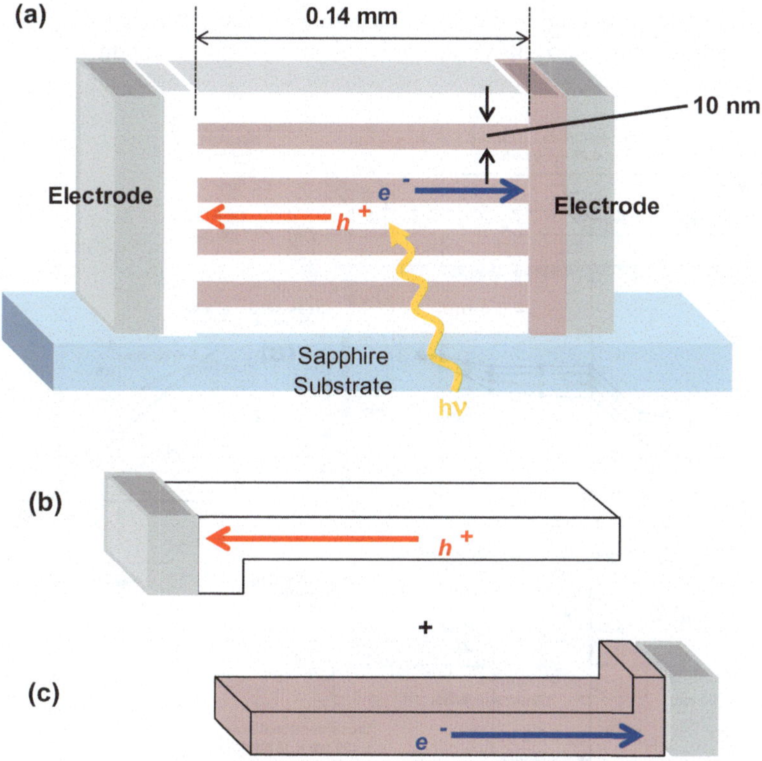

Fig. 3.15 **a** Lateral alternating multilayered junction. Present reported junction (Fig. 9) has layer thickness of 10 nm and lateral distance of 0.14 mm, i.e., aspect ratio reaches 1.4×10^4. **b** Hole pathway. **c** Electron pathway. **b** and **c** are simplest units for (**a**). Reproduced with permission from M. Kikuchi et al., *ACS Appl. Energy Mater.*, **2**, 2087 (2019). Copyright 2019, American Chemical Society

are the simplest units. The hole and electron pathways are selectively connected only to the hole- and electron-collecting electrodes by using the respective buffer layers. A photogenerated exciton diffuses to the nearest D/A interface and dissociates. The generated electrons and holes move laterally through the pathways and are collected by the electron (right side) and hole-collecting (left side) electrodes, respectively. Exciton collection is achieved through nanometer-scale control of the film thickness, and the long-distance lateral transport of carriers in high-carrier-mobility organic semiconductor films allows carrier collection. The present structure can be precisely designed via nanometer-scale control over the film thickness that is independent of percolation. Moreover, an unlimited vertical thickness is permitted, which allows for the absorption of the entire solar spectrum.

Recently, organic films with high mobilities for holes (reaching 43 $cm^2 \cdot V^{-1} \cdot s^{-1}$) [24–27] and electrons (reaching 1.7 $cm^2 \cdot V^{-1} \cdot s^{-1}$) [22] were reported. Because high mobility has been observed in the direction parallel to the substrate owing to strong π-π interactions (Fig. 3.15), this concept is suitable for the lateral device.

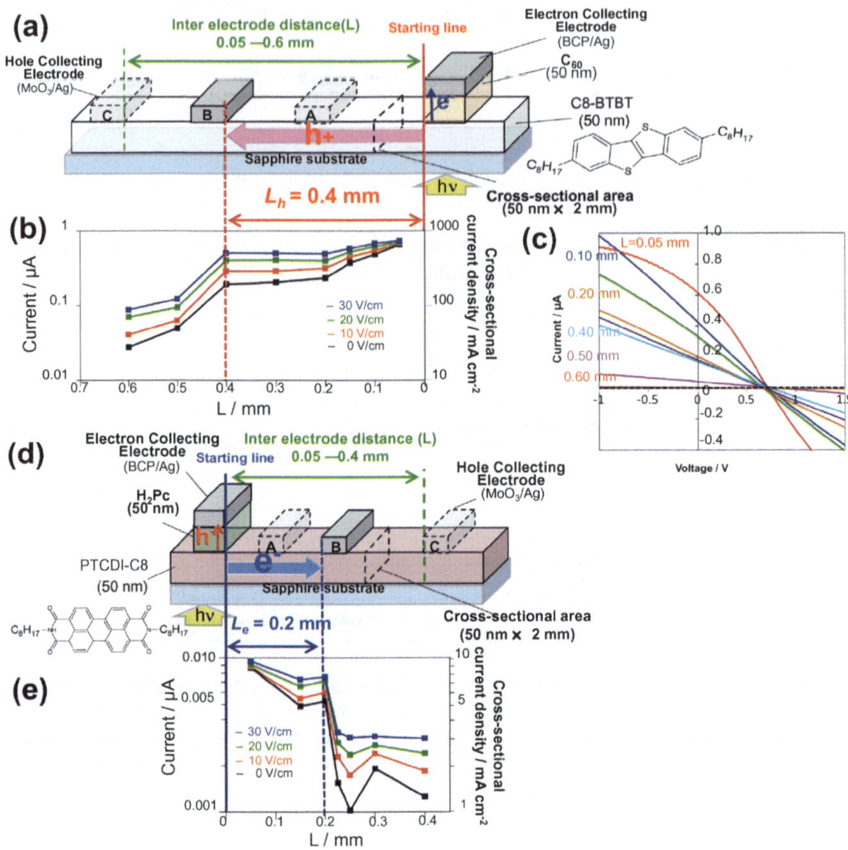

Fig. 3.16 **a** Hole pathway unit cells with electrode distance (L) from 0.05 to 0.6 mm. C-plane single-crystal sapphire substrates which were annealed at 1,200 °C for 1 h in air were used. **b** Dependence of photocurrent magnitude on L under constant electric fields of 0 (black), 10 (red), 20 (green), and 30 V·cm^{-1} (blue) between lateral electrodes. Current density calculated from area of cross section of C8-BTBT film is shown on right vertical axis. **c** Current-voltage (J-V) characteristics of hole pathway unit cells. **d** Electron pathway unit cells with L from 0.05 to 0.4 mm. **e** Dependence of photocurrent magnitude on L under constant electric field strengths of 0 (black), 10 (red), 20 (green), and 30 V·cm^{-1} (blue). Reproduced with permission from M. Kikuchi et al., *ACS Appl. Energy Mater.*, **2**, 2087 (2019). Copyright 2019, American Chemical Society

3.4.3 Carrier Pathway Unit Cells

Hole and electron pathway unit cells are shown in Figs. 3.16a and d, respectively. For the hole pathway unit cell (Fig. 3.16a), C8-BTBT, which shows a hole mobility (μ_h) of up to 43 $cm^2 \cdot V^{-1} \cdot s^{-1}$ [24, 28], was used as the hole transporter. A pair of electrodes separated by distance L was deposited on the film surface, and cells with an L of 0.05, 0.1, 0.2, 0.3, 0.4, 0.5, and 0.6 mm were fabricated. Fullerene (C_{60}), the acceptor (A), was inserted between the C8-BTBT, the donor (D), and the electron-collecting electrode. At the D/A interface, excitons dissociate into electrons and holes under light irradiation. Because the photogenerated electrons move vertically, the distance to the electron-collecting electrode is equal to the C_{60} film thickness (50 nm). Conversely, because the photogenerated holes move in the lateral direction, the distance to the hole-collecting electrode is equal to the inter-electrode distance (L), which is longer than 0.05 mm. Because the lateral distance is 1,000 times longer than the vertical distance, the characteristics of the hole pathway unit cells are dominated by the lateral hole transport. The left-hand edge of the electron-collecting electrode can be regarded as the starting line for the holes (solid red line). The range of the holes (L_h) is expressed by Eq. (3.2) [29–33],

$$L_h = \mu_h \tau_h E, \tag{3.2}$$

where μ_h, τ_h, and E are the hole drift mobility, the lifetime of holes, and the electric field, respectively. When L is shorter than the range L_h ($L_h > L$) (Fig. 3.16a, electrode A), the holes are collected by the hole-collecting electrode. However, when L is longer than the range L_h ($L_h < L$) (Fig. 3.16a, electrode C), the holes are not collected. Thus, the magnitude of the photocurrent decreases at approximately $L = L_h$(Fig. 3.16a, electrode B), and L_h can be determined by observing the sudden drop in the photocurrent. By using the electron pathway unit cells (Fig. 3.16d), the electron range (L_e) can be determined in a similar manner. PTCDI-C8 (Fig. 3.16d) which has a high electron mobility ($\mu_e = 1.7$ $cm^2 \cdot V^{-1} \cdot s^{-1}$ [22]) was used as the electron transporter. Phthalocyanine (H_2Pc), D, was inserted between the PTCDI-C8, A, and the hole-collecting electrode.

3.4.4 Carrier Ranges

Figure 3.16c shows the current–voltage (J-V) characteristics of the hole-transporting cells. At $L = 0.05$ mm, a clear photovoltaic curve was observed (red curve). Photovoltaic effects were observed for L values of 0.05, 0.1, 0.2, 0.3, 0.4, 0.5, and 0.6 mm. A sudden decrease in the photocurrent was observed between $L = 0.4$ and 0.5 mm, and the photocurrent disappeared at $L = 0.6$ mm.

The dependence of the photocurrent on the electrode distance (L) for constant electric field strengths (E) between the lateral electrodes of 0 (black), −10 (red), −

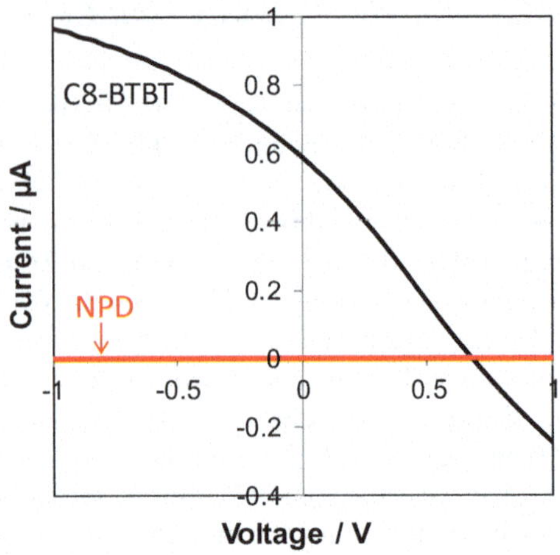

Fig. 3.17 Current-voltage (J-V) characteristics of cells shown in Fig. 5b with $L =$ 50 μm using C8-BTBT film (black curve) and NPD film (red curve). NPD cell shows no J_{SC} and no V_{OC}. Reproduced with permission from M. Kikuchi et al., *ACS Appl. Energy Mater.*, **2**, 2087 (2019). Copyright 2019, American Chemical Society

20 (green), and -30 V·cm^{-1} (blue), which were calculated by dividing the reverse applied voltage by L, are shown in Fig. 3.16b. Here, $E = 0$ V·cm^{-1} corresponds to the short-circuit condition. Sudden drops in the photocurrent at $L = 0.4$ mm were observed for all E values. Thus, we concluded that the hole range (L_h) is 0.4 mm. When a typical hole-transporting material [NPD; N,N-di(1-naphthyl)-N,N-diphenyl-(1,1-biphenyl)-4,4-diamine] with a low mobility of $\mu_h = 1.0 \times 10^{-4}$ cm^2·V^{-1}·s^{-1} was used, no photocurrent was observed, even at $L = 50$ μm (Fig. 3.17). Obviously, the high mobility of C8-BTBT is responsible for the long hole range (L_h) reaching 0.4 mm.

The cross-sectional area of the C8-BTBT film is 1×10^{-6} cm^2 because the film thickness and width are 50 nm and 2 mm, respectively (Fig. 3.16a, black broken rectangle). Thus, the cross-sectional current density through the hole-transporting layer reached 0.75 A·cm^{-2} with a very low electric field (-30 V·cm^{-1}) (see the right vertical axis in Fig. 3.16b). Therefore, the C8-BTBT film acts as an excellent lateral hole pathway.

The electron pathway unit cells exhibited clear photovoltaic curves. Sudden drops in the photocurrent for $L = 0.2$ mm were observed at $E = 0, -10, -20$, and -30 V·cm^{-1}(Fig. 3.16e). Thus, the electron range (L_e) was determined to be 0.2 mm. Moreover, the cross-sectional current density through the electron-transporting layer reached 10 mA·cm^{-2} with a very low electric field (30 V·cm^{-1}, see the right vertical axis in Fig. 3.16e). Therefore, the PTCDI-C8 film acts as an excellent lateral electron pathway.

3.4.5 Lateral Alternating Multilayered Junction

The hole range ($L_h = 0.4$ mm) and the electron range ($L_e = 0.2$ mm) were determined. Thus, we adjusted the width of the overlap between C8-BTBT and PTCDI-C8 in the lateral alternating multilayered junction to less than 0.4 mm and 0.2 mm, i.e., 0.14 mm for the total extractions of holes and electrons to the respective electrode (Fig. 3.15a). The hole and electron pathways were selectively connected to only the hole- and electron-collecting electrodes, respectively (Fig. 3.15a), by using a custom-made movable masking system (Fig. 3.18). The total number of pathway layers increased, while the total thickness of the alternating multilayered junction was maintained at 100 nm.

Superior photovoltaic characteristics were observed (Fig. 3.19a), with an open-circuit voltage (V_{OC}) between 0.88 and 1.0 V and a fill factor (FF) between 0.36 and 0.48, which is surprising considering the small electric fields at the short-circuit

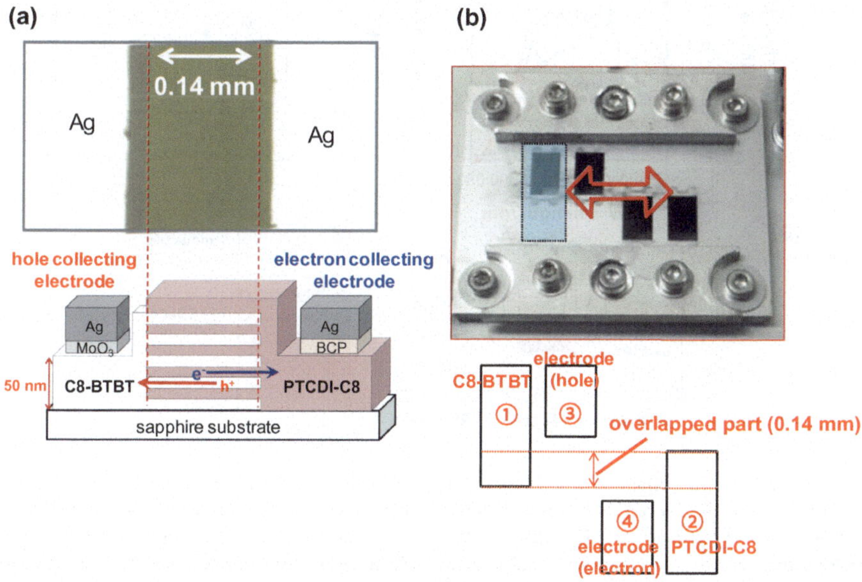

Fig. 3.18 **a** Photograph and corresponding structure of lateral alternating multilayered junction. **b** Photograph and illustration of metal mask. A sapphire substrate set on the mask (upper) was slid by a precise positioning system from side to side along the direction of the red double arrow. C8-BTBT, PTCDI-C8, the hole-collecting electrode (MoO_3/Ag), and the electron-collecting electrode (BCP/Ag) were deposited through the apertures in the order of aperture number, i.e., (1), (2), (3), and (4), respectively. To prepare the alternating multilayers, evaporation through apertures (1) and (2) was repeated. The cell area (7.7×10^{-3} cm^2) was defined by the length of the overlap of C8-BTBT and PTCDI-C8 (0.014 cm) and the electrode width (0.57 cm). Reproduced with permission from M. Kikuchi et al., *ACS Appl. Energy Mater.*, **2**, 2087 (2019). Copyright 2019, American Chemical Society

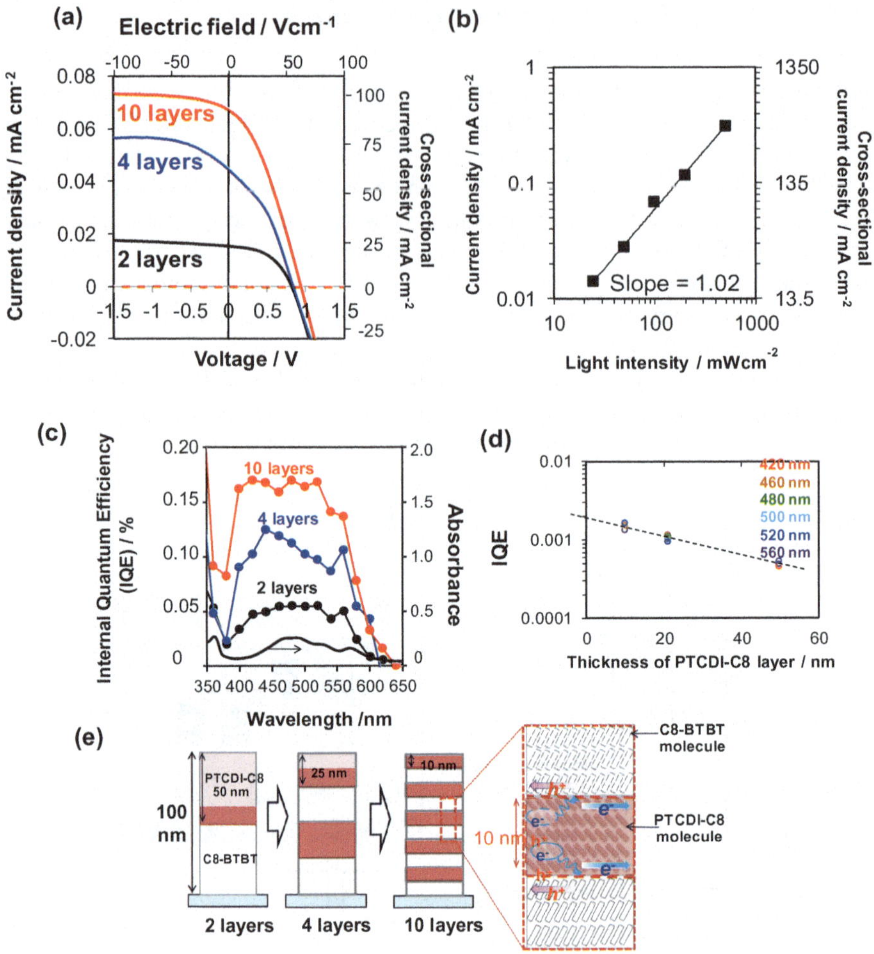

Fig. 3.19 **a** Current-voltage (J-V) characteristics of lateral alternating multilayered junctions of 2-layered (black), 4-layered (blue), and 10-layered (red) devices. Photocurrent density per cell area is shown on left vertical axis. Effective electric field in cell is shown on upper horizontal axis. Photocurrent density per cross-sectional area of multilayered film is shown on right vertical axis. Broken curve is dark current. **b** Dependence of short-circuit photocurrent density (J_{SC}) on simulated solar light intensity between 0.25 and 5 suns (from 25 to 500 mW·cm^{-2}) for 10-layered alternating multilayered cell. **c** Action spectra of internal quantum efficiency (IQE) of J_{SC} for each number of layers. Absorption spectrum of film with a total thickness of C8-BTBT (50 nm) and PTCDI-C8 (50 nm) is also shown (black curve). **d** Semilogarithmic plots of dependence of IQE at various wavelengths on PTCDI-C8 layer thickness. **e** Active region (red) and dead region (pink) of exciton collection in 2-, 4-, and 10-layered cells. Layer thickness of 10 nm corresponds to 5 and 3 molecules of PTCDI-C8 and C8-BTBT, respectively. Reproduced with permission from M. Kikuchi et al., *ACS Appl. Energy Mater.*, **2**, 2087 (2019). Copyright 2019, American Chemical Society

condition (70 V·cm^{-1}) and even smaller electric field near the open-circuit condition (<<70 V·cm^{-1}) (Fig. 3.19a). The photocurrent density was proportional to the simulated solar light intensity between 25 and 500 mW·cm^{-2} (Fig. 3.19b). The photocurrent density per cross-sectional area of the multilayered film reached 0.4 A·cm^{-2} ((Fig. 3.19b, right vertical axis). Thus, the photovoltaic operation of lateral alternating multilayered junctions with a macroscopic lateral electrode distance of 0.14 mm was demonstrated.

A corresponding vertical cell with an ITO/MoO_3/C8-BTBT(50 nm)/PTCDI-C8(50 nm)/BCP/Ag structure was usually operated with a built-in electric field under short-circuit conditions that are 10^4 times larger (10^5 V·cm^{-1}) when a cell thickness of 100 nm and V_{OC} of 1 V are considered.

By increasing the number of layers from 2 (black curve) to 4 (blue curve) and 10 (red curve), i.e., by decreasing the D/A interface distance from 50 to 25 and 10 nm, a dramatic increase in the photocurrent was observed. Photogenerated excitons within a diffusion length of the exciton from the D/A interface can reach the D/A interface (active layer) (Fig. 3.19e, red region). However, excitons photogenerated at distances greater than the exciton diffusion length cannot reach the D/A interface (dead layer) (Fig. 3.19e, pink region). Gradual elimination of the dead layer by decreasing the D/A interface distance (Fig. 3.19e, from left to right) increases the exciton-collection efficiency (η_{EC}), thus dramatically increasing the photocurrent.

The action spectra of the internal quantum efficiency (IQE) of the short-circuit photocurrent (J_{SC}) for these cells are shown in Fig. 3.19c. By increasing the number of layers from 2 to 4 and 10, i.e., by decreasing the D/A interface distance, dramatic increases in the IQE were observed. The exciton-collection efficiency (η_{EC}) could be quantitatively evaluated by using the IQE. The photocurrent owing to irradiation with visible wavelengths (400–600 nm) was generated by PTCDI-C8 because C8-BTBT is transparent. The semilogarithmic plots of the IQE at various wavelengths versus the PTCDI-C8 thickness are linear (Fig. 3.19d). The slope of the plot gives an exciton diffusion length of 12 nm, which is the distance at which 50% of the excitons can reach the D/A interface. By assuming that η_{EC} is 100% at 0 nm, η_{EC} for five monolayers of PTCDI-C8 with a thickness of 10 nm was 75% (Fig. 3.19e, right enlarged illustration). Thus, we concluded that the lateral alternating multilayered junction could collect most of the photogenerated excitons at the D/A interfaces. This analysis is identical to that for the vertical junctions mentioned in Sect. 3.3.

The clear saturation behavior of the photocurrent that is independent of the reverse bias (Fig. 3.19a) strongly suggests that most of the photogenerated electrons and holes are extracted laterally. Moreover, the proportional relationship between photocurrent and light intensity (Fig. 3.19b) suggests that the carrier recombination during lateral carrier transport was negligible. Thus, we concluded that the carrier-collection efficiency (η_{CC}) in the region of photocurrent saturation is unity. Therefore, the η_{CC} of 93%, which is determined by the ratio of the J_{SC} value to the photocurrent in the saturated region [34], was obtained (Fig. 3.19a, red curve).

Figure 3.20 shows the photocurrent–voltage (J-V) characteristics of the lateral alternating multilayered junction cell (red curve), vertical multilayered cell (blue curve), and bulk heterojunction (blended) cell (black curve). The total thickness of

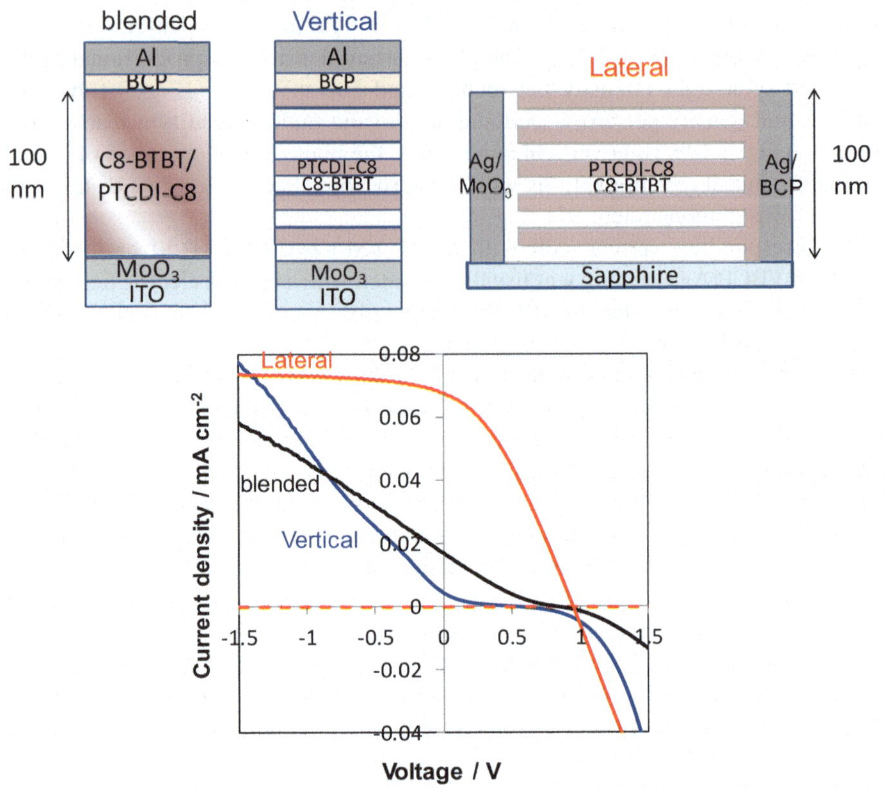

Fig. 3.20 Photocurrent-voltage (J-V) characteristics of the lateral multilayered junction cell (red curve), vertical multilayered cell (blue curve), and blended cell (black curve). Cell structures are also shown. Total thickness of all cells is 100 nm. Number of layers (10 layers) is the same for both vertical multilayered cell and lateral multilayered cell. Lateral alternating multilayered junction showed the superior characteristics. On the other hand, blended cell and vertical multilayered cell showed deteriorated characteristics caused by the absence of efficient routes for the hole and electron extraction. Reproduced with permission from M. Kikuchi et al., *ACS Appl. Energy Mater.*, **2**, 2087 (2019). Copyright 2019, American Chemical Society

all the cells is 100 nm, and the number of layers (10 layers) is the same for both the vertical multilayered cell and the lateral alternating multilayered junction cell. The lateral alternating multilayered junction showed superior performances compared with those of the bulk heterojunction cell and the vertical multilayered cell. For the vertical multilayered cell, the effects on exciton collection from increasing the number of D/A interfaces are obscured by the difficulty of hole and electron transport in the vertical direction across numerous barriers. The increase in the number of D/A interfaces clearly affected carrier transport only in the lateral direction.

With the present D/A combination of C8-BTBT and PTCDI-C8, the exciton dissociation efficiency (η_{ED}) was very small, calculated as 0.3% using Eq. (3.3).

$$IQE = \eta_{EC} \times \eta_{ED} \times \eta_{CC} \tag{3.3}$$

We presume that this low efficiency was caused by the long alkyl chains of C8-BTBT and PTCDI-C8 acting as 2-nm-thick insulators and inhibiting the initial charge separation. However, the lateral alternating multilayered junction (Fig. 3.15a) is a universal principle, and various kinds of organic semiconductors are suitable for use in these junctions. One of the advantages of a lateral carrier-collecting alternating multilayered cell is its unlimited thickness in the vertical direction (Fig. 3.21a). Therefore, tandem solar cells that can utilize the full solar spectrum can be freely designed. Owing to the millimeter-scale macroscopic lateral distance between the electrodes, modules with large surface areas can be fabricated by a simple masking system (Fig. 3.21b).

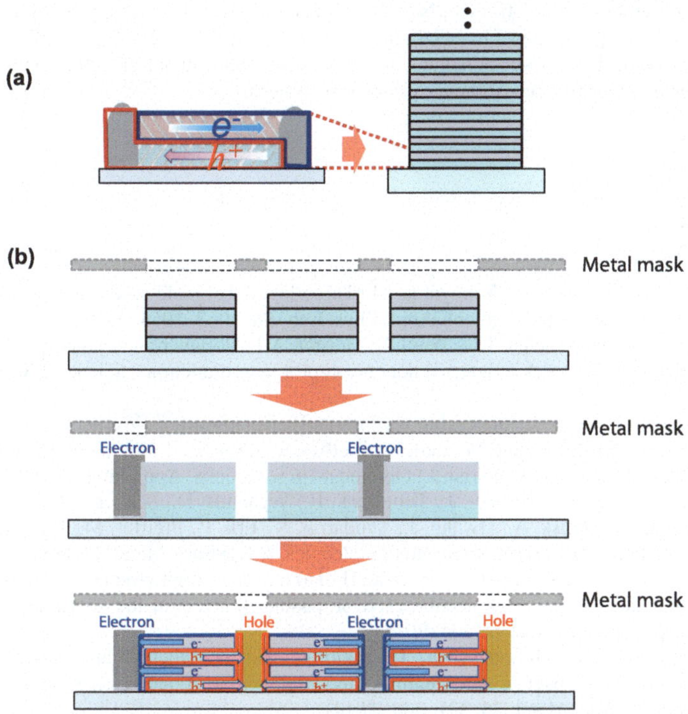

Fig. 3.21 **a** Unlimited thickness in the vertical direction. **b** Modules with large surface areas can be fabricated by a simple masking system owing to the millimeter-scale macroscopic lateral distance between the electrodes

3.5 Conclusion

Using percolation to form crystalline–amorphous nanocomposites of blended films between donor and acceptor molecules (which offers individual transport routes for electrons and holes) by the elevated substrate temperature during co-deposition enhances the photovoltaic performance.

The vertical superlattice junctions have ability to collect both excitons and carriers although the area of the vertical superlattice junction is very small.

The lateral multilayered junctions using high-mobility organic semiconductors are able to collect both excitons and carriers and have sufficient cell area. Therefore, the lateral junction can be regarded as an alternative blended junction for organic solar cells.

Acknowledgements Funding from the New Energy and Industrial Technology Development Organization (NEDO) is appreciated. Financial support from JSPS, KAKENHI (No. 17H02768) is gratefully acknowledged. The author appreciates A. Adachi and S. Ohashi of EpiTech Co. (Kyoto, Japan) for the design and construction of the movable mask system.

References

1. Hiramoto, M., Fujiwara, H., Yokoyama, M.: Three-layered organic solar cell with a photoactive interlayer of codeposited pigment. Appl. Phys. Lett. **58**, 1062–1064 (1991)
2. Yu, G., Gao, J., Hummelen, J.C., Wudl, F., Heeger, A.J.: Polymer photovoltaic cells enhanced efficiencies via a network of internal donor-acceptor heterojunctions. Science **270**, 1789–1791 (1995)
3. Maennig, B.,Drechsel, J., Gebeyehu, D., Simon, P., Kozlowski, F., Werner, A., Li, F., Grundmann, S., Sonntag, S., Koch, M., Leo, K., Pfeiffer, M., Hoppe, H., Meissner, D., Sariciftci, N. S.,Riedel, I., Dyakonov, V., Parisi, J.: Organic *p*-i-*n* solar cells. Appl. Phys. A,79, 1–14 (2004)
4. Riede, M., Uhrich, C., Widmer, J., Timmreck, R., Wynands, D., Schwartz, G., Gnehr, W.M., Hildebrandt, D., Weiss, A., Hwang, J., Sundarraj, S., Erk, P., Pfeiffer, M., Leo, K.: Efficient organic tandem solar cells based on small molecules. Adv. Funct. Mater. **21**, 3019–3028 (2011)
5. Li, W.N., Ye, L., Li, S. S., Yao, H. F., Ade, H., Hou, J. H.:A high-efficiency organic solar cell enabled by the strong intramolecular electron push-pull effect of the nonfullerene acceptor. Adv. Mater.,30, 1707170(8 pages) (2018)
6. Osaka, I., Saito, M., Koganezawa, T., Takimiya, K.: Thiophene-thiazolothiazole copolymers: significant impact of side chain composition on backbone orientation and solar cell performances. Adv. Mater. **26**, 331–338 (2014)
7. Kaji, T., Zhang, M., Nakao, S., Iketaki, K., Yokoyama, K., Tang, C.W., Hiramoto, M.: Coevaporant induced crystalline donor:acceptor blends in organic solar cells. Adv. Mater. **23**, 3320–3325 (2011)
8. Hiramoto, M.: Organic solar cells incorporating a p-i-n junction and a p-n homojunction. In: Sun, S.-S., Sariciftci, N.S. (eds.) Organic photovoltaics, Mechanisms, Materials and Devices, pp. 239–270. CRC Press, New York (2005)
9. Spanggaard, H., Krebs, F.C.: A brief history of the development of organic and polymeric photovoltaics. Sol. Energy Mater. Sol. Cells **83**, 125–146 (2004)
10. Tang, C.W.: Two-layer organic photovoltaic cell. Appl. Phys. Lett. **48**, 183–185 (1986)
11. Walzer, K., Maennig, B., Pfeiffer, M., Leo, K.: Highly efficient organic devices based on electrically doped transport layers. Chem. Rev. **107**, 1233–1271 (2007)

12. Sakai, K., Hiramoto, M.: Efficient organic *p-i-n* solar cells with very thick co-deposited i-layer consisting of highly purified organic semiconductors. Mol. Cryst. Liq. Cryst. **491**, 284–289 (2008)
13. Armin, A., Subbiah, J., Stolterfoht, M.,Shoaee, S., Xiao, Z., Lu, S., Jones, D. J., Meredith, P.:Reduced recombination in high efficiency molecular nematic liquid crystalline: fullerene solar cells. Adv. Energy Mater., 6, 1600939 (10 pages), (2016)
14. Jin, Y., Chen, Z., Dong, S., Zheng, N., Ying, L., Jiang, X.F., Liu, F., Huang, F., Cao, Y.: A novelnaphtho[1,2-c:5,6-c′]bis([1,2,5]thiadiazole)-based narrow-bandgap π-conjugated polymer with power conversion efficiency over 10%. Adv. Mater. **28**, 9811–9818 (2016)
15. Suemori, K., Miyata, T., Hiramoto, M., Yokoyama, M.: Enhanced photovoltaic performance in fullerene:phthalocyanine co-deposited films deposited on heated substrate. Jpn. J. Appl. Phys. **43**, L1014–L1016 (2004)
16. Suemori, K., Miyata, T., Yokoyama, M., Hiramoto, M.: Three-layered organic solar cells incorporating nanostructure-optimized phthalocyanine:fullerene co-deposited Interlayer. Appl. Phys. Lett., 86, 063509 (3 pages) (2005)
17. Suemori, K., Matsumura, Y., Yokoyama, M., Hiramoto, M.: Large Area Organic Solar Cells with Thick and Transparent Protection Layers. Jpn. J. Appl. Phys. **45**, L472–L474 (2006)
18. Hiramoto, M., Yamaga, T., Danno, M., Suemori, K., Matsumura, Y., Yokoyama,M.:Design of nanostructure for photo-electric conversion by organic vertical superlattice. Appl. Phys. Lett., 88, 213105 (3 pages)(2006)
19. Kikuchi, M., Takagi, K., Naito, H., Hiramoto, M.: Single crystal organic photovoltaic cells using lateral electron transport. Org. Electron. **41**, 118–121 (2017)
20. Laudise, R.A., Kloc, Ch., Simpkins, P.G., Siegrist, T.: Physical vapor growth of organic semiconductors. J. Cryst. Growth **187**, 449–454 (1998)
21. Jayaraman, A., Kaplan, M.L., Schmidt, P.H.: Effect of pressure on the Raman and electronic absorption spectra of naphthalene- and perylenetetracarboxylic dianhydrides. J. Chem. Phys. **82**, 1682–1687 (1985)
22. Chesterfield, R.J., McKeen, J.C., Newman, C.R., Ewbank, P.C., Filho, D.A.S., Brédas, J.-L., Miller, L.L., Mann, K.R., Frisbie, C.D.: Organic thin film transistors based on N-alkyl perylene diimides: charge transport kinetics as a function of gate voltage and temperature. J. Phys. Chem. B **108**, 19281–19292 (2004)
23. Kikuchi, M., Hirota, M., Kunawong, T., Shinmura, Y., Abe, M., Sadamitsu, Y., Moh, A.M., Izawa, S., Izaki, M., Naito, H., Hiramoto, M.: Lateral alternating donor/acceptor multilayered junction for organic solar cells. ACS Appl. Energy Mater. **2**, 2087–2093 (2019)
24. Yuan, Y.,Giri,G., Ayzner, A. L., Zoombelt, A. P., Mannsfeld, Chen, S. C. B. J., Nordlund, D., Toney, M. F., Huang, J., Bao, Z.,: Ultra-high mobility transparent organic thin film transistors grown by an off-centre spin-coating method. Nat. Commun., 5, 3005–3013 (2014)
25. Takeya, J., Yamagishi, M., Tominari, Y., Hirahara, R., Nakazawa, Y.: Very high-mobility organic single-crystal transistors with in-crystal conduction channels. Appl. Phys. Lett., 90, 102120(3 pages) (2007)
26. Minemawari, H., Yamada, T., Matsui, H., Tsutsumi, J., Haas, S., Chiba, R., Kumai, R., Hasegawa, T.: Inkjet printing of single-crystal films. Nature **475**, 364–367 (2011)
27. Haas, S., Takahashi, Y., Takimiya, K., Hasegawa, T.: High-performance dinaphtho-thienothiophene single crystal field-effect transistors. Appl. Phys. Lett.,95, 022111(3 pages) (2009)
28. Takimiya, K., Shinamura, S., Osaka, I., Miyazaki, E.: Thienoacene-based organic semiconductors. Adv. Mater. **23**, 4347–4370 (2011)
29. Naito, H., Kanemitsu, Y.: Relations between transient charge transport and the glass-transition temperature in amorphous chalcogenides. Phys. Rev. B **49**, 10131–10135 (1994)
30. Coropceanu, V., Cornil, J., da Silva Filho, D.A., Olivier, Y., Silbey, R., Bredas, J.-L.: Charge transport in organic semiconductors. Chem. Rev. **107**, 926–952 (2007)
31. Lecomber, P.G., Spear, W.E.: Electronic transport in amorphous silicon films. Phys. Rev. Lett. **25**, 509–511 (1970)

32. Schmidlin, F.W.: Theory of trap-controlled transient photoconduction. Phys. Rev. B **16**, 2362–2385 (1977)
33. Podzorov, V., Menard, E., Borissov, A., Kiryukhin, V., Rogers, J.A., Gershenson, M.E.: Intrinsic charge transport on the surface of organic semiconductors. Phys. Rev. Lett., **93**, 086602(4 pages) (2004)
34. Mori, D., Benten, H., Okada, I., Ohkita, H., Ito, S.: Highly efficient charge-carrier generation and collection in polymer/polymer blend solar cells with a power conversion efficiency of 5.7%. Energy Environ. Sci., **7**, 2939–2943 (2014)

Chapter 4
OPV with a Crystalline Organic Pigment Active Layer Up to 10 μm

Toshihiko Kaji

4.1 Introduction: Optical Design of Thickness of OPV and Its Photoelectric Conversion Layer

The photoelectric conversion efficiency of OPVs has reported greatly exceeded 10% in recent years, gradually approaching to a practical level of the solar cell performances [1–4]. It is expected that it can be manufactured with lightweight and flexible plastic substrates and that it has a low environmental impact. Since the invention of OPV for heterojunction and bulk heterojunction layers, various approaches such as material design and screening [1, 5, 6], thin-film growth process [7, 8], and new device structure [9, 10] have been developed to improve OPV photoelectric conversion efficiency. The optimization process often involves trade-offs between key cell designs such as active layer composition and thickness. For example, the light absorption of the active layer, that is, the photoelectric conversion layer, can be increased by increasing the film thickness of the layer, but beyond a certain thickness, the recombination of the photogenerated charges through the active layers suppresses the charge carrier transports and photocurrent generation.

Therefore, in order to increase the photocurrent at zero bias, that is, the short-circuit current J_{sc}, the thickness of the active layer has been designed in the certain optimal range where the photocurrent generation is not limited by charge recombination and transport [11, 12]. For example, in an OPV in which an organic pigment is vacuum-deposited, the 40-60-nm-thick active layer is typically used to absorb the first standing wave [11, 13–15]. The second and third standing waves can be used to further enhance absorption; thus, in the design of tandem OPVs, the bottom cell

The original version of this chapter was revised: Figure 4.6 (b) have been updated with new Figure. The correction to this chapter is available at https://doi.org/10.1007/978-981-15-9113-6_11

T. Kaji (✉)
Department of Applied Physics, Tokyo University of Agriculture and Technology, 2-24-16 Nakacho, Tokyo, Koganei 184-8588, Japan
e-mail: kaji-t@cc.tuat.ac.jp

M. Hiramoto and S. Izawa (eds.), *Organic Solar Cells*,
https://doi.org/10.1007/978-981-15-9113-6_4

is often placed about 200 nm from the metal electrode that acts as a mirror for the benefit from the second standing wave [1, 14, 16]. The thicker active layer is often used in OPV devices made of polymer, and the antinode of the third standing wave is often located at about 300 nm. However, beyond this thickness, the efficiency of polymer devices also tends to decrease, or peak, in both J_{sc} and fill factor, FF [16–19].

However, in some exceptional cases, it has been reported that OPV devices show high J_{sc} even in a very thick active layer of 1 μm [17, 20, 21]. This anomalous behavior has been reported both in the vacuum-deposited pigment devices [20] and in the solution spin-coated polymer devices [17, 21], although in these cases the relationship between the optical effects of standing waves has not been clarified as well, as a result of actual device operations.

4.2 Change in Relation Between Film Thickness of Active Layer and Solar Cell Characteristics of OPV by Crystallization

This section summarizes performance difference of thickness-dependent characteristics of crystallized and non-crystallized ZnPc:C_{60} OPVs. The thickness ranges from 40 nm to 10 μm [22].

4.2.1 Fabrication of Pigmented OPV Device with Crystallized Active Layer

In this study, in order to show the thickness dependence of solar cell characteristics of OPV devices, a typical combination of donor:acceptor mixed film, zinc phthalocyanine, and fullerene (ZnPc:C_{60}) was chosen [11–14, 23, 24]. The thickness range of the active layer is 40–10,000 nm, and the order, composition, and thickness of the layers of the fabricated OPV cell are as follows: [ITO /F_4TCNQ (0.6 nm)/CuI (3 nm)/ZnPc (2 nm)/C_{60} (2.5 nm)/ZnPc (1 nm)/ZnPc: C_{60} (volume ratio 1.5:1, 40–10000 nm)/C_{60} (5 nm)/Alq$_3$: C_{60} (1: 1, 25 nm)/LiF (2 nm)/Ag (100/600/1500 nm)]. ITO is a transparent electrode of indium-doped tin oxide, F_4TCNQ is 2,3,5,6-tetrafluoro-7,7,8,8-tetracyanoquino-dimethane, and Alq$_3$ is an abbreviation for tris(8-quinolinolato)aluminum. In order to cover the thick active layer uniformly, it was necessary to change the thickness of the Ag electrode.

For the vacuum evaporation of ZnPc, C_{60}, and ZnPc:C_{60} mixed layers, the "co-evaporant induced crystallization" method [7, 22, 23, 25] was used, which has been devised by the authors and is still under research and development. In this method, "co-evaporant" enables crystallization and growth control of organic pigment film during vacuum evaporation. "Co-evaporant" liquid molecules are simultaneously vaporized during the deposition of the organic pigment thin film. The role of this liquid molecule is to induce crystallization of the deposited film and precisely control the crystal grain size. Since the liquid molecule itself is more volatile than the pigment molecule, it can be removed from the deposited film if the substrate was heated during the growth of the deposited film. Polydimethylsiloxane (PDMS) with an appropriate chain length was used as the co-evaporant molecule, and the substrate temperature was kept at 70 °C during deposition. The other layers were deposited at a substrate temperature of 25 °C without co-evaporant molecules. The thin interfacial layers above and below the active layer were used for the purpose of extracting and blocking electrons and holes from the active layer, and as a template for growth of the active layer.

4.2.2 Relationship Between Active Layer Thickness and Solar Cell Characteristics

Figure 4.1 shows the relationship between the photoelectric conversion efficiency and the film thickness of the active layer of the crystallized OPV cell and the conventional OPV cell prepared by vacuum evaporation, measured under AM1.5G pseudo-sunlight irradiation. In Fig. 4.1a, the difference was clear. When crystallization by

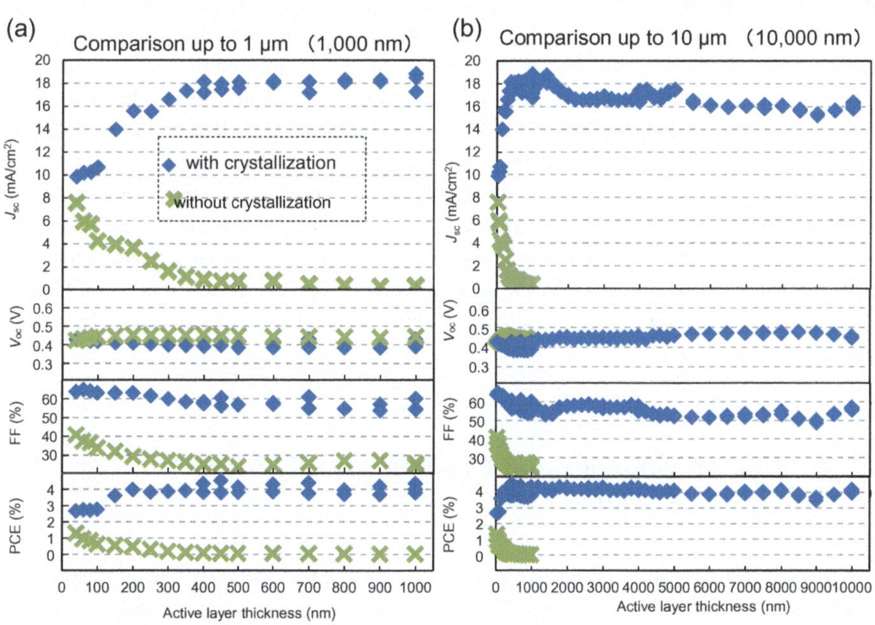

Fig. 4.1 Difference in the relationship between solar cell characteristics dependent on active layer thickness of ZnPc:C_{60} OPV cells with the presence or absence of crystallization **a** up to 1 μm and **b** up to 10 μm

PDMS was not induced and when the ZnPc:C_{60} layer became thicker than 40 nm, the short-circuit current density J_{sc} of the OPV cell drastically decreased to less than 1 mA/cm^2, and the photoelectric conversion efficiency also drastically decreased. This is as reported by many past research groups [10, 11]. The cause of the decrease is often thought to be an increase in electron–hole recombination in the ZnPc:C_{60} mixed layer. This ZnPc:C_{60} mixed layer was also vapor-deposited at a substrate temperature of 70 °C, but crystallization of the active layer could not be promoted only by heating the substrate. When crystallization was induced by co-evaporation of PDMS, J_{sc} increased dramatically with increasing active layer thickness and was maintained at about 18 mA/cm^2 between 400 and 1000 nm.

There was almost no effect of crystallization on the open-circuit voltage V_{oc}, and it varied between 0.40 and 0.45 V over the entire film thickness range. The fill factor (*FF*) was highly dependent on crystallization. Without crystallization, *FF* gradually decreased as the thickness of the mixed film increased, and leveled off at 25% above 400 nm, whereas with optimal crystallization, *FF* was maintained at more than 60% and in a thin-film thickness range and more than 55% in the 400–1000 nm range. As a result, the photoelectric conversion efficiency (PCE) of the crystallized device was around 4.0% even when the mixed layer thickness was 1,000 nm.

Here, the question arises as to how much thickness this characteristic can be maintained, and when we fabricated devices with successively increasing film thickness, we confirmed that the characteristic was almost maintained up to 10 μm, as shown in Fig. 4.1b. When the thickness reached 10 μm, the deposition time of the active layer reached three days and we were very tired to continue; then, this experiment was completed. The J_{sc} was about 16 mA/cm^2 at 2–10 μm, and the maximum value at 0.5–1.0 μm was 18 mA/cm^2. The *FF* varied at 2–10 μm and gradually decreased from 60 to 50%. In these devices, the Ag electrode surface had lost its metallic luster, so the surface became rougher as the active layer became thicker, and this is thought to cause problems with the contact of the electrodes. In particular, in order to measure the characteristics of devices with a very thick active layer (>4 μm), it was necessary to deposit a very thick Ag electrode (>1,500 nm). As a result of a gradual change in V_{oc} between 0.40 and 0.48 V, the photoelectric conversion efficiency (PCE) was able to maintain a value of about 4.0% in devices of all thicknesses up to 10 μm. It was a great surprise.

4.2.3 Film Morphology and Crystallinity of 10-μm Crystallized OPV

Using ultraviolet–visible absorption spectroscopy and X-ray diffraction (XRD), we confirmed the difference in the deposited film depending on the presence or absence of co-evaporant molecules in the 600-nm-thick ZnPc:C_{60} film. Comparing the absorption spectrum (Fig. 4.2a) with the XRD pattern (Fig. 4.2b), it can be seen that the absorption peak (about 445 nm) and the XRD peak (10.3° and 10.8°) due to the C_{60}

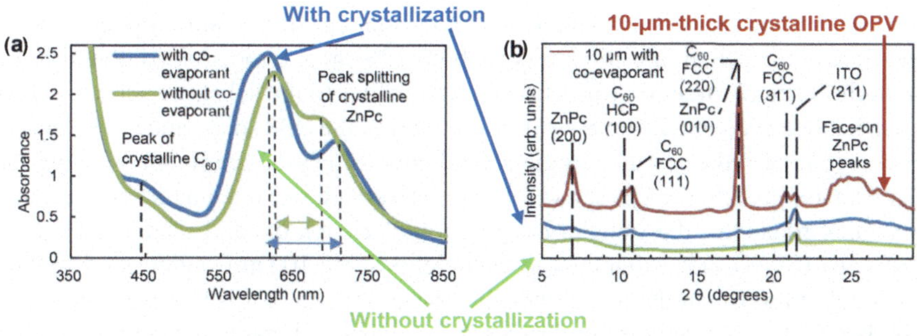

Fig. 4.2 Ultraviolet–visible absorption and X-ray diffraction change of ZnPc:C_{60} OPV device due to crystallization. **a** Ultraviolet–visible absorption and **b** X-ray diffraction

crystal only clearly appeared in the film co-evaporated with the co-evaporant. With respect to ZnPc, the absorption peak of the film deposited simultaneously with the co-evaporant molecules showed a broader splitting (620–710 nm), and the amorphous halo in the range of 7–10° disappeared in XRD. The changes in the XRD pattern observed in this thick crystalline film were also in agreement with those previously reported by other research groups in the thin (100 nm) ZnPc:C_{60} film [23].

The XRD peak of the 10-μm active layer is very sharp, but at the same time, both ZnPc and C_{60} clearly show peaks of various orientations [7, 11, 23, 24, 26, 27], which indicates that the orientations are not uniform. On the other hand, Fig. 4.3 shows an optical microscope image of a crystalline ZnPc:C_{60} device cross section with an

Fig. 4.3 Epi-illumination optical microscope image of crystalline ZnPc: C_{60} device cross section with film thickness over 10 μm. The top is Ag electrode, and the bottom is ITO glass substrate. Although the actual resolution of the microscope is around 300-400 nm, the presence of 100-nm rods and grains can be detected

assumed film thickness of 10 μm. This is a normal epi-illumination optical microscope image, the actual film thickness is about 12 μm, and rod-shaped domains are seen in the vertical direction (direction perpendicular to the substrate surface). This rod can be identified as a ZnPc crystal by comparing the bulk shape and color (purple red) of a general ZnPc acicular single crystal, and the crystal grains surrounding the rod can be assigned to the C_{60} crystal. The polarized microscope images suggest that many ZnPc rods have the same crystallographic orientation, and scanning electron microscopy (SEM) shows that the rod diameter is about 100 nm and the surrounding C_{60} crystal grains of similar size. Since the crystal growth of this film was optimized for higher J_{sc} by co-evaporant molecules, it is considered that it was optimized on a scale equivalent to the exciton diffusion length of ZnPc and C_{60} crystals.

Thus, it was found that the 10-μm active layer of the fabricated OPV consisted of ZnPc rod-shaped crystals surrounded by C_{60} crystal grains. Since the rod and granular crystal shapes of these crystals are very similar to those of their bulk crystals, other crystalline pigments with a needlelike or platelike crystal morphology in bulk, similar to ZnPc:C_{60}, are suggested to have the possibility of achieving an ultra-thick film that maintains the photoelectric conversion characteristics.

4.2.4 Relationship Between Active Layer Thickness and Absorption/Solar Cell Characteristics

Light absorption by the active layer of the OPV cell depends on several factors such as the film thickness, the standing wave due to the reflection of the upper metal electrode, and the roughness of the substrate, in addition to the absorption coefficient and spectrum of the material of the active layer. The solid line (-) and broken line (---) plots in Fig. 4.4 show the percentage of light absorbed by the ZnPc:C_{60} active layer deposited on the flat ITO glass substrate for organic EL while crystallizing with co-evaporant molecules. It is plotted against the film thickness of the active layer. In other words, these lines are corresponding to the optical efficiency of this solar cell for the standard solar spectrum in the 350–850 nm range, ignoring the optical losses used in the measurements and the optical losses of the electrical contacts. (Since it is not a practical module cell, it is strictly different from the definition of optical efficiency.)

Both the solid line (-) and the broken line (---) were calculated from the actual measurement data of the absorption spectrum of the actually prepared active layer based on the standard sunlight spectrum of AM1.5G. The optical efficiency was obtained from the "measurement result of actual OPV element containing Ag": solid line (-) and from the "measurement result of transmittance of the same active layer not containing Ag": broken line (---), respectively. Therefore, the solid line (-) is wavy reflecting the standing wave actually generated by the reflection of the Ag electrode, and the effect is canceled from the broken line (---). The broken line at the bottom of the graph shows the amplification of the standing wave due to the

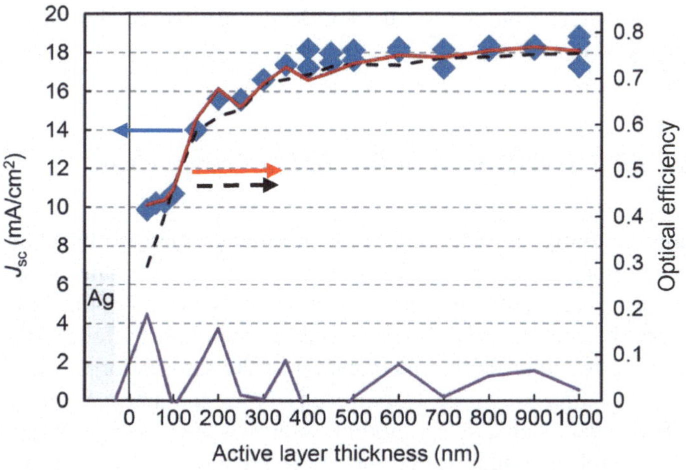

Fig. 4.4 Relationship between optical efficiency and short-circuit current density of the crystallized ZnPc:C_{60} OPV cell in reference to film thickness of active layer

reflection of Ag, calculated by the ratio of the absorption of the solid line (-) to the absorption of the broken line (---). We have also confirmed the effect of roughness of the transparent electrode substrate by using the textured fluorine-doped tin oxide (FTO) glass, where it is clear that the effect of texture is greater than it of the standing wave, so if you are interested, please refer to Ref. [22].

Here, in Fig. 4.4, the short-circuit current of the OPV element using this active layer is also plotted as a diamond (◆). The optical efficiency without Ag reflection (---) gradually increases with the thickness of the active layer, but the plots of both J_{sc} (◆) and optical efficiency with Ag reflection (-) show a wavy change with clear local maximums at less than 400 nm. This is because J_{sc} has a local maximum value when the film thickness is adjusted to the antinode of the standing wave, as in previous reports by various groups [14, 17, 18, 28, 29]. The effect of the standing wave in this graph is clear at 40, 200, and 350 nm, and the optical efficiency (-) and J_{sc} (◆) continue to increase until the third standing wave node (400 nm) and are almost saturated up to 1,000 nm. Such a behavior in which the optical efficiency and J_{sc} are strongly correlated is, of course, reasonable in principle, but as far as the author knows, the actual measured data of the OPV cell had never showed such a strong correlation up to 1,000 nm before this result.

4.3 Effect of Buffer Layer and Antireflection Film

The selection of buffer layer is important for stable comparison of the performances of ultra-thick OPVs. In this section, we disclose the difference of the effect of single and mixed buffer layers for the ultra-thick OPVs. The effect of antireflection film is also demonstrated in addition.

4.3.1 Buffer Layer Selection for Stable Comparison

In this study, the buffer layer is optimized in order to increase the film thickness while stabilizing J_{sc} and other characteristics in the device with an active layer whose thickness exceeds 400 nm. In this section, the selection of this buffer layer is explained. For this stabilization, LiF was used as an anode buffer layer just below the Ag electrode to suppress the intercalation of Ag into the organic film, and an Alq_3:C_{60} mixed buffer was used to uniformly cover the active layer. It has been reported [30–33] that such a buffer mixture improves the electronic conductivity of the buffer layer while maintaining the amorphous property such as the combination of BCP:PBD, BCP:C_{60}, and BPhen:C_{60}. Although the authors tried many other mixed buffers, Alq_3:C_{60} was the best for this purpose of the stabilization to the thickness.

Of course, if a single buffer layer of BCP or Alq_3 with a thickness of about 10 nm, which is used for ordinary OPVs, can be used without any problems in many cases. Usually between the single buffer and the mixed buffer, almost no difference in performance could be observed except the difference in loss due to the optical absorption of the buffers. The difference occurred when the optimum thickness of the single buffer varied due to the surface roughness of the active layer. This is because if the BCP or Alq_3 single buffer is too thick (in a few nanometers), the J_{sc} sharply decreases, but in the case of the mixed buffer, the J_{sc} only gradually decreases. Using the mixed buffer, now, it is possible to make a stable comparison. Because of this difference, in this study, Alq_3:C_{60} mixed buffer was used for the thickness-dependent characteristic investigation of the OPV using a crystalline organic mixed active layer with a rough surface.

The underlying interface buffer used a series of layers of F_4TCNQ (0.6 nm)/CuI (3 nm)/ZnPc (2 nm)/C_{60} (2.5 nm)/ZnPc (1 nm). The F_4TCNQ/CuI/ZnPc layer just above the ITO electrode acted as a hole transport layer to improve the anodic contact[11,27]. The C_{60}/ZnPc layer served as a template layer to reproduce the starting block of growth of the active ZnPc:C_{60} blend layer. This template contributed greatly to the reproducibility of repeated production of the device. The cell with all these buffer layers also showed good thermal stability during the two heating/cooling cycles: heating to 100 °C for 1 h and cooling to 25 °C without bias. They showed no deterioration in *J*–*V* characteristics.

To quantitatively distinguish the effect of these layers from the effect of co-evaporant-induced crystallization, we made a set of OPV cells with a 600-nm-thick ZnPc:C_{60} active layer with or without an underlying interfacial layer. The *J*–*V* characteristics of these cells and the incident photon-to-current efficiency (IPCE) spectra are compared in Fig. 4.5a and b. Crystallization with co-evaporant has the greatest effect on J_{sc}, but requires a lower interfacial layer to significantly improve J_{sc} and

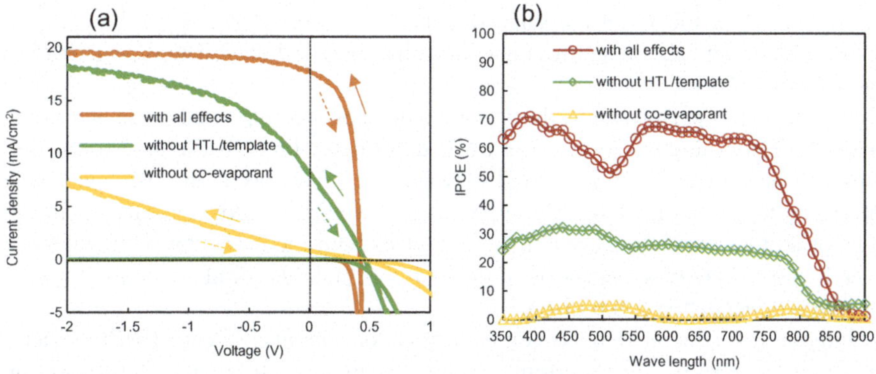

Fig. 4.5 Various effects on the photocurrent of ultra-thick pigment OPVs. **a** Current density–voltage (J–V) characteristics of a 600-nm-thick ZnPc:C_{60} cell, with or without crystallization effects due to co-evaporant, and with or without the underlying buffer layer. The solid/dashed style indicates the direction of the voltage sweep. **b** IPCE spectrum of the same set of OPV cells

FF. These buffers serve for both reproducibility for the stable comparison and the performance of the OPV cell.

On the other hand, it can be seen that the buffer selected to exhibit the most stable performance at various film thicknesses does not always maximize the photoelectric conversion efficiency of this crystalline OPV. For example, Fig. 4.6 shows the solar cell characteristics when the selection of the buffer is optimized in the active layer of 600 nm, and J–V curve in Fig. 4.6a shows 2.1 mA/cm^2 higher than the plots

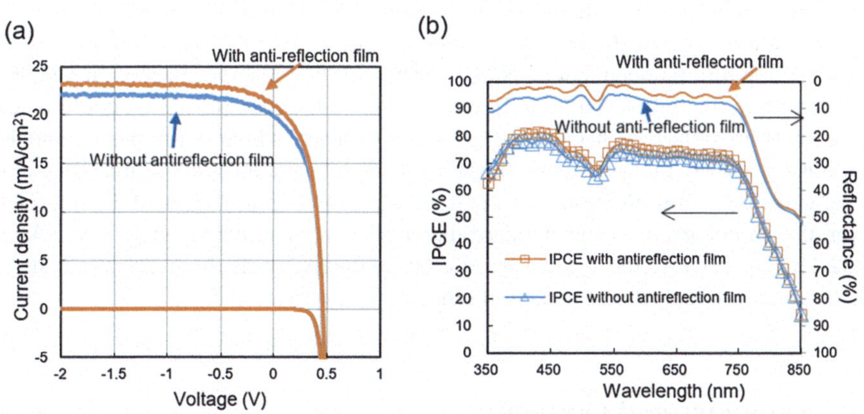

Fig. 4.6 Solar cell characteristics of crystalline ZnPc:C_{60} OPV cell with a thickness of 600 nm with buffer layer optimized for its performance. **a** J–V characteristics, **b** IPCE spectra and the reflectance of the same cell, with the presence or absence of antireflection film on ITO glass substrate

when comparing the J_{sc} at the same thicknesses in Figs. 4.1 and 4.4, here, J_{sc} of 20.04 mA/cm^2 and photoelectric conversion efficiency of 4.76%. The V_{oc} was 0.45 V, and FF was 52.6%.

For this better performance, the J_{sc} of the devices was improved by an additional layer of 40-nm rubrene:C_{60} (ratio = 1.5:1, deposited with PDMS co-evaporant and substrate heating at 70 °C) inserted between the active layer and the upper side interfacial layers in Fig. 4.1a. Although this insertion increased short-circuit current density up to around 2 mA/cm^2 for the devices with the active layer thickness of over 400 nm, there was dispersion of the increase, and this could not be used for the thickness dependence comparison.

Judging from the incident photon to current conversion efficiency (IPCE) spectra in Figs. 4.5b and 4.6b, these current increases are mainly attributed to the absorption of crystalline C_{60} around 450 nm of wave length [7]; it is considered that the charge carrier recombination at the upper interface has been suppressed by hole blocking because of the deeper highest occupied molecular orbital (HOMO) of rubrene: 5.4 eV [34], than it of ZnPc: 5.2 eV [11]. We note that almost the same phenomena were also observed when rubrene was replaced by the other donor materials, such as CBP: 6.0 eV [35], DTDCTB: 5.3 eV [36], and α-6T: 5.3 eV [37], but were not observed by the acceptor material of Alq_3: 5.7 eV [38]; even it has similar HOMO level to the donors above.

4.3.2 *Effect of Antireflection Film*

Finally, we confirmed the effect of the antireflection film to further improve J_{sc} in this device (Fig. 4.6a). Pasting an antireflection film onto the surface of a glass substrate has been known as one of the methods for improving J_{sc} and efficiency [39, 40]. This film reduces the loss due to the reflection of light on the outermost surface of the glass and increases the effective incident light to the OPV cell. Introducing of antireflection film reduced from 0.69% to 4.06% of the reflectance depending on the wavelength in the entire visible light range (Fig. 4.6b). As a result, J_{sc} could reach up to 21.16 mA/cm^2, even with ZnPc:C_{60} active layer, which is the most standard combination of organic pigments for OPVs made by vacuum evaporation. By using the effects of the antireflection film, crystallization, and optimization of the interface layer, the photocurrent of the standard materials was maximized, and FF was kept at 52.1%, but because the V_{oc} was restricted by the materials themselves to 0.46 V, the resultant photoelectric conversion efficiency was up to 5.03%.

4.4 Summary and Outlook

The authors have discovered that if an organic pigment is properly crystallized, it can be used as a photoelectric conversion layer of an OPV even if it is thickened to 10 μm. The material used in this study was a combination of the well-known organic pigment fullerene (C_{60}), a soccer-ball-shaped molecule, and phthalocyanine, which is also well known as a paint material, and these have been a standard donor:acceptor combination for vacuum-deposited OPVs for decades. Usually, the efficiency of organic solar cells using this combination of organic pigments decreases when the thickness exceeds 40–50 nm, but by growing the organic pigment crystals in an appropriate size (diameter of about 100 nm) (Fig. 4.3), the efficiency hardly decreased even when the thickness was increased to 10 μm (Fig. 4.1b). In addition, we demonstrated the relationship between the light absorbed by the organic pigment and the generated current in a wide range of thicknesses (40 nm–1 μm) in a corresponding manner (Fig. 4.4).

The thickness of 10 μm itself achieved in this study is much larger than the thickness that absorbs almost all visible light, so it cannot be directly applied to the OPV or organic light-emitting diode products. However, the potential of organic pigments, which are usually used in a thickness of several nm to several tens of nm, has been proved to work as a photoelectric conversion layer even with a thickness of 10 μm, which is an order of magnitude thicker than previously thought. We hope this fact will promote R&D of future organic photoelectric conversion devices using crystalline organic pigments.

Acknowledgments This section is written based on the results of collaboration with Mr. Mikimasa Katayama (at that time) at Tokyo University of Agriculture and Technology, and Professor Masahiro Hiramoto and Dr. Satoshi Nakao (at that time) at the Institute for Molecular Science. We would like to thank Professor C.W. Tang and Assistant Professor S. Dong at Hong Kong University of Science and Technology for their discussion and advices. In addition, part of the research introduced in this section was carried out with the support of JST-ALCA, JSPS Scientific Research Fund (17H04807), and Nanotechnology Platform Project (Molecular/Material Synthesis).

References

1. Meng, L. et al. Organic and solution-processed tandem solar cells with 17.3% efficiency. Science **361**, 1094–1098 (2018)
2. Kang, H., et al.: Bulk-heterojunction organic solar cells: five core technologies for their commercialization. Adv. Mater. **28**, 7821–7861 (2016)
3. Søndergaard, R., Hösel, M., Angmo, D., Larsen-Olsen, T.T., Krebs, F.C.: Roll-to-roll fabrication of polymer solar cells. Mater. Today **15**, 1–2 (2012)
4. Dou, L., et al.: 25th anniversary article: a decade of organic/polymeric photovoltaic research. Adv. Mater. **25**, 6642–6671 (2013)
5. Roncali, J., Leriche, P., Blanchard, P.: Molecular materials for organic photovoltaics: small is beautiful. Adv. Mater. **26**, 3821–3838 (2014)
6. Mishra, A., Bäuerle, P.: Small molecule organic semiconductors on the move: promises for future solar energy technology. Angew. Chem. Int. Ed. **51**, 2020–2067 (2012)
7. Kaji, T., et al.: Co-evaporant induced crystalline donor: acceptor blends in organic solar cells. Adv. Mater. **23**, 3320–3325 (2011)
8. Song, B., Rolin, C., Zimmerman, J.D., Forrest, S.R.: Effect of mixed layer crystallinity on the performance of mixed heterojunction organic photovoltaic cells. Adv. Mater. **26**, 2914–2918 (2014)

9. Luo, G., et al.: Recent advances in organic photovoltaics: device structure and optical engineering optimization on the nanoscale. Small **12**, 1547–1571 (2016)
10. Riede, M., et al.: Efficient organic tandem solar cells based on small molecules. Adv. Funct. Mater. **21**, 3019–3028 (2011)
11. Zhou, Y., et al.: Phase separation of co-evaporated ZnPc:C_{60} blend film for highly efficient organic photovoltaics. App. Phys. Lett. **100**, 233302 (2012)
12. Pfuetzner, S., Meiss, J., Petrich, A., Riede, M., Leo, K.: Thick C_{60}:ZnPc bulk heterojunction solar cells with improved performance by film deposition on heated substrates. App. Phys. Lett. **94**, 253303 (2009)
13. Wang, Z., Miyadera, T., Yamanari, T., Yoshida, Y.: Templating effects in molecular growth of blended films for efficient small-molecule photovoltaics. ACS Appl. Mater. Interf. **6**, 6369–6377 (2014)
14. Schueppel, R., et al.: Controlled current matching in small molecule organic tandem solar cells using doped spacer layers. J. Appl. Phys. **107**, 044503 (2010)
15. Xue, J., Rand, B.P., Uchida, S., Forrest, S.R.: A hybrid planar–mixed molecular heterojunction photovoltaic cell. Adv. Mater. **17**, 66–71 (2005)
16. Ko, S.-J., et al.: Photocurrent extraction efficiency near unity in a thick polymer bulk heterojunction. Adv. Funct. Mater. **26**, 3324–3330 (2016)
17. Osaka, I., Saito, M., Koganezawa, T., Takimiya, K.: Thiophene-thiazolothiazole copolymers: significant impact of side chain composition on backbone orientation and solar cell performances. Adv. Mater. **26**, 331–338 (2014)
18. Jung, J.W., Russell, T.P., Jo, W.H.: Highly crystalline low band gap polymer based on Thieno[3,4-c]pyrrole-4,6-dione for high-performance polymer solar cells with a >400 nm thick active layer. ACS Appl. Mater. Interf. **7**, 13666–13674 (2015)
19. Jin, Y., et al.: A Novel Naphtho[1,2-c:5,6-c′]Bis([1,2,5]Thiadiazole)-Based Narrow-Bandgap π-conjugated polymer with power conversion efficiency over 10%. Adv. Mater. **28**, 9811–9818 (2016)
20. Sakai, K., Hiramoto, M.: Efficient organic p-i-n solar cells having very thick co-deposited i-layer consisting of highly purified organic semiconductors. Mol. Cryst. Liq. Cryst. **491**, 284–289 (2008)
21. Price, S.C., Stuart, A.C., Yang, L., Zhou, H., You, W.: Fluorine substituted conjugated polymer of medium band gap yields 7% efficiency in polymer-fullerene solar cells. J. Am. Chem. Soc. **133**, 4625–4631 (2011)
22. Katayama, M., Kaji, T, Nakao, S., Hiramoto, M.: Ultra-thick organic pigment layer up to 10 μm activated by crystallization in organic photovoltaic cells. Front. Energy Res. **8**(4), 1–12 (2020)
23. Holzmueller, F., et al.: Co-evaporant induced crystallization of zinc phthalocyanine:C_{60} blends for solar cells. Org. Electron. **27**, 133–136 (2015)
24. Iketaki, K., Kaji, T., Nakao, S., Hiramoto, M.: Structural studies of the co-deposited i-layer of ZnPc:C_{60} p-i-n solar cells. Phys. Status Solidi C **8**, 637–639 (2011)
25. Kaji, T., Nakao, S., Hiramoto, M.: Effect of co-evaporant induced crystallization on needle growth of phthalocyanine thin films. Mol. Cryst. Liq. Cryst. **578**, 63–67 (2013)
26. Lee, J., Park, D., Heo, I., Yim, S.: Effect of cuprous halide interlayers on the device performance of ZnPc/C_{60} organic solar cells. Mater. Res. Bull. **58**, 132–135 (2014)
27. Cheng, C.H., et al.: Organic solar cells with remarkable enhanced efficiency by using a CuI buffer to control the molecular orientation and modify the anode. Appl. Phys. Lett. **97**, 083305 (2010)
28. Xue, J., Uchida, S., Rand, B.P., Forrest, S.R.: Asymmetric tandem organic photovoltaic cells with hybrid planar-mixed molecular heterojunctions. Appl. Phys. Lett. **85**, 5757–5759 (2004)
29. Maennig, B., et al.: Organic p-i-n solar cells. Appl. Phys. A **79**, 1–14 (2004)
30. Mori, T., Masumoto, Y.: Effect of organic alloy for suppression of polycrystallization in BCP thin film. J. Photopolym. Sci. Technol. **19**, 209–214 (2006)
31. Bartynski, A.N., et al.: A fullerene-based organic exciton blocking layer with high electron conductivity. Nano Lett. **13**, 3315–3320 (2013)

32. Xiao, X., Bergemann, K.J., Zimmerman, J.D., Lee, K., Forrest, S.R.: Small-molecule planar-mixed heterojunction photovoltaic cells with fullerene-based electron filtering buffers. Adv. Energy Mater. **4**, 1301557 (2014)
33. Kao, P.-C., Chu, S.-Y., Huang, H.-H., Tseng, Z.-L., Chen, Y.-C.: Improved efficiency of organic photovoltaic cells using tris (8-hydroxy-quinoline) aluminum as a doping material. Thin Solid Films **517**, 5301–5304 (2009)
34. Wang, D., et al.: Solution-processed organic films of multiple small-molecules and white light-emitting diodes. Org. Electron. **11**, 641–648 (2010)
35. Long, Z., et al.: The electroluminescence mechanism of non-doping PhOLEDs based on CBP/Ir(ppy)$_3$ investigated by delayed EL measurements. Org. Electron. **28**, 225–228 (2016)
36. Chen, Y.-H., et al.: Vacuum-deposited small-molecule organic solar cells with high power conversion efficiencies by judicious molecular design and device optimization. J. Am. Chem. Soc. **134**, 13616–13623 (2012)
37. Sakai, J., Taima, T., Yamanari, T., Saito, K.: Annealing effect in the sexithiophene:C_{70} small molecule bulk heterojunction organic photovoltaic cells. Sol. Energy Mater. Sol. Cells **93**, 1149–1153 (2009)
38. Wang, N., Yu, J., Zang, Y., Huang, J., Jiang, Y.: Effect of buffer layers on the performance of organic photovoltaic cells based on copper phthalocyanine and C_{60}. Sol. Energy Mater. Sol. Cells **94**, 263–266 (2010)
39. Sai, H., Matsui, T. and Matsubara, K. Stabilized 14.0%-efficient triple-junction thin-film silicon solar cell. Appl. Phys. Lett. **109**, 183506 (2016)
40. Chattopadhyay, S., et al.: Anti-reflecting and photonic nanostructures. Mat. Sci. Eng. R. **69**, 1–35 (2010)

Chapter 5
Polymer Solar Cells: Development of π-Conjugated Polymers with Controlled Energetics and Structural Orders

Itaru Osaka

5.1 Introduction

Organic solar cells have seen a dramatic progress in the last two decades by the development of new organic π-conjugated materials that are used as the photoactive materials [1–4]. Whereas, in the early stage of the area, the photoactive layer(s) were fabricated by vacuum evaporation of organic pigment dyes [5, 6], recently the layers are typically fabricated by solution processes of the blend of a polymeric π-conjugated material, as the p-type organic semiconductor, and a small-molecular π-conjugated material such as fullerenes and non-fullerenes, as the n-type organic semiconductor [7, 8]. In particular, π-conjugated polymers have been key materials for organic solar cells due to the advantage in film-forming property and film stability over small molecules.

One of the most common π-conjugated polymers is regioregular poly(3-hexylthiophene) (rrP3HT), in which 3-hexylthiophenes are linked to form a conjugated polymer backbone in a head-to-tail manner (Fig. 5.1) [9]. With a regio-controlled chemical structure, rrP3HT forms a crystalline lamellar structure with strongly π–π stacked backbones, leading to high charge carrier transport, while regiorandom polymer (raP3HT) is amorphous and thus has poor charge transport property. Although the absorption range only covers up to approximately 650 nm (optical bandgap of ~1.9 eV), rrP3HT, in combination with fullerenes, can give good power conversion efficiencies (PCEs) of 3–6% with external quantum efficiencies (EQEs) of ~80% [10, 11], which is believed to be as a consequence of the crystallinity. It is also important to note that rrP3HT affords better performance with a thick active layer film of ~200 nm, while most of the polymer-based solar cells show better performance with a thin layer of around 100 nm [12]. The thick layer

I. Osaka (✉)
Department of Applied Chemistry, Hiroshima University, Hiroshima, Japan
e-mail: iosaka@hiroshima-u.ac.jp

M. Hiramoto and S. Izawa (eds.), *Organic Solar Cells*,
https://doi.org/10.1007/978-981-15-9113-6_5

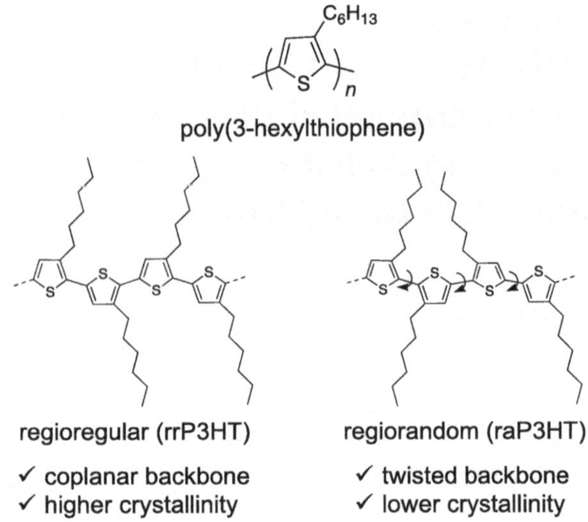

Fig. 5.1 Chemical structure of poly(3-hexylthiophene) (P3HT) (upper). Difference between regioregular and regiorandom P3HTs

is thought to be beneficial for large-area printing as well as for increasing the light absorption and thus the PCE.

The use of larger heteroaromatic fused rings as the building units is expected to provide better π–π interactions, and thus to enhance the crystallinity of π-conjugated polymers (Fig. 5.2). Further, "donor–acceptor (D–A)" polymers, where electron-rich (donor) and electron-poor (acceptor) heteroaromatic rings are alternatively incorporated in the backbone can also enhance the intermolecular π–π stacking, and thus the crystallinity, which likely originates in dipole–dipole interactions (Fig. 5.2) [13]. As a result, D–A polymers offer high charge carrier mobilities. In addition, such donor and acceptor arrangements result in molecular orbital mixing and thus intramolecular charge transfer (CT) interactions in the backbone, giving rise to narrow optical bandgaps, i.e., wide absorption ranges [14]. By choosing the strength of the donor

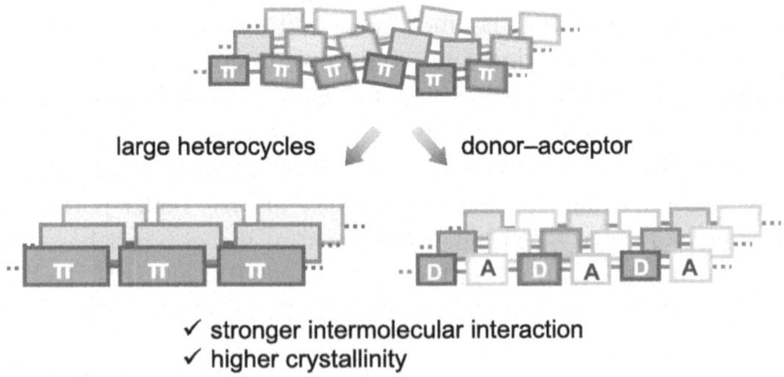

Fig. 5.2 Strategy to enhance the intermolecular interaction of the π-conjugated polymer backbone and thereby the crystallinity

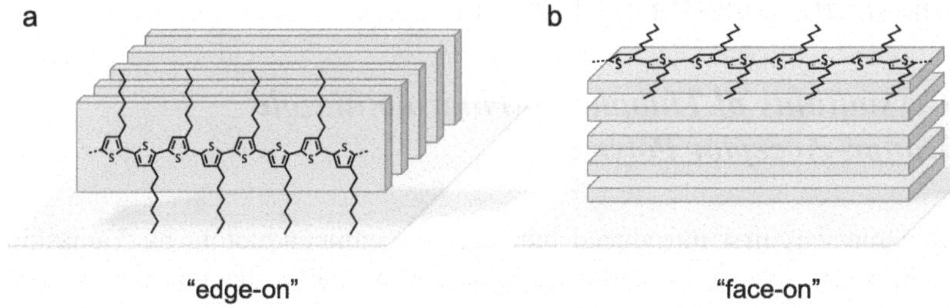

Fig. 5.3 Two different motifs of the backbone orientation **a** edge-on, **b** face-on

and acceptor characteristics, the energy levels of the highest occupied molecular orbital (HOMO) and the lowest unoccupied molecular orbital (LUMO) can also be tuned. Hence, the use of fused rings in a D–A motif is an important design strategy for creating high-performance polymers.

In parallel, the orientation of the polymer backbone must also be controlled to boost the performance of the solar cell. Typically, two orientation motifs can be formed in π-conjugated polymers with respect to the substrate plane. One with the backbone plane standing on the substrate is often called "edge-on" orientation, whereas one with the backbone lying flat on the substrate is often called "face-on" orientation (Fig. 5.3) [15]. As the solar cell requires higher charge transport in the out-of-plane direction with respect to the substrate (electrode) plane, face-on orientation is believed to be more desired.

This chapter will summarize our recent studies on the development of π-conjugated polymers with D–A motifs, specifically, based on thiazolothiazole and naphthobischalcogenadiazoles as the acceptor unit. By carefully designing the molecular structure, the crystallinity and the backbone orientation of the polymers can indeed be controlled. With the crystalline structure and favorable face-on backbone orientation, these polymers show high PCEs along with the maximized solar cell performance with thick active layers of around 300 nm. Importantly, a polymer system with naphthobisoxadiazole, with a deep HOMO and LUMO energy levels, allows us to have a significantly high V_{OC} of close to 1 V even with a narrow bandgap of ~1.5 eV, resulting in a markedly small photon energy loss. It is believed that this chapter will provide beneficial guidelines for the development of high-performance π-conjugated polymers for organic solar cells.

5.2 Thiazolothiazole-Based Polymers

5.2.1 Crystallinity of Thiophene–Thiazolothiazole Donor–Acceptor Polymers

Thiazolothiazole was first introduced into organic semiconductors by Yamashita and co-workers [16–18]. They synthesized a series of small-molecular compounds, which nicely functioned as p-type and n-type semiconductors in organic field-effect transistors (OFETs). On the other hand, thiazolothiazole-based π-conjugated polymers were first reported as luminescent materials [19]. We have then developed a series of donor–acceptor polymers, PTzQTs, bearing thiazolothiazole as the acceptor unit in the electron-rich polythiophene backbone (Fig. 5.4) [20]. PTzQTs had optical bandgaps (E_gs), determined by the onset of the absorption spectra in the film, of around 1.8 eV, which was about 0.1 eV narrower than rrP3HT, suggesting that there was a D–A interaction along the backbone. PTzQTs demonstrated charge carrier mobilities as high as 0.3 $cm^2\ V^{-1}\ s^{-1}$ in the OFET devices [21]. The main reason for the high mobility was ascribed to be their crystalline structure, with π–π stacking distances as close as 3.5 Å, along with the edge-on orientation, facilitating the in-plane charge carrier transport. This was evidenced by the two-dimensional grazing incidence wide-angle X-ray diffraction (2D GIXD) study, in which the lamellar and π–π stacking diffraction appeared along the q_z (out-of-plane) and q_{xy} (in-plane) axes, respectively (Fig. 5.5a). Such high crystallinity was believed to be brought by the enhanced intermolecular interactions owing to the D–A motif. In fact, PTzBT

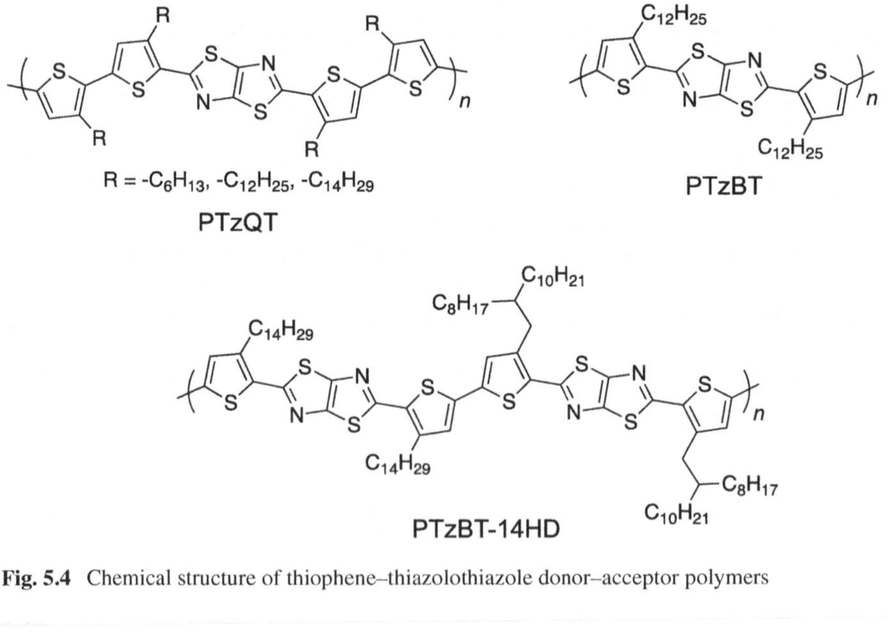

Fig. 5.4 Chemical structure of thiophene–thiazolothiazole donor–acceptor polymers

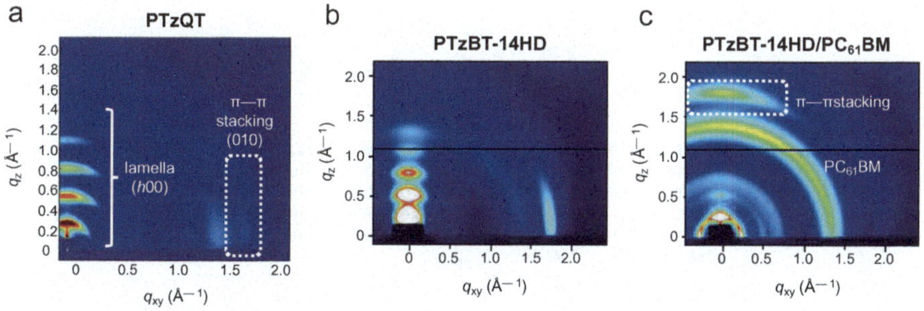

Fig. 5.5 2D GIXD patterns of the thiophene–thiazolothiazole polymers. **a** PTzQT film, **b** PTzBT-14HD film, and **c** PTzBT-14HD/$PC_{61}BM$ blend film

(Fig. 5.4) with the fewer alkylthiophene moiety compared to PTzQT was almost insoluble, while PBTTT [22] in which thiazolothiazole of PTzBT is replaced with thienothiophene thus being a donor–donor polymer, is known to soluble in toluene and chlorinated benzenes. This clearly demonstrates the impact of a D–A backbone.

PTzBT becomes soluble when a bulky branched alkyl group was partially introduced: PTzBT-14HD having tetradecyl and 2-hexyldecyl (HD) groups (Fig. 5.4) is soluble in chloroform. PTzBT-14HD formed a highly crystalline structure with the edge-on orientation in the thin film (Fig. 5.5b), leading to similarly high mobilities of ~0.4 $cm^2\ V^{-1}\ s^{-1}$ in OFETs [23]. Note that, however, the orientation was found to be "random" when the molecular weight was low, which resulted in a lower mobility of around 0.1 $cm^2\ V^{-1}\ s^{-1}$ [24]. More interestingly, PTzBT-14HD altered its backbone orientation into face-on orientation when it was blended with $PC_{61}BM$ while maintaining the crystalline structure (Fig. 5.5c). As a consequence, although the E_g of PTzBT-14HD was limited to ~1.8 eV, the polymer cell that combine with $PC_{61}BM$ showed a relatively high PCE of 5.7% [23].

5.2.2 *Control of Backbone Orientation in Thiophene–Thiazolothiazole Polymers*

The fact that the backbone orientation in PTzBT-14HD is sensitive to the molecular weight and blending with $PC_{61}BM$ inspired us to further study on the orientation using this simple polymer platform. We hypothesized that by changing the alkyl side chains on the PTzBT backbone in terms of the topology (linear or branched) and length can lead to different orientations, as they have different size and thus give different intermolecular interactions. We thus synthesized a series of PTzBTs by systematically tuning the side chains, where *n*-decyl (C10), *n*-dodecyl (C12), and *n*-tetradecyl (C14) linear alkyl groups and 2-ethylhexyl (EH), 2-butyloctyl (BO), HD, and 2-octyldodecyl (OD) branched alkyl groups were introduced as R^1 and R^1 (Fig. 5.6) [25].

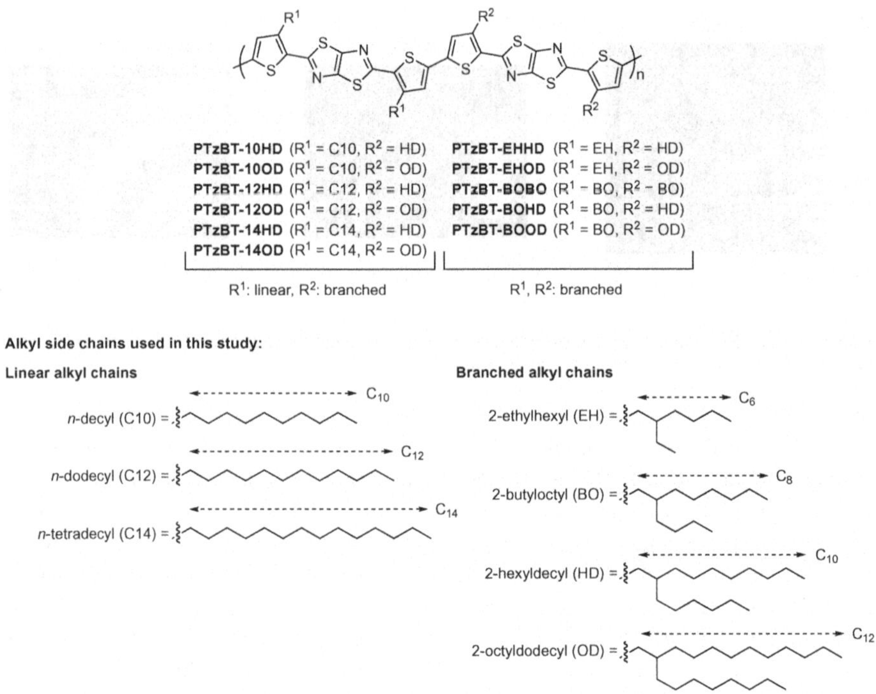

Fig. 5.6 Chemical structures of PTzBTs with different side chain compositions. Reproduced with permission [25]. Copyright (2014) Wiley

The polymers afforded several different textures in the 2D GIXD image that are characteristics of certain orientational motifs depending on the side chain composition (Fig. 5.7). PTzBT-14HD (R^1 = C14 and R^2 = HD) exhibited a texture corresponding to a predominant edge-on orientation as also shown in Fig. 5.5b. When R^1 was shorter such as C12 and C10 (PTzBT-12HD, -10HD), the diffraction texture, in particular, for π–π stacking structure, became bimodal and face-on, respectively. Similar change in the texture was also observed when the longer branched side chain (OD) was used as R^2 (PTzBT-10OD, -12OD, and -14OD). PTzBT-14OD gave the texture similar to that of PTzBT-12HD, where both the polymers have the length difference of two carbons between R^1 and R^2. By shortening the linear side chain, PTzBT-12OD and 10OD, the face-on fraction became more dominant. These results suggest that PTzBTs can form the face-on orientation when the size of the branched side chain is larger. In addition, as PTzBT-10HD and -12OD showed strong diffractions corresponding to the face-on orientation, introduction of the linear and branched side chains with the same length could be a key to direct the backbone more face-on with higher crystallinity at the same time.

When the linear side chain (R^1) of PTzBT-14HD was replaced with the branched EH or BO group (PTzBT-EHHD or -BOHD), the polymers exhibited the texture corresponding to the face-on orientation. In particular, in PTzBT-BOHD, where the

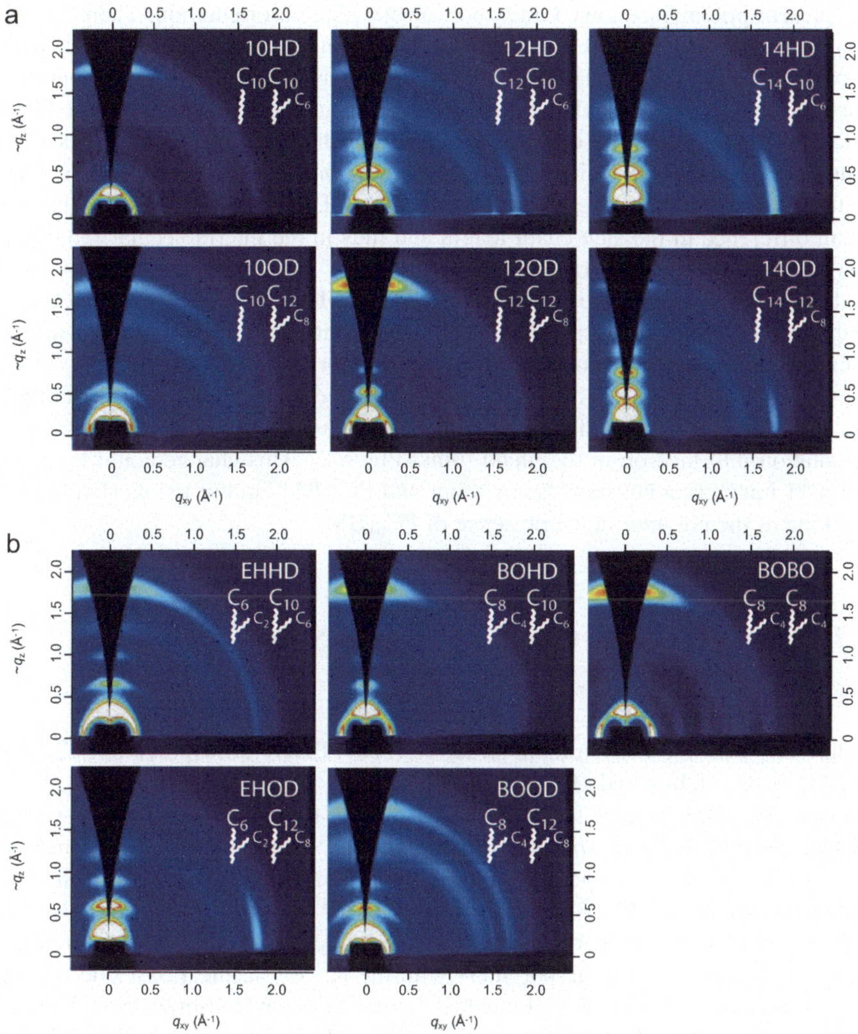

Fig. 5.7 2D GIXD patterns of polymer neat films for PTzBTs with **a** linear and branched side chains and **b** with all-branched side chains. Reproduced with permission [25]. Copyright (2014) Wiley

length difference between R^1 and R^2 is smaller than in PTzBT-EHHD, it was more preferentially oriented in the face-on manner. The texture of PTzBT-BOOD, where the length difference between R^1 and R^2 is the same as PTzBT-EHHD, appeared similar to that of PTzBT-EHHD. When both R^1 and R^2 is BO, PTzBT-BOBO, the face-on orientation was more enhanced, which could be understood by the more regular structure than the other PTzBTs. As an exception in the all-branched polymers, PTzBT-EHOD, with the largest length difference between R^1 and R^2 among the present polymers, showed a texture for the edge-on orientation. Therefore, the length difference between R^1 and R^2, in which the length of the trunk part is taken into account in case of branched side chains, seems to play an important role to determine the backbone orientation. When the difference is larger the polymers form an edge-on orientation: note that in the case of the linear–branched system, the linear side chain must be longer. When the difference is smaller, on the other hand, the polymers tend to form a well-ordered face-on orientation. Overall, it is very interesting that such small difference in the side chain length and thus in the intermolecular interaction can impact the backbone orientation.

Interestingly, however, in the polymer/$PC_{61}BM$ blend films, all the polymers displayed the texture corresponding to face-on orientation. In other words, while the polymers that form face-on orientation in the polymer neat films, "face-on-polymers," preserved their orientation, the polymers that form edge-on or bimodal orientation in the polymer-only films, "edge-on-polymers" or "bimodal-polymers," changed their orientation into face-on in the blend films (Fig. 5.8). This change could be due to the π–π interaction between the polymer and $PC_{61}BM$, and/or the weakened π–π stacking of the polymer in the presence of $PC_{61}BM$.

5.2.3 Impact of Side Chain Topology in Solar Cell Performance

Solar cells with the conventional structure (ITO/PEDOT:PSS/(polymer/$PC_{61}BM$)/Ca/Al) were fabricated by spin coating the chloroform solution of the polymer/$PC_{61}BM$ blend. Figures 5.9a and b show the current density (J)–voltage (V) curves and the EQE spectra of the solar cells with the active layer thickness of roughly 200 nm. The optimal polymer to $PC_{61}BM$ ratio was 1:2 for most of the polymers, and 1:1 for PTzBT-BOOD, 1:3 for-BOBO, and 1:4 for-EHHD. The solar cell parameters are summarized in Table 5.1. Interestingly, a notable difference in V_{OC} was found between the polymers with the R^1, R^2 = linear, branched groups (linear–branched) and R^1, R^2 = branched, branched groups (all-branched). The cells with the linear–branched system gave V_{OC} of 0.81–0.82 V, whereas the cells with all-branched system gave 0.88–0.90 V, despite the fact that the HOMO energy level of the polymers is almost the same. The difference in V_{OC} on the side chain composition is seemingly independent of the orientation. The V_{OC} difference is possibly due to the different intermolecular interaction between the polymer and $PC_{61}BM$ [25]. The weak π–π stacking crystallinity of the all-branched system compared to the linear–branched system in the presence of $PC_{61}BM$ as described above, which can also be counted as weaker polymer–$PC_{61}BM$ interactions, might reduce the charge recombination, in turn leading to the high V_{OC}.

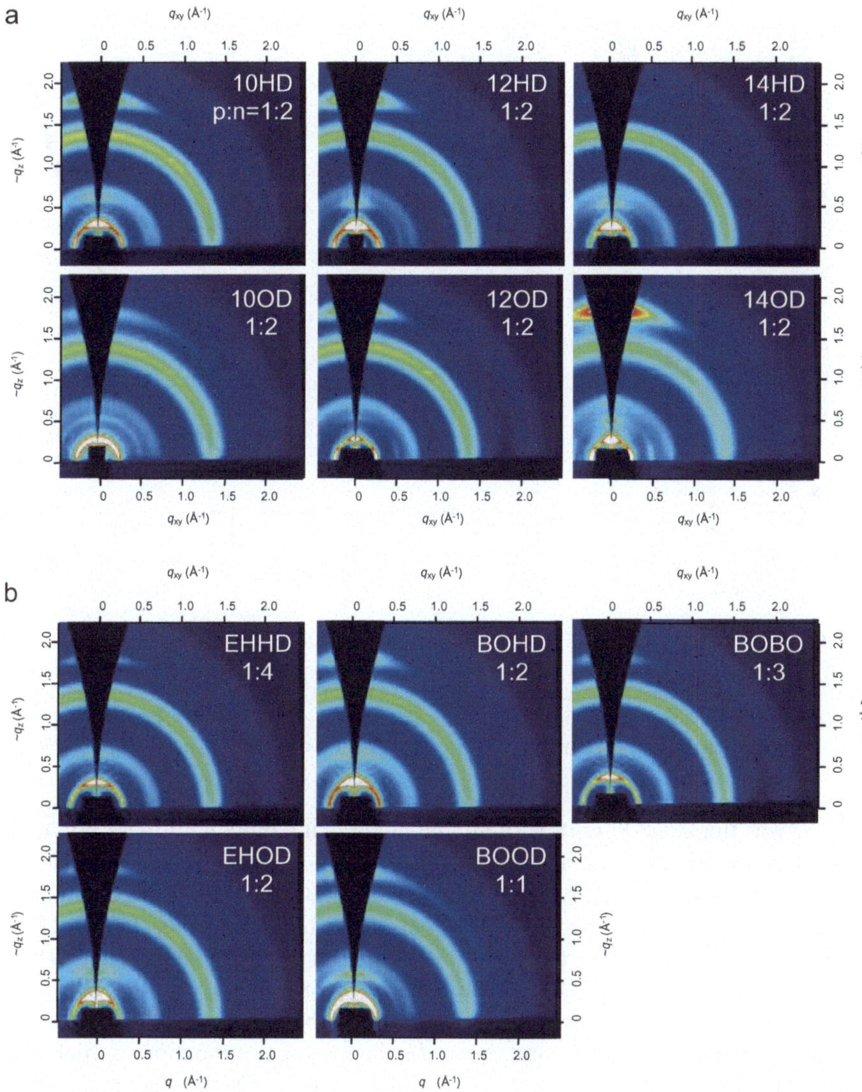

Fig. 5.8 2D GIXD patterns of PTzBT/$PC_{61}BM$ blend films. PTzBTs **a** with linear and branched side chains and **b** with all-branched side chains. Reproduced with permission [25]. Copyright (2014) Wiley

5.2.4 Impact of Backbone Orientation in Solar Cell Performance

The performance of the cells based on PTzBTs was insensitive to the backbone orientation when the active layer was thin (<200 nm), whereas it was quite sensitive when

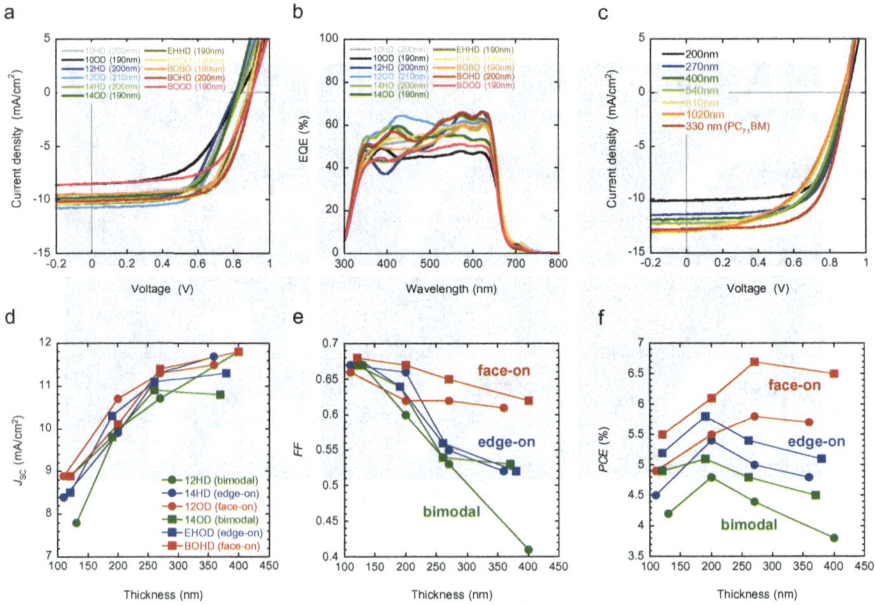

Fig. 5.9 **a, b** *J–V* curves and EQE spectra of the solar cells using PTzBTs/PC_{61}BM as the active layer with the thickness of ca. 200 nm. **c** *J–V* curves of the solar cell with PTzBT-BOHD/PC_{61}BM with different active layer thickness, and with PTzBT-BOHD/PC_{71}BM with the 330 nm thickness. (d-f) Thickness dependence of J_{SC}, FF, and PCE of the solar cells using PTzBT-12HD, -14HD, -12OD, -14OD, EHOD, and -BOHD. Reproduced with permission [25]. Copyright (2014) Wiley

the layer was thick (>200 nm). It should be noted that although all these polymers formed face-on orientation in the blend films as mentioned above, and the solar cell performance was quite sensitive to the primary orientation in the polymer-only film. Figure 5.9d–f displays the short-circuit current density (J_{SC}), fill factor (FF), and PCE of the cells as a function of the active layer thickness, in which PTzBT-14HD,-EHOD (edge-on-polymers), -12HD, -14OD (bimodal-polymers), and -12OD, -BOHD (face-on-polymers) were chosen as representative of the polymers with three different orientations. Clearly, for all the polymers, J_{SC} increased as a function of thickness, reflecting the increased volume of the light-absorbing layer. On the other hand, while in face-on-polymers, FF was mostly preserved above 0.6 even with the 400 nm thickness, in bimodal-polymers and edge-on-polymers, FF dropped to below 0.6 when the thickness was above 200 nm. As a result, PCEs increased for face-on-polymers and decreased for bimodal-polymers and edge-on-polymers when the active layer was thickened to above 200 nm. One can, in part, explain this different trend in the cell performance as a function of the active layer thickness in terms of the charge transport properties. The face-on-polymers afford higher out-of-plane mobilities, and thereby the charge carrier can travel further through the bulk film, which would presumably contribute to high FF even in the thicker films. On the other hand, in the edge-on-polymers and bimodal-polymers, the charge carrier can be trapped or can

Table 5.1 Backbone orientation, charge carrier mobility, and photovoltaic parameters of polymer-based BHJ cells for PTzBT-12HD, -12OD, -14HD, -14OD, -EHOD, and -BOHD

Side chains	Orientation[a]	Photovoltaic parameters					
		p:n ratio[b]	Thickness [nm]	J_{SC} [mA/cm^2]	V_{OC} [V]	FF	PCE [%]
12HD	Bimodal	1:2	130	7.8	0.81	0.67	4.2
			200	10.0	0.81	0.60	4.8
			270	10.7	0.81	0.53	4.4
			350	11.8	0.80	0.41	3.8
12OD	Face-on	1:2	130	8.9	0.83	0.66	4.9
			210	10.7	0.82	0.62	5.5
			260	11.3	0.82	0.62	5.8
			370	11.5	0.82	0.61	5.7
14HD	Edge-on	1:2	110	8.4	0.81	0.67	4.5
			200	9.9	0.82	0.66	5.4
			270	11.3	0.81	0.55	5.0
			360	11.7	0.78	0.52	4.8
14OD	Bimodal	1:2	120	8.9	0.82	0.67	4.9
			190	9.8	0.82	0.64	5.1
			260	10.9	0.81	0.54	4.8
			370	10.8	0.79	0.53	4.5
EHOD	Edge-on	1:2	120	8.5	0.89	0.68	5.2
			190	10.3	0.88	0.64	5.8
			260	11.1	0.87	0.56	5.4
			380	11.3	0.88	0.52	5.1
BOHD	Face-on	1:2	120	8.9	0.91	0.68	5.5
			200	10.1	0.90	0.67	6.1
			270	11.4	0.89	0.65	6.7
			400	11.8	0.89	0.62	6.5
			540	12.2	0.88	0.60	6.4
			810	12.8	0.87	0.53	5.9
			1020	12.9	0.87	0.51	5.8
		1:2[c]	330	12.7	0.90	0.65	7.5

[a]Backbone orientation in the polymer-only film. [b]Polymer (p) to PCBM (n) weight ratio. [c]$PC_{71}BM$ was used as the n-type material

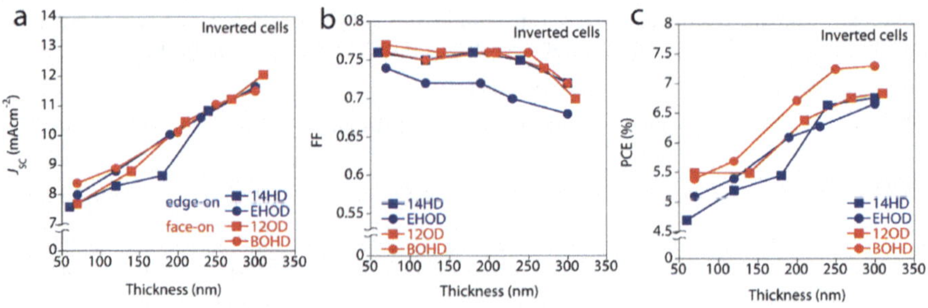

Fig. 5.10 Thickness dependence of the photovoltaic parameters for the PTzBT-based inverted cells: **a** J_{SC}, **b** FF, **c** PCE. Reproduced with permission [26]. Copyright (2018) American Chemical Society

recombine before reaching the electrode, which gives rise to the loss of FF due to the relatively lower mobility.

Having well-ordered face-on structure, with better solubility than the rest of polymers, PTzBT-BOHD turned out to be the best polymer for the use in solar cells in this series (Fig. 5.9c). The PCE of the cell that used PTzBT-BOHD was 5.5% (J_{SC} = 8.9 mA cm^{-2}, V_{OC} = 0.91 V, FF = 0.68) at the thickness of 120 nm, which increased to 6.7% (J_{SC} = 11.4 mA cm^{-2}, V_{OC} = 0.89 V, FF = 0.65) at the thickness of 270 nm. Notably, the PTzBT-BOHD-based cell still demonstrated a PCE of 5.8% even at an extraordinarily thick active layer of 1 μm. In addition, with the use of $PC_{71}BM$ as the n-type material and the optimization of the active layer thickness, the solar cells of PTzBT-BOHD demonstrated a maximum PCE of 7.5% (J_{SC} = 12.7 mA cm^{-2}, V_{OC} = 0.90 V, FF = 0.65) at the active layer thickness of 330 nm.

Interestingly, however, we found that the trend of the thickness dependence in the inverted cells (ITO/ZnO/(PTzBT/$PC_{61}BM$)/MoO_x/Ag) was in sharp contrast to that in the conventional cells. Figure 5.10a–c depicts the thickness dependence of J_{SC}, FF, and PCE of the inverted cells. In all cases, J_{SC} increased from ca. 8–12 mA cm^{-2} as the photoactive layers thickened, reflecting the increased volume of the photoactive layer. The FF for both the edge-on-polymers and face-on-polymers were almost the same (~0.75) when the photoactive layers were very thin (ca. 70 nm), which similarly decreased gently to around 0.7 when the layers were thickened to ca. 300 nm. As a result, PCEs increased as a function of the photoactive layer thickness for both the edge-on- and face-on-polymers.

5.2.5 *Distribution of Backbone Orientation Through Film Thickness*

To discuss the difference in the trend of the thickness dependence between the inverted and conventional cells, we conducted in-depth GIXD studies for the blend films of PTzBT-14HD and -EHOD, and PTzBT-12OD and BOHD, as representatives of the edge-on- and face-on-polymers, respectively, which were fabricated on both the ZnO surface (ZnO coated ITO glass) and the PEDOT:PSS surface (PEDOT:PSS coated ITO glass) with different thicknesses [26]. We then performed a pole figure analysis for the lamellar diffraction of each film [27, 28]. Figure 5.11 depicts the line-cut profiles of the lamellar diffraction along the azimuth angle (χ) for PTzBT-14HD and-12OD on the ZnO and PEDOT:PSS surfaces. The face-on/edge-on ratio can be quantified by dividing the peak area for the face-on crystallite (A_{xy}: 0–45° and 135–180°) by that for the edge-on crystallite (A_z: 55–125°) (see Supporting Information). Note that A_{xy}/A_z does not necessarily indicate the real face-on/edge-on ratio, since the scattering from the direct beam overlaps with the first order lamellar diffraction, especially along the q_z axis, which overestimates the A_z value.

Figure 5.12a and b show the plots of A_{xy}/A_z as a function of the thickness for the blend PTzBT/$PC_{61}BM$ films fabricated on ZnO and PEDOT:PSS, respectively. In all the polymers, the A_{xy}/A_z values were slightly higher for the films on ZnO than for the films on PEDOT:PSS, indicating that the polymers have a higher tendency to form face-on orientation on ZnO than PEDOT:PSS. Further, in both surfaces, the

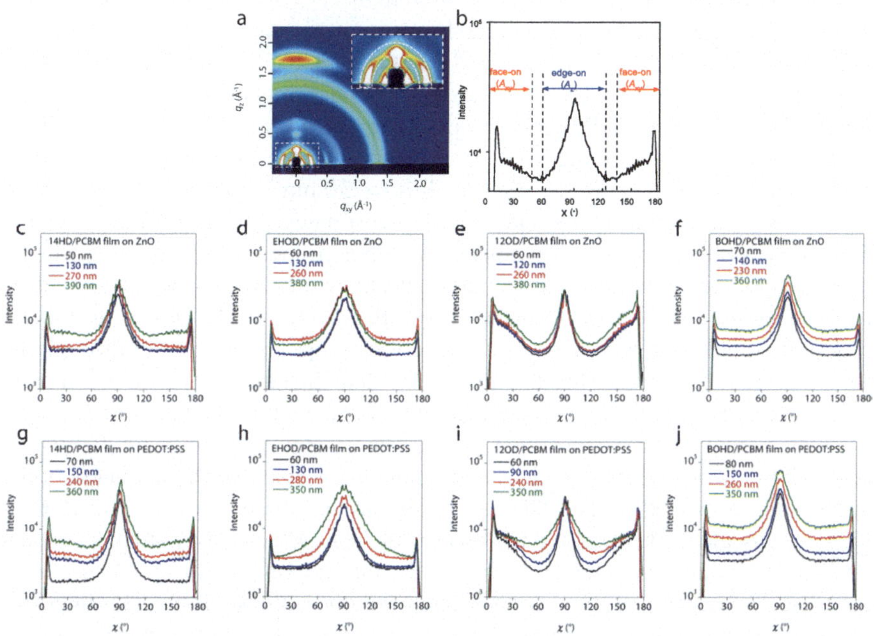

Fig. 5.11 **a, b** Simple model for pole figure analysis. **a** 2D GIXD pattern showing how to take line-cut profiles along the azimuth angle (χ). **b** Pole figure plot for the lamella diffraction cut along the χ angle. **c–j** Pole figure plots extracted from the lamellar diffraction for PTzBT/$PC_{61}BM$ blend films on the ITO/ZnO **c–f** and ITO/ZnO **g–j** substrates: **c, g** PTzBT-14HD, **d, h** PTzBT-EHOD, **e, i** PTzBT-12OD, **f, j** PTzBT-BOHD Reproduced with permission [26]. Copyright (2018) American Chemical Society

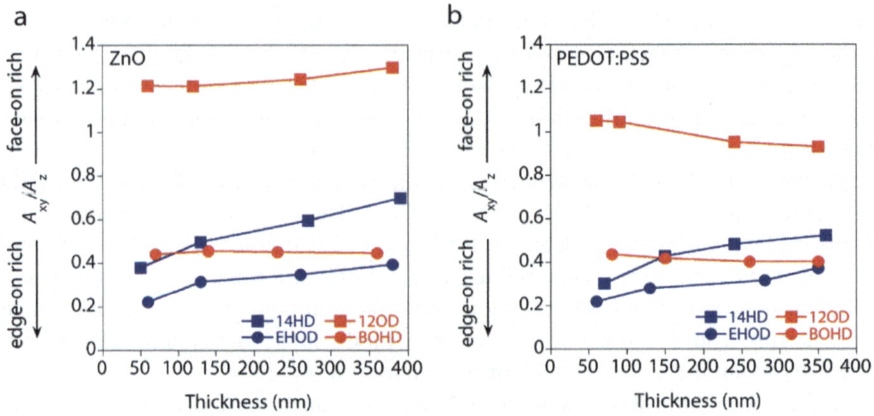

Fig. 5.12 Dependence of A_{xy}/A_z on film thickness, where A_{xy}/A_z corresponds to the ratio of the face-on to edge-on orientation. **a** Films on the ZnO surface, **b** films on the PEDOT:PSS surface. Reproduced with permission [26]. Copyright (2018) American Chemical Society

face-on polymers (12OD and BOHD) showed higher A_{xy}/A_z values than the edge-on polymers (14HD and EHOD), reflecting the primary orientation, in the thinner films. Interestingly, whereas the A_{xy}/A_z value did not change by the film thickness for the face-on polymers, it gently increased as the film thickness increased for the edge-on polymers. Notably, 14HD (edge-on polymer) were found to show even higher A_{xy}/A_z values, thus possessing a larger population of the face-on crystallite, than BOHD (face-on polymer) in the thicker films (>250 nm). However, the difference in A_{xy}/A_z by the thickness does not directly correlate with the difference in the photovoltaic performances by the thickness: although the edge-on polymers had a larger population of the face-on crystallite at larger thicknesses, which should be favorable for the solar cell performance, they afforded lower FF. This is most likely because the thickness, which is the distance the charge carriers travel through the cell, is a more critical factor for determining FF than the ratio of face-on orientation.

The increase in the face-on crystallite with increasing film thickness in the edge-on-polymers may rather suggest that the face-on to edge-on ratio is not distributed evenly along the film thickness. It is reasonable to assume that the thickness of the "interfacial" layer of the PTzBT/PC_{61}BM film around the bottom surface (ZnO or PEDOT:PSS) and the face-on to edge-on ratio in the interfacial layer are independent of the total film thickness. Hence, the increase in the face-on fraction mainly occurs in the bulk and arises from the increase in the bulk volume in thicker films. This indicates that the face-on fractions are more abundant in the bulk than at the interface of the film-bottom surface. On the other hand, the fact that the face-on/edge-on ratio did not change with the film thickness in the face-on-polymers suggests that the face-on to edge-on ratio is distributed evenly along the film thickness.

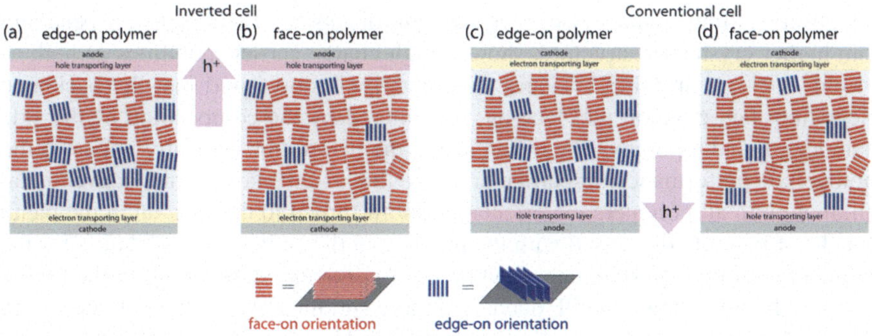

Fig. 5.13 Schematic images of the blend films of the edge-on-polymer (**a**, **c**) and the face-on-polymer (**b**, **d**), showing the distribution of the orientation in the inverted (**a**, **b**) and conventional cells (**c**, **d**). Note that $PC_{61}BM$ molecules/aggregates are not illustrated for the sake of simplicity. Reproduced with permission [26]. Copyright (2018) American Chemical Society

5.2.6 *Correlation Between Thickness Dependence in FF and Backbone Orientation*

The distribution of the backbone orientation through the film thickness correlates with the difference in the thickness dependence of FF between the edge-on-polymers and the face-on-polymers. In the edge-on-polymers, the population of the face-on crystallite is relatively larger in the bulk, and the population of the edge-on crystallites is relatively larger at the film-bottom interlayer interface. This distribution could facilitate the vertical hole transport in the inverted cell, where the generated holes flow toward the top interlayer (MoO_x) through the face-on rich region (Fig. 5.13a). In contrast, this distribution of the orientation should be detrimental to the conventional cell, where the generated holes flow toward the bottom interlayer through the edge-on rich region (Fig. 5.13c). Note that we did not observe the segregation of $PC_{61}BM$ at the substrate interface, which would also affect the device performance [29, 30]. It is thus quite reasonable that, with the better matching between orientation distribution and the cell structure, the edge-on polymers showed a gentler decrease in FF with increasing film thickness in the inverted cell than in the conventional cell. With respect to the face-on polymers, the face-on to edge-on ratio is distributed evenly through the film thickness (Fig. 5.13b and d). It is thus assumed that the generated holes flow similarly in both the upper and lower directions, and thereby the thickness dependence in FF is insensitive to the cell structure.

5.2.7 *Summary*

A series of thiophene–thiazolothiazole polymers was summarized. Due to the donor–acceptor backbone, these polymers had strong intermolecular interactions, leading to high crystallinity with close π–π stacking distances. The polymers, benefited from higher crystalline structure, showed high charge carrier mobilities as high as 0.4 $cm^2\ V^{-1}\ s^{-1}$ in OFET devices. Importantly, it was found that the backbone orientation, i.e., edge-on and face-on, was altered by the composition of the side chain length and topology. The difference in the orientation critically impacted the photovoltaic performance, in particular, at the thick active layer in the conventional cell, though in all polymers the orientation motif turned to the face-on in the blend film. The FF gently decreased with the increase in the active layer thickness for the polymers with face-on orientation, while the J_{SC} increased, resulting in the higher PCE. On the other hand, the FF decreased more significantly for the polymers with edge-on orientation, giving rise to the lower PCE. However, this was not the case in the inverted cell. Both the edge-on-polymers and face-on-polymers showed similar dependence in the photovoltaic performance by the active layer thickness, where the FF gently decreased with increasing thickness. We found that, interestingly, there is a clear difference in the distribution of the backbone orientation through the film thickness between the edge-on-polymers and face-on-polymers in the blend film, which seemingly correlated well with the difference in the thickness dependence in the photovoltaic performance.

5.3 Naphthobischalcogenadiazole-Based Polymers

5.3.1 Naphthobischalcogenadiazoles

Among the many acceptor building units, benzochalcogenadiazoles (BXz) such as 2,1,3-benzoxadiazole (BOz) [31], 2,1,3-benzothiadiazole (BTz) [14], and 2,1,3-benzoselenadiazole (BSz) [32] (Fig. 5.14) including their derivatives have been widely investigated and have afforded a large number of π-conjugated polymers that show great performances in polymer solar cells. As their doubly fused system,

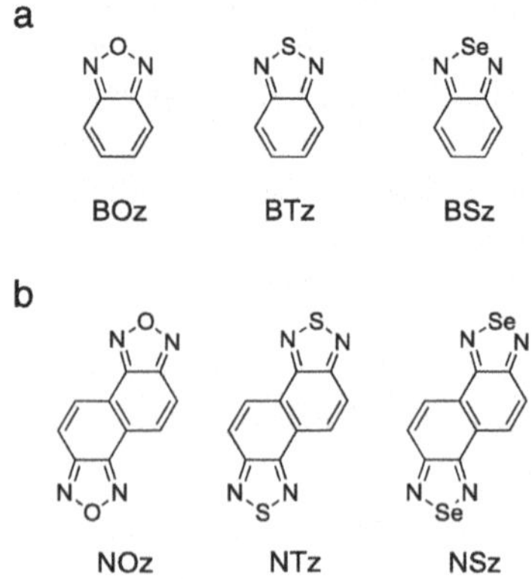

Fig. 5.14 Chemical structures of **a** benzochalcogenadiazoles and **b** naphthobischalcogenadiazoles

naphthobischalcogenadiazoles (NXz) such as naphtho1,2-*c*:5,6-*c*′]bis[1, 2, 5]oxadiazole (NOz) [33, 34], naphtho1,2-*c*:5,6-*c*′]bis[1, 2, 5]thiadiazole (NTz) [35, 36], and naphtho1,2-*c*:5,6-*c*′]bis[1, 2, 5]selenadizaole (NSz) [33] (Fig. 5.14) are emerging acceptor units that can be used for semiconducting polymers. The advantages of NXz over BXz are the extended π-electron system that enhances the intermolecular interactions and thus the ordering of the polymers, and the enhanced electron deficiency that deepens the HOMO and LUMO energy levels. Further, with its centrosymmetric structure, NXz is beneficial for designing more regularly structured polymers, which would contribute to the polymer ordering, relative to BXz with an axisymmetric structure. In particular, NOz and NTz are found to be promising building units for π-conjugated polymers.

5.3.2 *Quaterthiophene–Naphthobisthiadiazole Polymer (PNTz4T): Comparison with Benzothiadiazole Analog*

NTz is the most widely studied unit among the family of NXz [37, 38]. While the group of Cao and Huang has reported a series of benzodithiophene–NTz polymers and their application to solar cell [35], we have been studying a series of quaterthiophene–NTz polymers, such as PNTz4T (Fig. 5.15a) [36]. In these polymers, the 2-decyltetradecyl (DT) long branched alkyl group was used as the substituent to ensure sufficient solubility in solvents, which was attached to the thiophene rings neighboring the NTz moiety at β-positions heading the unsubstituted bithiophene moiety to avoid steric hindrance between NTz and alkylthiophene. The absorption of PNTz4T covered whole visible range with the absorption maximum (λ_{max}) of 725 nm and the absorption edge (λ_{edge}) reaching 805 nm, which were red-shifted for approximately 50 nm relative to those of PBTz4T (Fig. 5.15b). As a consequence, the E_g was determined to be 1.54 eV, which was 0.1 eV smaller than that of PBTz4T. Both the HOMO and LUMO energy levels were deeper for PNTz4T than PBTz4T. These results are most likely due to the strong electron-poor nature of NTz as compared to BTz.

The hole mobility, evaluated by using OFET devices, for PNTz4T was as high as 0.56 $cm^2\ V^{-1}\ s^{-1}$ and approximately one order of magnitude higher than that for PBTz4T (~0.05 $cm^2\ V^{-1}\ s^{-1}$). In solar cells with a conventional structure (ITO/PEDOT:PSS/(polymer/$PC_{61}BM$)/LiF/Al), whereas the PBTz4T cell showed a PCE of 2.6% (J_{SC} = 5.6 mA cm^{-2}, V_{OC} = 0.74 V, FF = 0.63), the PNTz4T cell showed an approximately twofold higher PCE of 6.3% (J_{SC} = 12.0 mA cm^{-2}, V_{OC} = 0.76 V, FF = 0.69) (Fig. 5.15c).

The higher photovoltaic performance of PNTz4T relative to PBTz4T was understood by investigating the thin-film structure on the basis of 2D GIXD measurements. Figure 5.16a–d shows the 2D GIXD patterns of PNTz4T and PBTz4T in the neat and blend film. In the PNTz4T neat film, diffractions assignable to the lamellar structure, up to the fifth order, and a diffraction assignable to the π–π stacking structure,

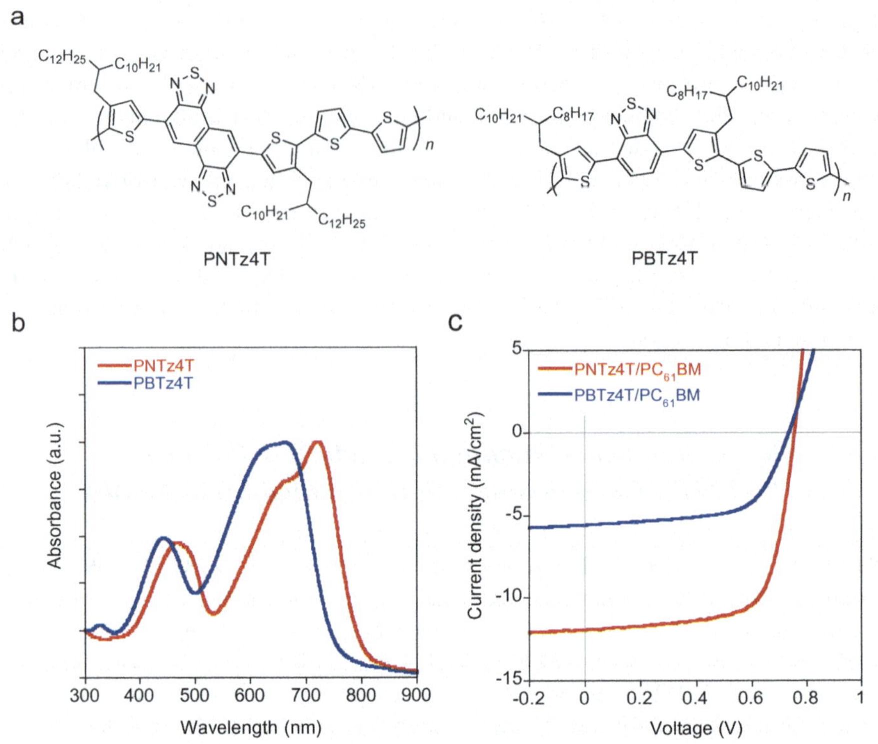

Fig. 5.15 **a** Chemical structure of PNTz4T and PBTz4T, **b** UV-vis absorption spectra of PNTz4T and PBTz4T in thin films, and **c** *J*–*V* curves of the solar cells based on PNTz4T/$PC_{61}BM$ and PBTz4T/$PC_{61}BM$ blends

appeared on the quasi q_z (~$q_z \approx 0.25$ Å^{-1}) and q_{xy} ($q_{xy} \approx 1.7$ Å^{-1}) axes, respectively. This suggested that the polymer formed a high crystalline structure, with the backbones predominantly oriented in the edge-on manner. The π–π stacking distance was determined to be 3.5 Å, which was quite small for π-conjugated polymers. In contrast, PBTz4T showed lamellar and π–π stacking diffractions along the q_z and ~q_{xy} axes, indicating a face-on orientation. Further, the lamellar diffraction appeared only for the first order, indicating a less ordered structure. Interestingly, in the polymer/$PC_{61}BM$ blend films, PNTz4T mainly oriented in a face-on manner as the π–π stacking diffraction appeared on the ~q_z axis. Further, π–π stacking distance unchanged by blending, suggesting that the crystallinity was maintained. On the other hand, PBTz4T provided a much less crystalline feature in the blend film as there was no π–π stacking diffraction.

NTz is a more highly extended fused ring as compared to BTz, and thus it is fairly reasonable that PNTz4T forms a more highly ordered structure as compared to PBTz4T. However, the difference in crystallinity could be too large. We speculated that this marked difference originated in the difference of symmetry between the NTz

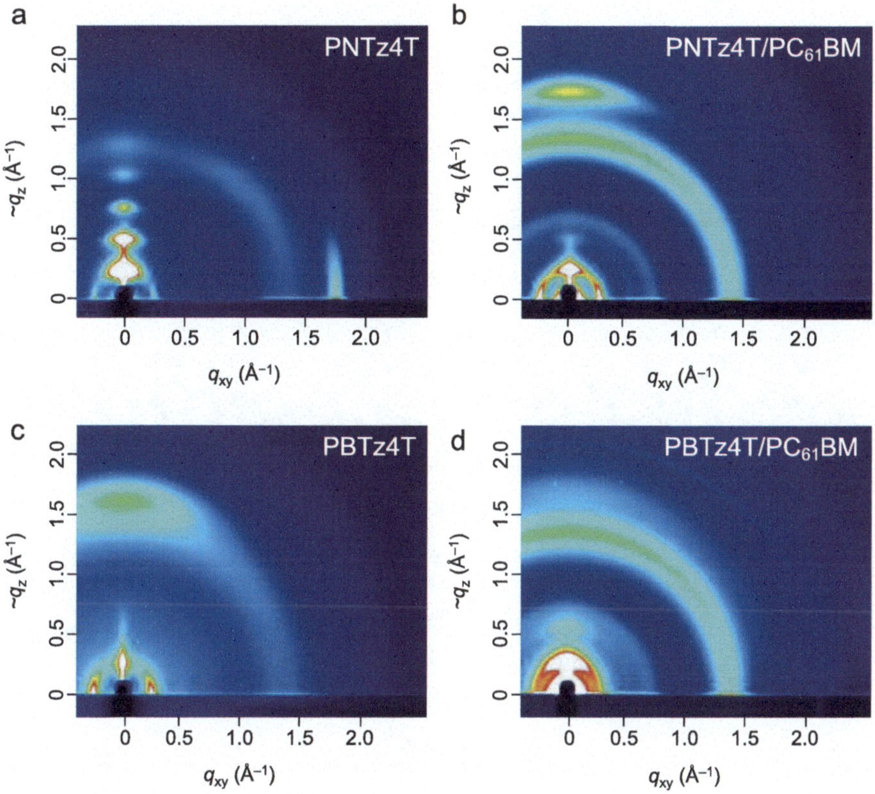

Fig. 5.16 2D GIXD patterns of the thin films. **a** PNTz4T, **b** PNTz4T/$PC_{61}BM$, **c** PBTz4T, **d** PBTz4T/$PC_{61}BM$. Reproduced with permission [38]. Copyright (2017) Wiley

and BTz unit. As shown by the single-crystal X-ray analysis of the model compounds, NTz2T and BTz2T, the NTz (or BTz)–thiophene linkage is fixed in one configuration. Thus, NTz, with a centrosymmetrical structure, affords an *anti*-arrangement of the thiophene rings that sandwich NTz, whereas BTz, with an axisymmetrical structure, affords a *syn* arrangement of the neighboring thiophenes (Fig. 5.17a). Based on these arrangements, PNTz4T gives a more linear-shaped backbone as compared to PBTz4T, which gives a "wavy" shape (Fig. 5.17b). The backbone shape might largely affect the packing structure [39, 40], and this relatively linear backbone shape, together with the rigidity of NTz, could lead to the highly ordered packing structure in the thin film in PNTz4T compared to PBTz4T.

In addition, the photovoltaic performance of PNTz4T was further improved by the device optimization (Fig. 5.18) [3]. As PNTz4T formed a crystalline structure with the favorable face-on orientation, as similar to PTzBTs, the performance of the PNTz4T cell was maximized at thicker active layers. When the active layer thickness was increased to approximately 300 nm, J_{SC} improved markedly, resulting in the PCE of 8.7%. Replacing $PC_{61}BM$ by $PC_{71}BM$ slightly improved PCE to 8.9%

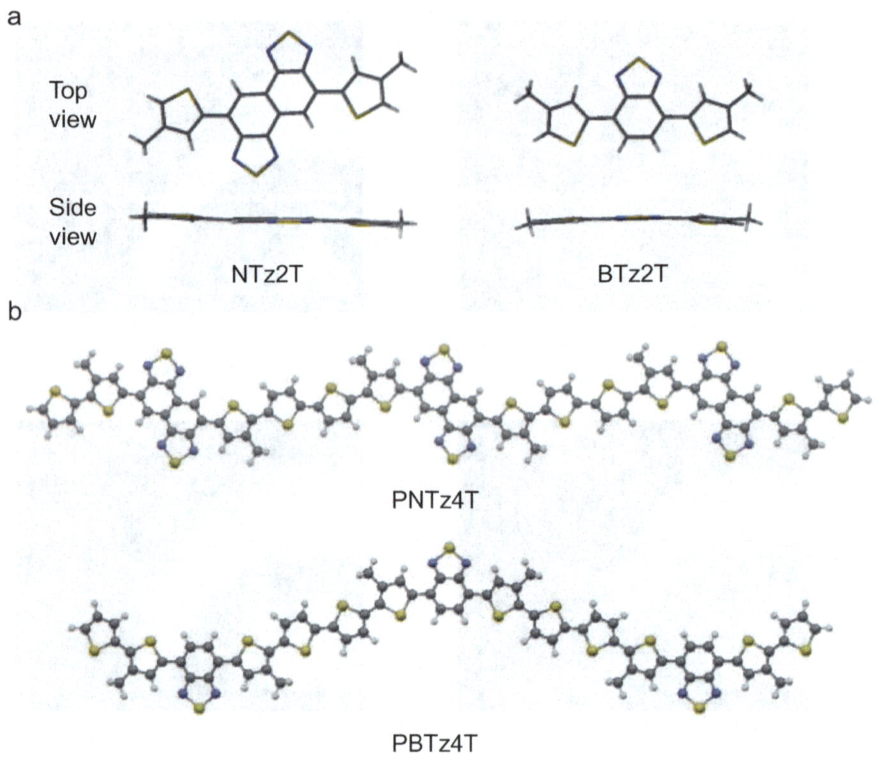

Fig. 5.17 **a** Crystal structures of NTz2T and BTz2T determined by single-crystal X-ray analysis, **b** optimized backbone structures for PNTz4T and PBTz4T. Reproduced with permission [36]. Copyright (2012) American Chemical Society

owing to the increased absorption at wavelengths of around 400–600 nm. Further, the use of the inverted structure (ITO/ZnO/(PNTz4T:PCBM)/MoO_x/Ag) resulted in a higher J_{SC} and thus the PCEs of 9.8% with $PC_{61}BM$ and 10.1% with $PC_{71}BM$. The improved performance in the inverted cell can also be explained by the distribution of the orientation through the film thickness, as is the case in PTzBT.

5.3.3 Molecular Design of Polymers Using NTz and NOz Toward Reduction of Photon Energy Loss

5.3.3.1 Importance of Photon Energy Loss

In order to improve the efficiency further, it is necessary to realize a high J_{SC} and a high V_{OC} at the same time. However, it is a well-known fact that there is a trade-off

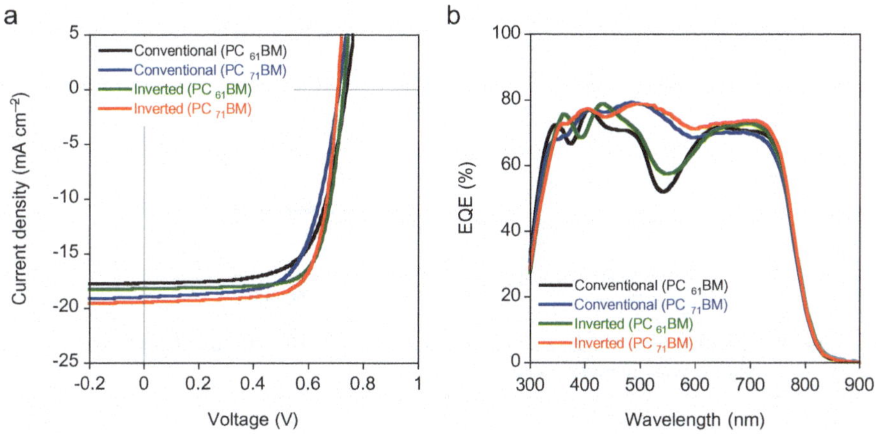

Fig. 5.18 **a** J–V curves and **b** EQE spectra of the optimized PNTz4T-based cells with conventional and inverted architectures. Reproduced with permission [3]. Copyright (2015) Nature Publishing group

between those two parameters, and thus minimizing the trade-off is one of the most crucial issues. A key to resolving the issue in polymer solar cells is to reduce the relatively large photon energy loss (E_{loss}), defined by $E_g - eV_{\text{OC}}$, which is also called voltage loss [41, 42]. The E_{loss} in organic solar cells is typically 0.7–1.0 eV [41, 43], which is larger than that in inorganic solar cells and perovskite solar cells, which are around 0.4–0.5 eV [42, 44]. Thus, the V_{OC} for organic solar cells is essentially lower than that for inorganic and perovskite solar cells.

The E_{loss} as well as the trade-off is strongly related to the match of the molecular orbital energy levels between the polymer and fullerene. Given the fundamental working mechanism of the organic solar cell, the realization of both high J_{SC} and V_{OC} requires that the polymer must have both a narrower E_g and a deeper HOMO energy level. This inevitably results in a deeper LUMO energy level, diminishing the energy offset of the LUMOs between the polymer and fullerene (ΔE_L) (Fig. 5.19). In other words, reducing the ΔE_L results in the reduction of E_{loss} and thus a high V_{OC}. However, as ΔE_L is considered to be a driving force for the photoinduced charge separation [45, 46], a reduced ΔE_L causes a loss of the driving force, in turn giving rise to a low J_{SC} and thereby a low PCE, even though it can bring about a high V_{OC}. In this regard, managing the energetics between polymers and fullerenes and thus to minimize the trade-off is a crucial issue.

5.3.3.2 Fluorination on the PNTz4T Backbone

With the strong electron-withdrawing nature, the introduction of fluorine into the polymer backbone can deepen the energy levels while minimally changing E_g [47]. Therefore, the fluorine atom can be a powerful functional group for reducing the

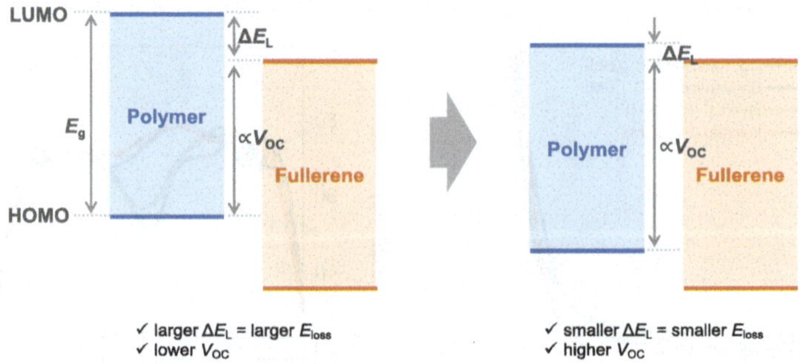

Fig. 5.19 Relationship between the energetics of the materials (HOMO and LUMO energy levels, ΔE_L) and the photovoltaic parameters (V_{OC}, E_{loss})

E_{loss}. It is also believed that the fluorine atom offers noncovalent attractive interactions between the hydrogen or sulfur atoms (F···H or F···S), which may contribute to enhancing the coplanarity of the polymer backbone and hence the crystallinity. Thus, we have synthesized a series of PNTz4T derivatives with the fluorine atoms: PNTz4TF2 and PNTz4TF4 with two and four fluorine groups on the bithiophene moiety [48], PNF4T with two fluorine groups on the NTz moiety, and PNF4TF2 with two fluorine groups on each NTz and bithiophene moiety (Fig. 5.20) [49]. To better distinguish the chemical structure, PNTz4T, PNTz4TF2, PNTz4TF4, PFN4T, and PFN4TF2 will be hereinafter called F0–F0, F0–F2, F0–F4, F2–F0, and F2–F2, respectively.

Figure 5.21a depicts the energy diagrams of the polymers, in which HOMO and LUMO energy levels were determined by the photoelectron yield spectroscopy (PYS) and the low-energy inverse photoelectron spectroscopy (LEIPS) [50]. The HOMO energy levels of F0–F2 and F0–F4 were –5.27 and –5.46 eV, respectively, which were deeper than that of F0–F0 (–5.09 eV) by 0.18 and 0.37 eV. The LUMO energy levels of F0–F2 and F0-F4 were –3.18 and –3.30 eV, respectively, which were also deeper than that of F0-F0 (–3.12 eV), but only by 0.06 and 0.18 eV. Thus, the electronic effect of the fluorine atom was larger on HOMO than on LUMO when fluorine atoms were introduced on the bithiophene moiety. When fluorine was introduced on the NTz moiety (F2–F0), while the HOMO energy level was slightly deeper by

Fig. 5.20 Chemical structures of fluorinated PNTz4T derivatives

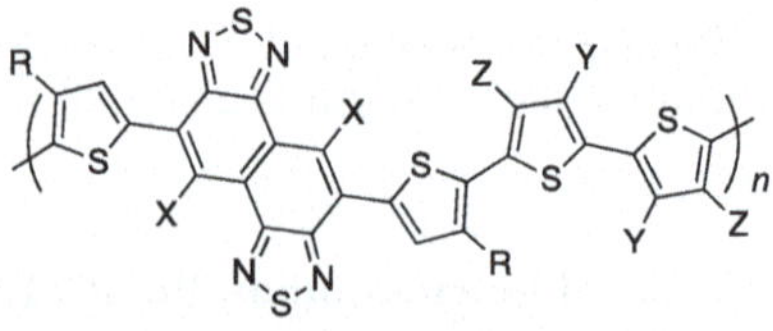

F0-F0 / PNTz4T (X = H, Y = H, Z = H)
F0-F2 / PNTz4TF2 (X = H, Y = F, Z = H)
F0-F4 / PNTz4TF4 (X = H, Y = F, Z = F)
F2-F0 / PNF4T (X = F, Y = H, Z = H)
F2-F2 / PNF4TF2 (X = F, Y = F, Z = H)

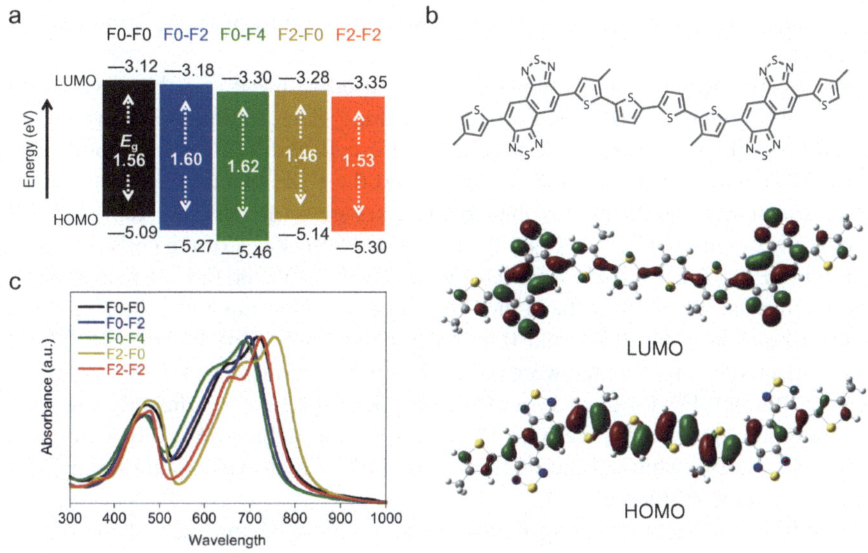

Fig. 5.21 **a** Energy diagrams of PNTz4T (F0–F0) and its fluorinated derivatives. The HOMO and LUMO energy levels were determined by the photoelectron yield spectroscopy and the low-energy photoelectron spectroscopy, respectively. **b** Geometry of HOMO and LUMO for the model compound of F0–F0 calculated by the DFT method (B3LYP/6-31 g(d)). **c** UV-vis absorption spectra of F0–F0 and its fluorinated derivatives in thin films

0.05 eV, i.e., –5.14 eV, than that of F0–F0, the LUMO energy level was deeper by 0.16 eV, i.e., –3.28 eV, than that of F0–F0, indicating that the electronic effect was larger on LUMO than on HOMO. Further, the HOMO and LUMO energy levels of F2–F2 were found to equally down shifted compared to F0–F0, resulting in –5.30 and –3.35 eV, respectively. These shifts of energy levels can be understood by the HOMO and LUMO geometry of the backbone, as evidenced by the DFT calculation (B3LYP/6-31 g(d)), where HOMOs are mainly localized on the oligothiophene unit, whereas LUMOs are mainly localized on the NTz moieties (Fig. 5.21b).

Figure 5.21c shows the UV-vis absorption spectra of the polymers in thin films. F0–F2 and F0–F4 gave absorption maximum (λ_{max}) at 697 and 688 nm, which were blue-shifted by 24 and 33 nm relative to F0–F0 that had λ_{max} of 721 nm, respectively. The E_gs of F0–F2 and F0–F4 were calculated to be 1.60 and 1.62 eV from the λ_{edge} (774 and 765 nm), respectively, and were slightly wider than that of F0–F0 (E_g = 1.56 eV, λ_{edge} = 793 nm). On the other hand, the λ_{max} and λ_{edge} of F2–F0 were 760 and 850 nm, respectively, both of which were red-shifted by 40–50 nm from those of F0–F0. Accordingly, E_g of F2–F0 was calculated to be 1.46 eV, which was reduced by ≈0.1 eV relative to that of F0–F0. F2–F2 gave an absorption band, with a λ_{max} of 727 nm and a λ_{edge} of 810 nm that was similar to that of F0–F0, resulting in an E_g of 1.53 eV. Thus, it is concluded that the fluorination on the NTz moiety reduces the bandgap of the polymer, whereas the fluorination on the bithiophene moiety enlarges the bandgap. This trend is consistent with results in the shift of HOMO and LUMO energy levels.

We fabricated solar cells that used these polymers with ITO/ZnO/(polymer/$PC_{71}BM$)/MoO_x/Ag stacking. The optimum polymer to $PC_{71}BM$ weight ratio was 1:1.5 for the F2–F2 cell and 1:2 for the other cells. Figure 5.22a and b displays the J–V curves and the EQE spectra of the optimized cells, respectively, and the photovoltaic parameters are summarized in Table 5.2. All the polymers exhibited higher V_{OC}s than F0–F0, where V_{OC}s of the cells that used F0–F2, F0–F4, F2–F0, F2–F2 were 0.82 V, 0.93 V, 0.73 V, and 0.84 V, respectively, mostly reflecting the shift of the HOMO energy level. Consequently, the E_{loss}s was found to be 0.78, 0.69, 0.73, and 0.69 eV for the F0–F2, F0–F4, F2–F0, F2–F2 cells, respectively, all of which were reduced from that of the F0–F0 (0.84 eV) cells. Further, although F0–F4 gave a low PCE of 6.5%, due to a significantly low J_{SC}, other polymers provided quite high PCEs despite the reduced E_{loss}. In particular, F0–F2 and F2–F2 exhibited the PCE close to 11%, which was among the highest values for polymer/fullerene cells.

In addition, it was interesting that there was a clear dependence of the photovoltaic properties on the fluorination position when the active layer thickness was changed. Figure 5.22c–e depicts the dependence of J_{SC}, FF, and PCE on the active layer thickness. In F0–F0 and F0–F2, J_{SC} increased as the active layer thickness increased to above 300 nm, most likely due to the increased photon absorption, whereas in F0–F2, F0–F4, and F2–F2, with the fluorine atoms on the bithiophene

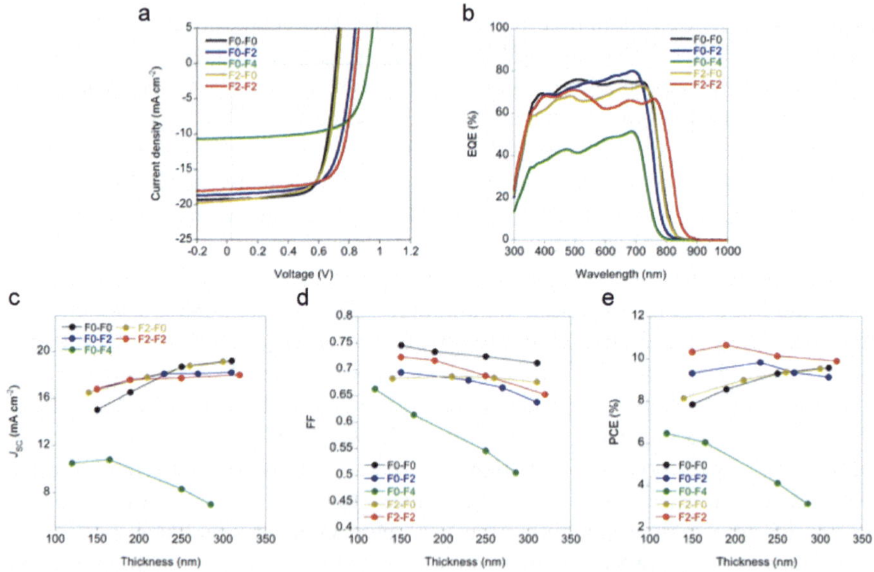

Fig. 5.22 Photovoltaic performances of the cells that used PNTz4T (F0–F0) and its fluorinated derivatives in combination with $PC_{71}BM$. **a** J–V curves, **b** EQE spectra, thickness dependence of **c** J_{SC}, **d** FF, and **e** PCE

Table 5.2 Photovoltaic parameters of the optimized polymer/$PC_{71}BM$ cells

Polymer	Thickness (nm)[a]	J_{SC} (mA cm^{-2})	V_{OC} (V)	FF	PCE (%)[b]	E_{loss} (eV)[c]
F0–F0	320	19.4	0.71	0.71	9.6 [9.2]	0.84
F0–F2	240	19.3	0.82	0.68	10.5 [10.1]	0.78
F0–F4	120	10.5	0.93	0.66	6.5 [5.9]	0.69
F2–F0	300	19.2	0.73	0.68	9.6 [9.2]	0.73
F2–F2	190	17.8	0.84	0.72	10.8 [10.4]	0.69

[a]Thickness of the active layer. [b]Maximum power conversion efficiency. The brackets are average power conversion efficiencies obtained from more than 10 devices. [c]Photon energy loss defined by $E_g - eV_{OC}$

moiety, J_{SC} increased at first but almost plateaued at around 200 nm thickness or even decreased for F0–F4 at more than 150 nm. With respect to FF, in F0–F0 and F0–F2, it decreased very gently as the thickness increased, whereas in other polymers, it decreased more steeply. As a result, F0–F0 and F0–F2 showed gradual increases in PCE with increasing thickness, whereas F0–F2, F0–F4, and F2–F2 showed gradual decreases above 200 nm thickness. Thus, the optimum thickness for the F2–F0 and F0–F0 cells was >300 nm, whereas that for the F2–F2 and F0–F2 cells was around 200 nm or less. The trend was correlated well with the bimolecular recombination [48, 49]. The differences in the thickness dependence and recombination were supported by the structural order of the polymers in the blend film, in which F0–F2, F0–F4, and F2–F2 having fluorine atoms on the bithiophene moiety showed lower crystallinity as revealed by the 2D GIXD study [48, 49]. These results were somewhat interesting because the polymers with the fluorinated bithiophene moieties (F0–F2, F0–F4, and F2–F2) should have more coplanar backbones due to the noncovalent F···S interactions. A plausible reason is that these polymers have relatively lower solubility due to the more coplanar backbones originating in the F···S interactions, which makes them solidify more quickly before self-organizing to pack in order during the spin coating, resulting in the lower crystallinity.

5.3.3.3 Quaterthiophene–NOz Polymer

NOz, which is an oxygen analog of NTz, has significantly more electron-poor nature than NTz, most likely as a result of the stronger electron negativity of oxygen than sulfur [33]. Thus, the incorporation of NOz can further deepen both the HOMO and LUMO energy levels while maintaining the E_g. In fact, a quaterthiophene–NOz polymer (PNOz4T) [33, 51], in which NTz in PNTz4T was replaced with NOz (Fig. 5.23a), had an absorption spectrum slightly red-shifted from PNTz4T, thus even narrowed E_g of 1.52 eV (Fig. 5.23b). The HOMO and LUMO energy levels, estimated by the cyclic voltammetry (Fig. 5.23c), were –5.48 (±0.007) eV and –3.65 (±0.014) eV, which were 0.34 and 0.19 eV deeper than those of PNTz4T with the HOMO and LUMO energy levels of –5.14 (±0.022) eV and –3.46 (±0.022) eV

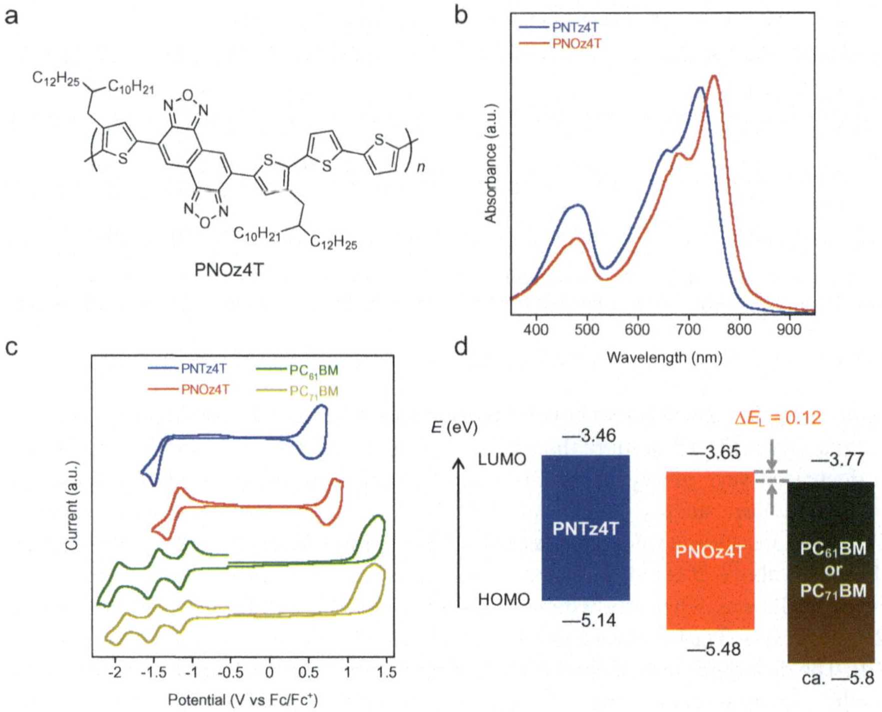

Fig. 5.23 **a** Chemical structure of PNOz4T. **b** UV-vis absorption spectra of PNOz4T in the thin film in comparison with PNTz4T. **c** Cyclic voltammograms of PNTz4T, PNOz4T, $PC_{61}BM$, and $PC_{71}BM$. **d** Energy diagrams of PNTz4T, PNOz4T, $PC_{61}BM$, and $PC_{71}BM$. The HOMO and LUMO energy levels were determined by onset redox potentials in the cyclic voltammograms. Reproduced with permission [51]. Copyright (2015) Nature Publishing Group

(Fig. 5.23d). Importantly, whereas ΔE_L for PNTz4T (between $PC_{71}BM$) was 0.31 eV, that for PNOz4T was 0.12 eV, which is much smaller than the widely referenced empirical threshold value of 0.3 eV [45]. This indicates that the driving force of the photoinduced charge separation is reduced in the PNOz4T system compared with the PNTz4T system. Note that, however, the estimation of the energetics based on the LUMO energy level evaluated by the measurements on the isolated polymer and PCBMs is considered to neglect the influence of exciton binding energy and other interfacial effects [52, 53]. Alternatively, the energy offset can be estimated by the difference of the excited state energy of the polymer and the charge transfer (CT) state energy (E_{CT}) of the polymer/PCBM blend [41, 54]. Thus, we attempted to determine E_{CT} of the blend films by means of the electroluminescence measurement. However, for the PNOz4T/$PC_{71}BM$ system, the emission band of the pure PNOz4T and the CT emission band of the PNOz4T/$PC_{71}BM$ blend system measured by the electroluminescence cannot be distinguished, and thus E_{CT} cannot be evaluated. This suggests, however, that the excited state of the polymer and the CT state of the blend are energetically very close, and thereby the energy offset is very small [55]. This is consistent with the small ΔE_L for the PNOz4T/$PC_{71}BM$ blend system measured by cyclic voltammetry.

The solar cells were fabricated using an inverted architecture, ITO/ZnO/(PNOz4T/$PC_{61}BM$ or $PC_{71}BM$)/MoO_x/Ag. Figure 5.24a and b displays the J–V curves and EQE spectra of the PNOz4T cells. Although ΔE_L was smaller than the empirical threshold, the PNOz4T cells exhibited fairly high PCEs, with the maximum PCE of 8.9% (J_{SC} = 14.5 mA cm^{-2}, V_{OC} = 0.96 V, FF = 0.64) when combined with $PC_{71}BM$. The observed V_{OC}s of 0.96–1.0 V are higher by more than 0.2 V, even close to 0.3 V, than those observed for the PNTz4T cells, consistent with the difference in the HOMO energy level. Importantly, as E_g of PNOz4T was 1.52 eV, E_{loss} of the PNOz4T/PCBM system were found to be 0.52–0.56 eV, which

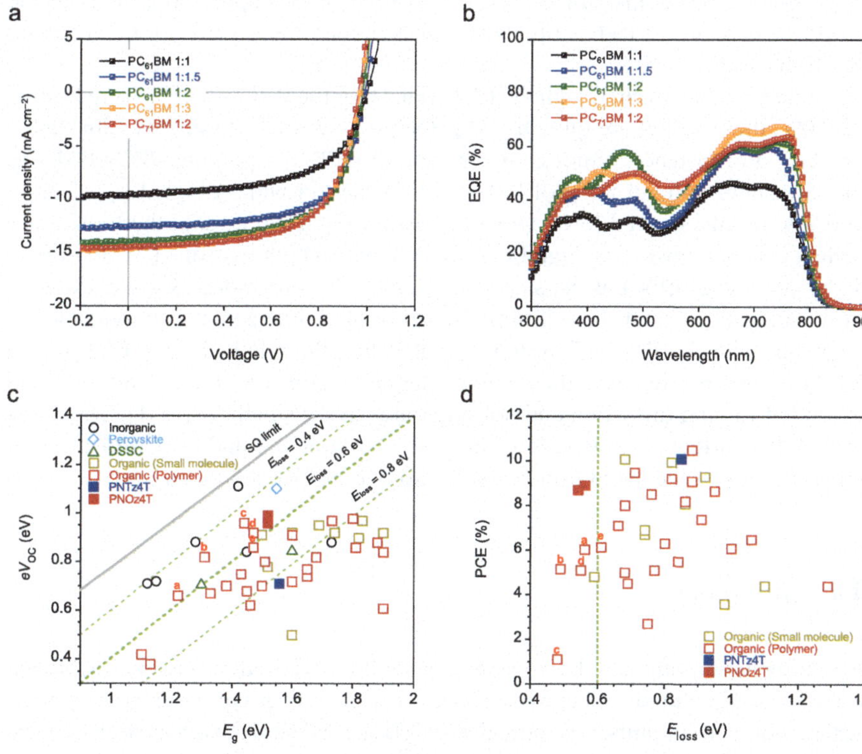

Fig. 5.24 Photovoltaic performance of the PNOz4T-based cells. **a** J–V curves of PNOz4T/$PC_{61}BM$ cells with different polymer to $PC_{61}BM$ weight (p/n) ratios and a PNOz4T/$PC_{71}BM$ cell with a p/n ratio of 1:2. **b** EQE spectra of PNOz4T/$PC_{61}BM$ cells with different p/n ratios and a PNOz4T/$PC_{71}BM$ cell with a p/n ratio of 1:2. **c** Plots of eV_{OC} against E_g, which is determined from the absorption onset, for various solar cell systems. The gray line and green dotted line show E_{loss} calculated from the Shockley–Queisser (SQ) theory and E_{loss} = 0.4, 0.6, and 0.8 eV. **d** Plots of PCE against E_{loss} for organic solar cells (small molecules and polymers) extracted from c. The red open squares a–e indicate the polymer systems with $E_{loss} \le 0.6$ eV. Reproduced with permission [51]. Copyright (2015) Nature Publishing Group

are significantly reduced from that of the PNTz4T/PCBM system (~0.85 eV) and are even smaller than 0.6 eV that has been referenced as the empirical limit for polymer solar cells [41].

Figure 5.24c depicts the plots of eV_{OC} against E_g for the PNTz4T and PNOz4T systems in comparison with various solar cell systems such as inorganic, perovskite, dye-sensitized (DSSC), small molecule-based organic, and polymer solar cells, in which the lines of E_{loss} estimated by the Shockley–Queisser theory [56] and the empirically estimated limit (0.6 eV) are drawn. Figure 5.24d shows the plots of PCE against E_{loss} for the PNTz4T and PNOz4T systems along with other organic systems plotted in Fig. 5.24c [51]. The red squares of a–e indicate the polymer systems with $E_{loss} \leq 0.6$ eV. It is clear that the E_{loss} of the PNOz4T system are among the lowest values reported so far for polymers solar cells, and even approach the values for inorganic or perovskite solar cells, 0.4–0.5 eV. More importantly, PCEs of the PNOz4T system are in fact the highest values among the organic systems having $E_{loss} \leq 0.6$ eV.

We note here that, even though the EQE values for the PNOz4T cells were limited to ca. 60%, these values are quite high for the polymer/PCBM systems with such a small ΔE_L. For instance, although the system of DPP-2Tz polymers/$PC_{71}BM$ was reported to have a small ΔE_L of 0.09–0.21 eV and provide a high V_{OC} of more than 0.9 V, it only showed low EQEs of 5–40% [57]. Therefore, to the best of our knowledge, in demonstrating high PCEs of ~9% with a high V_{OC} of ~1 V, small E_{loss} and a small energy offset at the same time, PNOz4T is regarded as quite a unique narrow bandgap polymer. In addition, the study of charge generation dynamics in comparison with the PNTz4T system revealed that the relatively low EQE of the PNOz4T system originates in the relatively large domain size of the blend films and thus limited exciton diffusion, and not originates in the small ΔE_L [51]. Thus, it is expected that further modification of the polymer structure and/or optimization of the morphology of the blend film can still improve EQE and thus J_{SC}.

5.3.4 Summary

This section has highlighted the π-conjugated polymers based on naphthobisthiadiazole (NTz) and naphthobisoxadiazole (NOz). The advantages of these building units over their counterpart units, benzothiadiazole (BTz) and benzooxadiazole (BOz), are their stronger electron-poor nature, extended π-electron system with rigid structure, and structural symmetry, all of which are important for developing high-performance polymers. In fact, polymers based on NTz showed deeper HOMO and LUMO energy levels, narrower bandgap, better ordering structures, and better device performances in both OFETs and OPVs compared to the polymer based on BTz. In particular, a quaterthiophene-NTz polymer, PNTz4T, exhibited efficiencies reaching 10% when blended with $PC_{71}BM$ in the solar cell. The introduction of the fluorine atoms successfully reduced the photon energy loss, resulting in the enhanced V_{OC}, and thereby the improved efficiencies that are close to 11%, which are among the highest reported so far for polymer/fullerene cells. Further, an NOz-based polymer, PNOz4T, was found to be able to significantly reduce the photon energy loss of the cell close to that for inorganic solar cells. Importantly, PNOz4T has allowed efficient photoinduced charge separation even with a very small driving force energy of nearly 0 eV. As a result, PNOz4T exhibited the maximum PCE of 8.9%, which was the highest value for polymer/fullerene systems with small E_{loss} such as lower than 0.6 eV.

5.4 Conclusion and Outlook

This chapter summarized our recent studies on the development of π-conjugated polymers based on thiazolothiazole and naphthobischalcogenadiazoles. The electron-poor nature of the building units resulted in donor–acceptor (D–A) backbones when incorporated into a polythiophene backbone. With the large π-electron system in addition to the D–A motif, the polymers formed highly ordered crystalline structures with close π–π stacking, leading to high charge carrier mobilities, which greatly benefits the photovoltaic performance. We discovered that the backbone orientation can be controlled by the careful design of the side chain composition. Further, we found that there is distribution of the orientation motifs through the film thickness. Thus, the match between the orientation including its distribution and the cell stack is crucial for maximizing the efficiency. The use of such polymers having high crystallinity and favorable orientation afforded high-efficiency polymer solar cells with thick active layers. In parallel, we also discovered that the photon energy loss, which is one of the most critical issues in organic solar cells, can be reduced as small as that for inorganic solar cells by molecular design, as evidenced by PNOz4T. We believe that these findings will be important guidelines for the development of new π-conjugated polymers and for the further improvement of the polymer solar cells.

A crucial issue would be that the reduction of the photon energy loss is always accompanied by a loss of driving force energy for charge separation, giving rise to a lower photocurrent (J_{SC}). Although PNOz4T showed relatively high photocurrent with the EQE more than 60% even under a small photon energy loss, such trade-off is still remained unsolved. Importantly, however, recent studies in this area have shown that the use of non-fullerene n-type materials [58–60] can realize small photon energy loss as small as ~0.5 eV and high photocurrent with EQEs reaching 80% at the same time, leading to significantly high efficiencies of ~18% [61]. Further, studies for understanding the mechanism underneath such unconventional phenomenon in PNOz4T as well as those non-fullerene materials would lead to new strategy in designing novel high-performance materials and thereby higher efficiencies.

Acknowledgements The author is grateful to KAKENHI (24685030 and 16H04196) from Japan Society for the Promotion of Science and Precursory Research for Embryonic Science and Technology (PRESTO) and the Advanced Low Carbon Technology Research and Development (ALCA) Programs from Japan Science and Technology Agency for funding. The author thanks Dr. M. Saito, Dr. K. Kawashima, Dr. Y. Tamai, Dr. V. Vohra, Dr. T. Koganezawa, Prof. K. Takimiya, Prof. H. Murata, Prof. H. Yoshida, Prof. Y. Ie, and Prof. H. Ohkita for their collaborative works.

References

1. Brabec, C., Scherf, U., Doyakonov, V.: Organic Photovoltaics: Materials, Device Physics, and Manufacturing Technologies, Wiley-VCH (2008)
2. Liang, Y., Xu, Z., Xia, J., Tsai, S.-T., Wu, Y., Li, G., Ray, C., Yu, L.: For the bright future—bulk heterojunction polymer solar cells with power conversion efficiency of 7.4%. Adv. Mater. **22**, E135–E138 (2010)
3. Vohra, V., Kawashima, K., Kakara, T., Koganezawa, T., Osaka, I., Takimiya, K., Murata, H.: Efficient inverted polymer solar cells employing favourable molecular orientation. Nat. Photon **9**, 403–408 (2015)
4. Yuan, J., Zhang, Y., Zhou, L., Zhang, G., Yip, H.-L., Lau, T.K., Lu, X., Zhu, C., Peng, H., Johnson, P.A., Leclerc, M., Cao, Y., Ulanski, J., Li, Y., Zou, Y.: Single-junction organic solar cell with over 15% efficiency using fused-ring acceptor with electron-deficient core. Joule **3**, 1140 (2019)
5. Tang, C.W.: Two-layer organic photovoltaic cell. Appl. Phys. Lett. **48**, 183–185 (1986)
6. Hiramoto, M., Fujiwara, H., Yokoyama, M.: Three-layered organic solar cell with a photoactive interlayer of codeposited pigments. Appl. Phys. Lett. **58**, 1062–1064 (1991)
7. Yu, G., Gao, J., Hummelen, J.C., Wudl, F., Heeger, A.J.: Polymer photovoltaic cells: enhanced efficiencies via a network of internal donor-acceptor heterojunctions. Science **270**, 1789–1790 (1995)
8. Gnes, S., Neugebauer, H., Sariciftci, N.S.: Conjugated polymer-based organic solar cells. Chem. Rev. **107**, 1324–1338 (2007)
9. Osaka, I., McCullough, R.D.: Advances in molecular design and synthesis of regioregular polythiophenes. Acc. Chem. Res. **41**, 1202–1214 (2008)
10. Li, G., Shrotriya, V., Huang, J., Yao, Y., Moriarty, T., Emery, K., Yang, Y.: High-efficiency solution processable polymer photovoltaic cells by self-organization of polymer blends. Nat. Mater. **4**, 864–868 (2005)
11. Zhao, G., He, Y., Li, Y.: 6.5% Efficiency of polymer solar cells based on poly(3-hexylthiophene) and indene-C60 bisadduct by device optimization. Adv. Mater. **22**, 4355–4358 (2010)
12. Dennler, G., Scharber, M.C., Brabec, C.J.: Polymer-fullerene bulk-heterojunction solar cells. Adv. Mater. **21**, 1323–1338 (2009)
13. Biniek, L., Schroeder, B.C., Nielsen, C.B., McCulloch, I.: Recent advances in high mobility donor–acceptor semiconducting polymers. J. Mater. Chem. **22**, 14803–14813 (2012)
14. Mühlbacher, D., Scharber, M., Morana, M., Zhu, Z., Waller, D., Gaudiana, R., Brabec, C.: High photovoltaic performance of a low-bandgap polymer. Adv. Mater. **18**, 2884–2889 (2006)
15. Osaka, I., Takimiya, K.: Backbone orientation in semiconducting polymers. Polymer **59**, A1–A15 (2015)
16. Ando, S., Nishida, J., Inoue, Y., Tokito, S., Yamashita, Y.: Synthesis, physical properties, and field-effect transistors of novel thiophene/thiazolothiazole co-oligomers. J. Mater. Chem. **14**, 1787–1790 (2004)
17. Ando, S., Nishida, J., Fujiwara, E., Tada, H., Inoue, Y., Tokito, S., Yamashita, Y.: Characterization and field-effect transistor performance of heterocyclic oligomers containing a thiazolothiazole unit. Chem. Lett. **33**, 1170–1171 (2004)

18. Ando, S., Nishida, J., Tada, H., Inoue, Y., Y, S., Yamashita, S.: High performance n-type organic field-effect transistors based on π-electronic systems with trifluoromethylphenyl groups. J. Am. Chem. Soc. **127**, 5336–5337 (2005)
19. Peng, Q., Peng, J.-B., Kang, E.T., Neoh, K.G., Cao, Y.: Synthesis and electroluminescent properties of copolymers based on fluorene and 2, 5-di(2-hexyloxyphenyl) thiazolothiazole. Macromolecules **38**, 7272–7298 (2005)
20. Osaka, I., Sauvé, G., Zhang, R., Kowalewski, T., McCullough, R.D.: Novel thiophene-thiazolothiazole copolymers for organic field-effect transistors. Adv. Mater. **19**, 4160–4165 (2007)
21. Osaka, I., Zhang, R., Sauvé, G., Smilgies, D.M., Kowalewski, T., McCullough, R.D.: High-lamellar ordering and amorphous-like π-network in short-chain thiazolothiazole-thiophene copolymers lead to high mobilities. J. Am. Chem. Soc. **131**, 2521–2529 (2009)
22. McCulloch, I., Heeney, M., Bailey, C., Genevicius, K., MacDonald, I., Shkunov, M., Sparrowe, D., Tierney, S., Wagner, R., Zhang, W., Chabinyc, M.L., Kline, R.J., McGehee, M.D., Toney, M.F.: Liquid-crystalline semiconducting polymers with high charge-carrier mobility. Nat. Mater. **5**, 328–333 (2006)
23. Osaka, I., Saito, M., Mori, H., Koganezawa, T., Takimiya, K.: Drastic change of molecular orientation in a thiazolothiazole copolymer by molecular-weight control and blending with $PC_{61}BM$ leads to high efficiencies in solar cells. Adv. Mater. **24**, 425–430 (2012)
24. Osaka, I., Zhang, R., Liu, J., Smilgies, D.M., Kowalewski, T., McCullough, R.D.: Highly stable semiconducting polymers based on thiazolothiazole. Chem. Mater. **22**, 4191–4196 (2010)
25. Osaka, I., Saito, M., Koganezawa, T., Takimiya, K.: Thiophene–thiazolothiazole copolymers: significant impact of side chain composition on backbone orientation and solar cell performances. Adv. Mater. **26**, 331–338 (2014)
26. Saito, M., Koganezawa, T., Osaka, I.: Correlation between distribution of polymer orientation and cell structure in organic photovoltaics. ACS Appl. Mater. Interfaces. **10**, 32420–32425 (2018)
27. Rivnay, J., Mannsfeld, S.C.B., Miller, C.E., Salleo, A., Toney, M.F.: Quantitative determination of organic semiconductor microstructure from the molecular to device scale. Chem. Rev. **112**, 5488–5519 (2012)
28. Duong, D.T., Toney, M.F., Salleo, A.: Role of confinement and aggregation in charge transport in semicrystalline polythiophene thin films. Phys. Rev. B **86**, 205205 (2012)
29. Xu, Z., Chen, L.-M., Yang, G., Huang, C.-H., Hou, J., Wu, J., Li, G., Hsu, C.-S., Yang, Y.: Vertical phase separation in poly(3-hexylthiophene): fullerene derivative blends and its advantage for inverted structure solar cells. Adv. Funct. Mater. **19**, 1227–1234 (2009)
30. Guo, S., Zhou, N., Lou, S.J., Smith, J., Tice, D.B., Hennek, J.W., Ortiz, R.P., Navarrete, J.T.L., Li, S., Strzalka, J., Chen, L.X., Chang, R.P.H., Facchetti, A., Marks, T.J.: Polymer solar cells with enhanced fill factors. Nat. Photon **7**, 825–833 (2013)
31. Blouin, N., Michaud, A., Gendron, D., Wakim, S., Blair, E., Neagu-Plesu, R., Belletetê, M., Durocher, G., Tao, Y., Leclerc, M.: Toward a rational design of poly(2,7-carbazole) derivatives for solar cells. J. Am. Chem. Soc. **130**, 732–742 (2008)
32. Zhou, E., Cong, J., Hashimoto, K., Tajima, K.: A benzoselenadiazole-based low band gap polymer: synthesis and photovoltaic application. Macromolecules **46**, 763–768 (2013)
33. Kawashima, K., Osaka, I., Takimiya, K.: Effect of chalcogen Aatom on the properties of naphthobischalcogenadiazole-based π-conjugated polymers. Chem. Mater. **27**, 6558–6570 (2015)
34. Zhang, Z., Lin, F., Chen, H.C., Wu, H.C., Chung, C.L., Lu, C., Liu, S.H., Tung, S.H., Chen, W.C., Wong, K.T., Chou, P.T.: A Silole copolymer containing a ladder-type Heptacylic Arene and Naphthobisoxadiazole Moieties for highly efficient polymer solar cells. Energy Environ. Sci. **8**, 552–557 (2015)
35. Wang, M., Hu, X., Liu, P., Li, W., Gong, X., Huang, F., Cao, Y.: Donor–acceptor conjugated polymer based on naphtho[1,2-*c*:5,6-*c*]bis[1,2,5]thiadiazole for high-performance polymer solar cells. J. Am. Chem. Soc. **133**, 9638–9641 (2011)

36. Osaka, I., Shimawaki, M., Mori, H., Doi, I., Miyazaki, E., Koganezawa, T., Takimiya, K.: Synthesis, characterization, and transistor and solar Cell applications of a naphthobisthiadiazole-based semiconducting polymer. J. Am. Chem. Soc. **134**, 3498–3507 (2012)
37. Mataka, S., Takahashi, K., Ikezaki, Y., Hatta, T., Tori-i, A., Tashiro, M.: Sulfur nitride in organic chemsitry. XIV, selective formation of benzo-and benzobis [1,2,5]thiadiazole skeleton in the reaction of tetrasulfur tetranitride with naphthalenols and related compounds. Bull. Chem. Soc. Jpn. **64**, 68–73 (1991)
38. Osaka, I., Takimiya, K.: Naphthobischalcogenadiazole conjugated polymers: emerging materials for organic electronics. Adv. Mater. **39**, 1605218 (2017)
39. Rieger, R., Beckmann, D., Mavrinskiy, A., Kastler, M., Müllen, K.: Backbone curvature in polythiophenes. Chem. Mater. **22**, 5314–5318 (2010)
40. Osaka, I., Abe, T., Shinamura, S., Takimiy, K.: Impact of isomeric structures on transistor performances in naphthodithiophene semiconducting polymers. J. Am. Chem. Soc. **133**, 6852–6860 (2011)
41. Veldman, D., Meskers, S.C.J., Janssen, R.A.J.: The energy of charge-transfer states in electron donor–acceptor blends: insight into the energy losses in organic solar cells. Adv. Funct. Mater. **19**, 1939–1948 (2009)
42. King, R.R., Bhusari, D., Boca, A., Larrabee, D., Liu, X.Q., Hong, W., Fetzer, C.M., Law, D.C., Karam, N.H.: Band gap-voltage offset and energy production in next-generation multijunction solar cells. Prog. Photovolt **19**, 797–812 (2011)
43. Wang, M., Wang, H., Yokoyama, T., Liu, X., Huang, Y., Zhang, Y., Nguyen, T.Q., Aramaki, S., Bazan, G.C.: High open circuit voltage in regioregular narrow band gap polymer solar cells. J. Am. Chem. Soc. **136**, 12576–12579 (2014)
44. Green, M.A., Ho-Baillie, A., Snaith, H.J.: The emergence of perovskite solar cells. Nat. Photon **8**, 506–514 (2014)
45. Scharber, M.C., Mühlbacher, D., Koppe, M., Denk, P., Waldauf, C., Heeger, A.J., Brabec, C.J.: Design rules for donors in bulk-heterojunction solar cells –towards 10% energy-conversion efficiency. Adv. Mater. **18**, 789–794 (2006)
46. Dimitrov, S.D., Durrant, J.R.: Materials design considerations for charge generation in organic solar cells. Chem. Mater. **26**, 616–630 (2014)
47. Zhou, H., Yang, L., Stuart, A.C., Price, S.C., Liu, S., You, W.: Development of fluorinated benzothiadiazole as a structural unit for a polymer solar cell of 7% efficiency. Angew. Chem. Int. Ed. **50**, 2995–2998 (2011)
48. Kawashima, K., Fukuhara, T., Suda, Y., Suzuki, Y., Koganezawa, T., Yoshida, H., Ohkita, H., Osaka, I., Takimiya, K.: Implication of Fluorine Atom on Electronic Properties, Ordering Structures, and Photovoltaic Performance in Naphthobisthiadiazole-Based Semiconducting Polymers. J. Am. Chem. Soc. **138**, 10265–10275 (2016)
49. Saito, M., Fukuhara, T., Kamimura, S., Ichikawa, H., Yoshida, H., Koganezawa, T., Ie, Y., Tamai, Y., Kim, H.-D., Ohkita, H., Osaka, I.: Impact of noncovalent sulfur–fluorine interaction position on properties, structures, and photovoltaic performance in naphthobisthiadiazole-based semiconducting polymers. Adv Energy Mater **10**, 1903278 (2020)
50. Yoshida, H.: Principle and application of low energy inverse photoemission spectroscopy: a new method for measuring unoccupied states of organic semiconductors. J. Electron Spectrosc. Relat. Phenom. **204**, 116–124 (2015)
51. Kawashima, K., Tamai, Y., Ohkita, H., Osaka, I., Takimiya, K.: High-efficiency polymer solar cells with small photon energy loss. Nat. Commun. **6**, 10085 (2015)
52. Brédas, J.L., Norton, J.E., Cornil, J., Coropceanu, V.: Molecular understanding of organic solar cells: the challenges. Acc. Chem. Res. **42**, 1691–1699 (2009)
53. Brédas, J.L.: Mind the gap! Mater. Horiz. **1**, 17–19 (2014)
54. Vandewal, K., Tvingstedt, K., Gadisa, A., Inganäs, O., Manca, J.V.: On the origin of the open-circuit voltage of polymer-fullerene solar cells. Nat. Mater. **8**, 904–909 (2009)
55. Vandewal, K., Ma, Z., Bergqvist, J., Tang, Z., Wang, E., Henriksson, P., Tvingstedt, K., Andersson, M.R., Zhang, F., Inganäs, O.: Quantification of quantum efficiency and energy losses in

low bandga polymer:fullerene solar cells with high open-circuit voltage. Adv. Funct. Mater. **22**, 3480–3490 (2012)
56. Shockley, W., Queisser, H.J.: Detailed balance limit of efficiency of p–n junction solar cells. J. Appl. Phys. **32**, 510–519 (1961)
57. Li, W., Hendriks, K.H., Furlan, A., Wienk, M.M., Janssen, R.A.J.: High quantum efficiencies in polymer solar cells at energy losses below 0.6 eV. J. Am. Chem. Soc. **137**, 2231–2234 (2015)
58. Nielsen, C.B., Holliday, S., Chen, H.-Y., Cryer, S.J., McCulloch, I.: Non-fullerene electron acceptors for use in organic solar cells. Acc. Chem. Res. **48**, 2803–2812 (2015)
59. Hou, J., Inganäs, O., Friend, R.H., Gao, F.: Organic solar cells based on non-fullerene acceptors. Nat. Mater. **17**, 119–128 (2018)
60. Zhang, J., Tan, H.S., Guo, X., Facchetti, A., Yan, H.: Material insights and challenges for non-fullerene organic solar cells based on small molecular acceptors. Nat. Energy **3**, 720–731 (2018)
61. Liu, Q., Jiang, Y., Jin, K., Qin, J., Xu, J., Li, W., Xiong, J., Liu, J., Xiao, Z., Sun, K., Yang, S., Zhang, X., Ding, L.: 18% Efficiency organic solar cells. Sci. Bull. **65**, 272–275 (2020)

Chapter 6
Charge Carrier Dynamics in Polymer Solar Cells

Hideo Ohkita

6.1 Introduction

Polymer solar cells have attracted increasing interest as a next-generation solar cell because they have excellent advantages of colorful, flexible, lightweight, and especially high-throughput and low-cost production by printing techniques. Thus, they have been intensively studied over the world for a long time. As a result, the power conversion efficiency (PCE) has been steadily improved year by year in the last two decades. In 1995, the PCE was less than 1% [1, 2]. Currently, the PCE is more than 17%, which was reported for tandem solar cells in 2018 [3] and is more than 18%, which was reported for single-junction cells in 2020 [4]. This efficiency is even higher than that of amorphous silicon solar cells.

These improvements are partly due to remarkable developments of various photovoltaic materials such as soluble fullerene derivatives like a [6,6]-phenyl-C_{61}-butyric acid methyl ester (PCBM) [5], crystalline conjugated polymers like a regioregular poly(3-hexylthiphene) (RR-P3HT) [6], low-bandgap conjugated polymers like a poly[2,6-(4,4-bis-(2-ethylhexyl)-4*H*-cyclopenta[2,1-*b*;3,4-*b*′]dithiophene)-*alt*-4,7(2,1,3-benzothiadiazole)] (PCPDTBT) [7], and non-fullerene acceptors like a 6,6,12,12-tetrakis(4-hexylphenyl)-6,12-dihydro-dithieno[2,3-*d*:2′,3′-*d*′]-*s*-indaceno[1,2-*b*:5,6-*b*′]*dithiophene* (ITIC) [8]. In other words, such diversity of organic materials is one of the great advantages over inorganic solar cells. For fullerene-based polymer solar cells, as described in Sect. 6.4, charge dissociation is likely to be more efficient in polymer solar cells based on highly crystalline conjugated polymers.

The original version of this chapter was revised: Greek symbol have been updated in the Equation 6.1. The correction to this chapter is available at https://doi.org/10.1007/978-981-15-9113-6_11

H. Ohkita (✉)
Department of Polymer Chemistry, Graduate School of Engineering, Kyoto University, Katsura Campus, Kyoto, Nishikyo-Ku 615-8510, Japan
e-mail: ohkita@photo.polym.kyoto-u.ac.jp

M. Hiramoto and S. Izawa (eds.), *Organic Solar Cells*,
https://doi.org/10.1007/978-981-15-9113-6_6

In addition, new approaches recently proposed have effectively boosted the PCE of polymer solar cells furthermore. Ternary blend solar cells are promising approaches to expand the light-harvesting wavelength range from visible to near-IR region. For example, we have demonstrated that the short-circuit current density (J_{SC}) can be effectively enhanced by addition of near-IR dye molecules into RR-P3HT/PCBM binary blend solar cells [9, 10]. The key to success of this approach is location of dye molecules in ternary blend films: Dye molecules should be located at a RR-P3HT/PCBM interface in blend films so that they can contribute to photocurrent generation effectively [11, 12]. Interestingly, most dye molecules are spontaneously located at the RR-P3HT/PCBM interface in ternary blend films, which is revealed by transient absorption spectroscopy [13], as described in Sect. 6.4.4.

In order to improve photovoltaic performance rationally, it is of particular importance to gain in-depth understanding of photovoltaic conversion mechanism in polymer solar cells. Thus, we need to directly observe dynamics of transient species such as exciton and charge carriers generated in polymer solar cells. In polymer solar cells, as described in Sect. 6.2, the exciton and charge carrier dynamics ranges from femtoseconds up to microseconds on a temporal scale [14–16]. Thus, time-resolved measurement techniques are essential for understanding photovoltaic conversion mechanism. In particular, transient absorption spectroscopy is one of the most powerful methods for studying the exciton and charge carrier dynamics directly. This chapter describes recent progress in understanding of photovoltaic conversion mechanism in polymer solar cells studied by time-resolved optoelectronic measurements.

6.2 Photovoltaic Conversion in Polymer Solar Cells

This section describes elementary processes of photovoltaic conversion in polymer solar cells. Figure 6.1 shows a schematic illustration of the photovoltaic conversion process in bilayered polymer solar cells based on hole-transporting (donor) and electron-transporting (acceptor) materials. Under the solar illumination, a photon is absorbed by either donor or acceptor material. The photon absorption efficiency (η_A) is defined as the ratio of the number of absorbed photons to the number of incident photons at a wavelength. Here, it is considered that a photon is absorbed by the donor material. Roughly speaking, the photon absorption excites an electron in the highest occupied molecular orbital (HOMO) of the donor material into the lowest unoccupied molecular orbital (LUMO), resulting in a hole in the HOMO and an electron in the LUMO. This electron–hole pair is tightly bound because of large Coulomb interaction in organic materials. This bound electron–hole pair is called a singlet exciton. This state is also called singlet excited state, which consists of several different electron configurations more strictly. In polymer solar cells, excitons cannot dissociate into free charge carriers at room temperature because the exciton binding energy is much larger than thermal energy k_BT. This is partly because relative permittivity ε_r is as small as 3–4 in most organic materials, and hence a critical distance r_C at which the

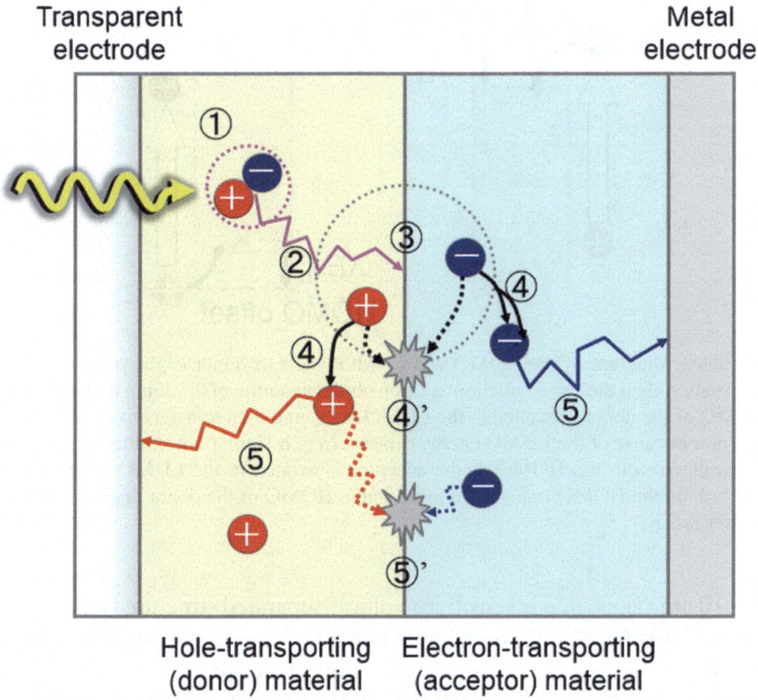

Fig. 6.1 Elementary processes of photovoltaic conversion in polymer solar cells with a bilayered device structure based on donor (light yellow part) and acceptor (light blue part) materials. (1) exciton generation upon photon absorption by the donor material, (2) exciton diffusion into a donor/acceptor interface, (3) charge transfer from the exciton arriving at the interface, (4) charge dissociation into free charge carriers, (4)' geminate (monomolecular) recombination of an electron–hole pair at the charge transfer state, (5) charge collection of free charge carriers to each electrode, and (5)' non-geminate (bimolecular) recombination of free charge carriers

thermal energy of a charge carrier is equal to the Coulomb attractive potential energy $e^2/4\pi\varepsilon_r\varepsilon_0 r_C$ would be as long as 14–19 nm at room temperature. Here, k_B is the Boltzmann constant, T is the absolute temperature, e is the elementary charge, and ε_0 is the vacuum permittivity. In contrast, excitons easily dissociate into free carriers in inorganic solar cells like silicon even at room temperature. This is partly due to large ε_r of inorganic semiconductors. For example, ε_r of crystalline silicon is as large as 11.9 and hence r_C would be as short as 5 nm at room temperature. This is the most critical difference in photovoltaic conversion mechanism between organic and inorganic solar cells. As such the exciton cannot dissociate into free carriers in polymer solar cells but can migrate randomly in polymer domains. Thus, some excitons can arrive at a donor/acceptor interface and the others cannot before deactivating to the ground state radiatively or non-radiatively. The exciton diffusion efficiency (η_{ED}) is defined as the ratio of the number of excitons arriving at the interface to the number of excitons generated. In most conjugated polymers, the exciton diffusion length is typically as

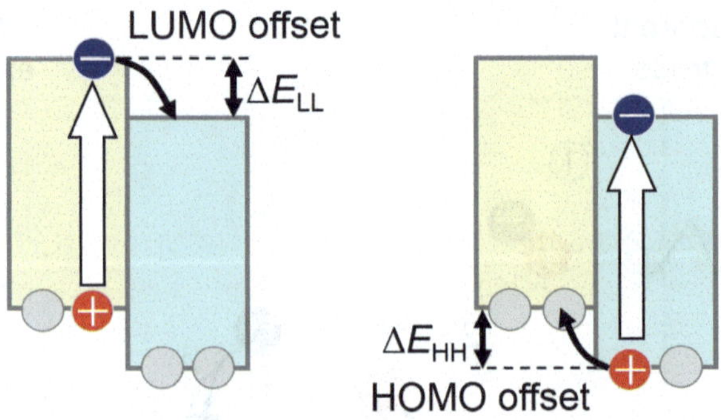

Fig. 6.2 Energy diagram of the HOMO and LUMO levels of donor (light yellow) and acceptor (light blue) materials at the heterojunction: **a** Upon photoexcitation of the donor material, an electron in the HOMO of the donor is excited to the LUMO level, and then transfers to more stable LUMO of the acceptor because of the LUMO energy offset ΔE_{LL}. **b** Upon photoexcitation of the acceptor material, an electron in the HOMO of the acceptor is excited to the LUMO level, and then the remaining hole in the HOMO transfers to more stable HOMO of the donor because of the HOMO energy offset ΔE_{HH}

short as ~10 nm. Thus, η_{ED} is sensitive to phase-separated structures in blend films: It would be small in largely phase-separated blend structures and large in finely mixed blend structures. For excitons arriving at the interface, an electron in the LUMO of the donor can transfer to a more stable LUMO of the neighboring acceptor because of the energy offset ΔE_{LL} between the two LUMO levels as shown in Fig. 6.2a. Upon photoexcitation of the acceptor instead of the donor, the same is equally true of hole transfer from the HOMO of the acceptor to that of the neighboring donor because of the energy offset ΔE_{HH} between the two HOMO levels as shown in Fig. 6.2b. It has been believed that these energy offsets should be large enough to break the Coulomb binding energy of electron–hole pairs in excitons. As a result, the charge transfer (CT) state is formed at the donor/acceptor interface. The charge transfer efficiency (η_{CT}) is defined as the ratio of the number of CT states formed at the interface to the number of excitons arriving at the interface. Some CT states recombine to the ground state radiatively or non-radiatively. This monomolecular recombination is called geminate recombination. The other CT states dissociate into free charge carriers. The charge dissociation efficiency (η_{CD}) is defined as the ratio of the number of CT states dissociating to free charge carriers to the number of CT states generated. After the charge dissociation, hole carriers can migrate in donor domains and electron carriers can migrate in acceptor domains. During the charge migration, some electron carriers encounter hole carriers at the donor/acceptor interface and recombine to the ground state radiatively or non-radiatively. This bimolecular recombination is called non-geminate recombination. Finally, electron and hole carriers are collected to cathode and anode electrodes, respectively. The charge collection efficiency (η_{CC}) is defined as the ratio of the number of charge carriers collected to the number of charge carriers generated. In summary, photocurrent generation is a final product of a series of elementary processes, which include (1) photon absorption to generate exciton, (2) exciton diffusion into a donor/acceptor interface, (3) charge transfer at the interface, (4) charge dissociation into free charge carriers, and (5) charge collection into each electrode. The charge dissociation is in competition with (4)' geminate (monomolecular) recombination to the ground state. The charge collection is in competition with (5)' non-geminate (bimolecular) recombination to the ground state, as shown in Fig. 6.1. Thus, the external quantum efficiency (EQE) of polymer solar cells is equal to the product of the efficiency of each elementary process: EQE $= \eta_{\mathrm{A}} \times \eta_{\mathrm{ED}} \times \eta_{\mathrm{CT}} \times \eta_{\mathrm{CD}} \times \eta_{\mathrm{CC}}$. The internal quantum efficiency (IQE) is given by IQE $= \eta_{\mathrm{ED}} \times \eta_{\mathrm{CT}} \times \eta_{\mathrm{CD}} \times \eta_{\mathrm{CC}}$.

These elementary processes in photovoltaic conversion range over a wide temporal scale as described below. The photon absorption is an electronic transition and hence occurs on a time scale of femtoseconds (~10^{-15} s). The exciton diffusion typically occurs on a timescale of picoseconds to sub-nanoseconds (10^{-12}–10^{-10} s) though it depends on phase-separated structures. The charge transfer has been reported to be promptly completed in the order of tens of femtoseconds (~10^{-14} s). The geminate recombination typically ranges from picoseconds to nanoseconds (10^{-12}–10^{-9} s). The charge collection time typically ranges from sub-microseconds to microseconds (10^{-7}–10^{-6} s), though it depends on the charge mobility, the thickness of the active layer, and the electric field applied to the active layer. In other words, the elementary processes in photovoltaic conversion range from femtoseconds to microseconds (over nine orders of magnitude on a temporal scale). This chapter focuses on these rapid photovoltaic conversion events studied by time-resolved optoelectronic measurements such as transient absorption spectroscopy and discusses the recent findings obtained from the kinetics analyses.

6.3 Optoelectronic Measurements

6.3.1 Transient Absorption Spectroscopy

Transient absorption spectroscopy is one of the most powerful tools for studying the dynamics of short-lived transient species such as excitons and charge carriers. Historically, Norrish and Porter developed the flash photolysis technique around 1950 before the invention of lasers [17, 18]. Their pioneering work enabled us to directly observe such short-lived transient products and therefore opened a new research field. As a result, they were awarded the Nobel Prize in Chemistry 1967 with Eigen for their studies of extremely fast chemical reactions, effected by disturbing the equilibrium by means of very short pulses of energy [19, 20]. Owing to the emergence of short-pulsed lasers, this technique has been further improved in temporal resolution. Currently, we can directly observe ultrafast phenomena with a resolution of femtoseconds. In 1999, Zeweil was awarded the Nobel Prize in Chemistry for his studies of the transition states of chemical reactions using femtosecond spectroscopy [21]. These two typical measurement methods are described below.

For ultrafast measurements up to tens of nanoseconds, pump and probe techniques are typically employed. In this technique, ultrashort laser pulses are split into two pulsed light sources: one is employed as a pump light source for exciting the sample and the other is employed as a probe light source for measuring transmittance of the sample. Figure 6.3 shows a pump and probe femtosecond transient absorption measurement system we employed. As shown in the figure, the probe light passes through the additional path length in the optical delay line and hence arrives at the sample following the pump light. The delay time can be tuned by controlling the optical delay length. For example, when the optical decay length is set at 30 cm, the probe light arrives at the sample 1 ns after the pump light excites the sample. In this case, we can measure the change in optical density ΔOD of the sample 1 ns after the laser excitation. By changing the optical delay length systematically, we can trace the time evolution of transient absorption spectra.

For measurements after nanoseconds, laser photolysis techniques are widely employed. In this technique, a short-pulsed laser is employed as a pump light source for exciting the sample and a stable white light such as tungsten or xenon lamp is employed as a probe light source for measuring transmittance of the sample. Figure 6.4 shows the highly sensitive microsecond transient absorption measurement system we employed. At such a later time stage, most transient species have already decayed. A highly sensitive transient absorption measurement system is required to detect small signals. As shown in the figure, a tungsten lamp with a stabilized power source is employed as a probe light to reduce fluctuation of the probe light intensity. To reduce scattering light, stray light, and emission from the sample, two monochromators and appropriate optical filters are placed before and after the sample in the probe light line. The probe light passing through the sample is detected with a Si or

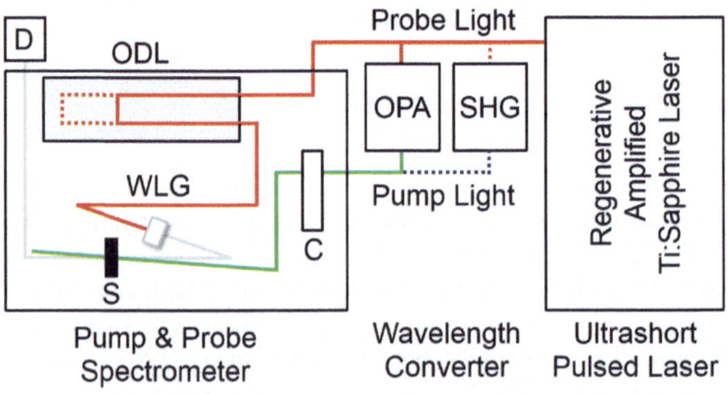

Fig. 6.3 Block diagram of a pump and probe femtosecond transient absorption measurement systems. This system consists of an ultrashort pulsed laser (regenerative amplified Ti:sapphire laser), wavelength converters, and a pump and probe spectrometer: SHG second harmonic generator, OPA optical parametric amplifier, C chopper, ODL optical delay line, WLG white light generator, S sample, and D detector. As a detector, a CMOS linear sensor is employed for the visible wavelength range and an InGaAs linear diode array sensor is employed for near-IR wavelength range. ©[2016] IEEE, Reprinted, with permission, from [16]

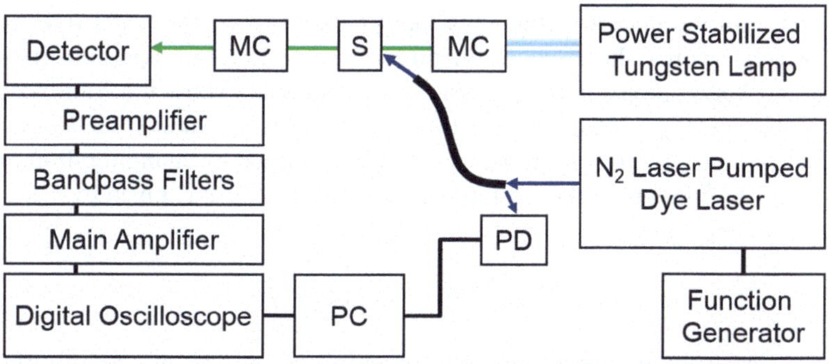

Fig. 6.4 Block diagram of a highly sensitive microsecond transient absorption measurement system. This system consists of a pump light source (nanosecond dye laser pumped by N_2 laser), a probe light source (power-stabilized tungsten lamp), and a detection system: MC monochromator, S sample, PD p-i-n photodiode to detect a part of the pump laser light as a trigger signal, and PC computer. As a detector for the probe light, a Si p-i-n photodiode is employed for the visible wavelength range and an InGaAs p-i-n photodiode is employed for near-IR wavelength range. ©[2016] IEEE, Reprinted, with permission, from [16]

InGaAs p-i-n photodiode depending on the measuring wavelength. The signal from the photodiode is pre-amplified, sent to electronic band-pass filters to improve the signal-to-noise ratio, and amplified again with the main amplification system. The amplified signal is collected with a digital oscilloscope, which is synchronized with a trigger signal of the laser pulse from another photodiode. By using this system, the detectable ΔOD is as small as 10^{-5}–10^{-6} depending on the measuring time domain and the accumulation times.

6.3.2 *Transient Photovoltage and Photocurrent Measurements*

Transient photovoltage (TPV) and transient photocurrent (TPC) measurements have been widely employed to study charge carriers dynamics and charge carrier density in solar cells under operation conditions. For polymer solar cells, Durrant and his co-workers were the first to apply TPV and TPC techniques in order to discuss charge carriers dynamics in polymer solar cells as reported previously [22]. Details have been described in their papers [23, 24]. Here, it will be briefly explained how charge carrier lifetime and density are evaluated by TPV and TPC measurements.

In TPV measurements, polymer solar cells are operated at the open-circuit under white light (simulated solar) illumination and then are excited by a small perturbation pulsed laser light. As a result, photovoltage is slightly increased from the open-circuit voltage (V_{OC}) because of the slight increase in the charge density Δn due to minority excess charges generated by the small perturbation excitation. The minority charges recombine with the majority charges steadily generated by the white light illumination. In other words, the decay of the minority charges follows pseudo-first-order kinetics and hence is given by an exponential function. Thus, the lifetime of the minority charges $\tau_{\Delta n}$ is evaluated as a time constant of the exponential decay. Similarly, the lifetimes $\tau_{\Delta n}$ for different V_{OC}s can be obtained by changing illumination intensities. The following relationship between $\tau_{\Delta n}$ and V_{OC} is given by

$$\tau_{\Delta n} = \tau_{\Delta n_0} \exp\left(-\frac{eV_{\mathrm{OC}}}{\nu k_{\mathrm{B}}T}\right) \tag{6.1}$$

where $\tau_{\Delta n}$ and ν are obtained from an intersection and a slope in Logarithmic plots of $\tau_{\Delta n}$ against V_{OC}, respectively.

In TPC measurements, polymer solar cells are operated at the short-circuit under white light (simulated solar) illumination and then are excited by the same small perturbation pulsed laser light as TPV. Under the short-circuit condition, transient photocurrent decay is ascribed to charge collection to electrodes. As described previously, the total charge carrier density n can be evaluated from TPC analyses [23, 24]. The following relationship between n and V_{OC} is given by

$$n = n_0 \exp\left(\frac{eV_{\mathrm{OC}}}{mk_{\mathrm{B}}T}\right) \tag{6.2}$$

where n_0 and m are obtained from an intersection and a slope in Logarithmic plots of n against V_{OC}, respectively. Finally, the lifetime of the total charge carriers τ_n is given by $\tau_n = (1 + \lambda)\tau_{\Delta n} = \delta\tau_{\Delta n}$ where $\lambda = m/\nu$ and $\delta = 1 + \lambda$. Here, δ is the empirical reaction order that describes how the recombination rate scales with all charge carrier density in a device including both free and trapped charges [25]. By using Eqs. (6.1) and (6.2), τ_n can be expressed as a function of the carrier density n.

$$\begin{aligned}
\tau_n &= (1 + \lambda)\tau_{\Delta n} = (1 + \lambda)\tau_{\Delta n_0} \exp\left(-\frac{eV_{\mathrm{OC}}}{\nu k_{\mathrm{B}}T}\right) \\
&= (1 + \lambda)\tau_{\Delta n_0} n_0^{\frac{m}{\nu}} n_0^{-\frac{m}{\nu}} \exp\left(\frac{eV_{\mathrm{OC}}}{mk_{\mathrm{B}}T}\right)^{-\frac{m}{\nu}} \\
&= (1 + \lambda)\tau_{\Delta n_0} n_0^{\lambda} n^{-\lambda} = \tau_0 n^{-\lambda}
\end{aligned} \tag{6.3}$$

6.4 Charge Generation Dynamics

This section describes the charge generation dynamics in various polymer/PCBM blend films studied by transient absorption spectroscopy [26–38]. Figure 6.5 shows the photovoltaic materials studied in this chapter. As shown in the figure, conjugated polymers have similar structures but different crystallinity. For example, RRa-P3HT exhibits the same backbone as RR-P3HT but exhibits totally different crystallinity from RR-P3HT: RRa-P3HT is a typical amorphous conjugated polymer but RR-P3HT is a typical crystalline conjugated polymer. PCPDTBT and PSBTBT have almost the same backbone except for one atom: carbon for PCPDTBT and sulfur for PSBTBT. PSBTBT exhibits higher crystallinity than PCPDTBT. PNTz4T and PNOz4T also have almost the same backbone except for one atom: sulfur for PNTz4T and oxygen for PNOz4T. PNOz4T exhibits higher crystallinity than PNTz4T although both polymers are highly crystalline polymers. Acceptor materials are fixed to fullerene derivatives of PCBM or $PC_{71}BM$, which have almost the same LUMO level of –3.7 eV. Note that there are wide variations in ΔE_{LL} for these polymer/fullerene blends. It will be discussed later how polymer crystallinity impacts on the charge generation efficiency in Sect. 6.6.

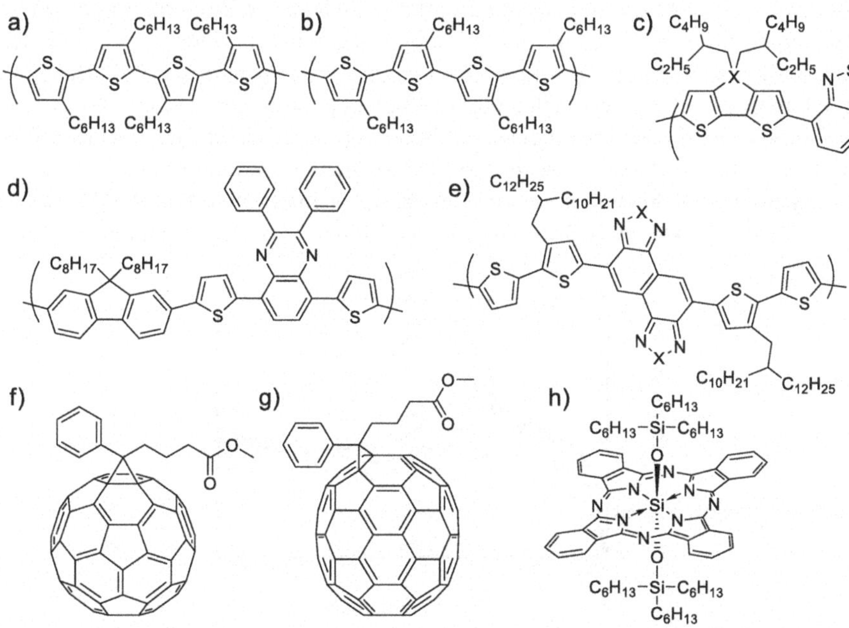

Fig. 6.5 Chemical structures of photovoltaic materials studied in this chapter: **a** RRa-P3HT, **b** RR-P3HT, **c** PCPDTBT (X = C), PSBTBT (X = Si), **d** NP-7, **e** PNTz4T (X = S), PNOz4T (X = O), **f** PCBM, **g** $PC_{71}BM$, and **h** SiPc6

6.4.1 Amorphous Polymer Solar Cells

Figure 6.6 shows transient absorption spectra of RRa-P3HT pristine and RRa-P3HT/PCBM blend films excited at 400 nm [27]. For the RRa-P3HT pristine film, a large absorption band was observed at around 1000 nm immediately after the laser excitation. This band is ascribed to polymer singlet excitons, which decays with a time constant of 270 ps under a low excitation intensity, as reported previously [26]. The negative absorption signal is observed at around 500 nm, which is consistent with RRa-P3HT absorption in the ground state and hence is ascribed to ground-state photobleaching (GSB) signals. For the RRa-P3HT/PCBM blend films, on the other hand, the singlet exciton band is reduced to almost half of that observed for the pristine film even at 0 ps. Instead, new absorption bands are observed at around 800 and 1600 nm, which are ascribed to polymer polarons. In other words, polymer polarons are promptly generated even at 0 ps in the blend films. Subsequently, the broadband of singlet excitons completely disappears at 0.2 ps. Instead, a small and sharp absorption band is clearly observed at around 1050 nm, which is ascribed to PCBM radical anion. As described in [27], the singlet exciton signals decay with a time constant of ~0.2 ps while the polaron signals increase with a time constant of ~0.2 ps. This agreement suggests that polymer polarons are rapidly generated from singlet excitons with a time constant of ~0.2 ps, which are more than two orders of magnitude faster than the lifetime of singlet excitons. In other words, polymer polarons are assumed to be generated from singlet excitons with a 100% efficiency from a kinetic point of view. Such a highly efficient polaron generation is consistent with almost 100% quenching efficiency of photoluminescence of RRa-P3HT/PCBM blend films. This is probably because PCBM molecules are likely to be distributed relatively homogeneously in an amorphous polymer matrix like RRa-P3HT. In the case of 50 wt% PCBM, PCBM would be located at intervals of a few nanometers and

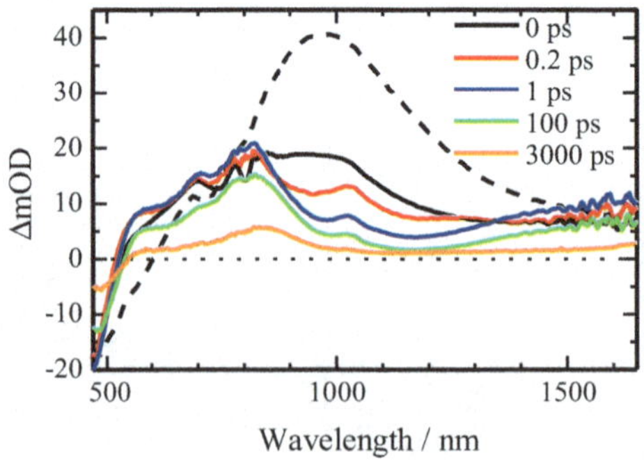

Fig. 6.6 Transient absorption spectra of RRa-P3HT pristine (broken line) and RRa-P3HT/PCBM blend (solid lines) films measured at 0 (black), 0.2 (red), 1, (blue), 100 (light green), and 3000 (orange) ps after the laser excitation. The excitation wavelength was 400 nm (~30 μJ cm^{-2}). The ΔOD was corrected for variations in the absorption at the excitation wavelength. Adapted with the permission from [27]. Copyright 2010 American Chemical Society

therefore can quench efficiently singlet excitons before the exciton diffusion. This would be true of other amorphous polymer blends. Indeed, such a prompt polaron generation in a few picoseconds is observed for other amorphous polymer/PCBM blend films like PCPDTBT/PCBM [29] and N-P7/PCBM blends [30].

As shown in Fig. 6.7, the PCBM anion band decays exponentially with a time constant of 813 ps (~70%) and a constant fraction (~30%). On the other hand, the GSB signals recover with the same time constant. This agreement suggests that PCBM anions recombine with polymer polarons to the ground state. In addition, this rise and decay time constant is independent of the excitation intensities ranging from 6 to 120 $\mu J\ cm^{-2}$, suggesting that the recombination is geminate recombination of CT pairs formed at the RRa-P3HT/PCBM interface. Thus, 70% of polymer polarons and PCBM anions form interfacial CT states and then geminately recombine to the ground state, and 30% of them dissociate into free charge carriers. In other words, the charge dissociation efficiency η_{CD} is as low as ~30% in RRa-P3HT/PCBM blend films. Similarly, the geminate recombination is one of the major losses in PCPDTBT/PCBM and NP-7/PCBM blend films. As a result, the charge dissociation efficiency is 0.5 and 0.7 for PCPDTBT/PCBM blends without and with additive [29], respectively, and 0.65 for NP-7/PCBM blends [30]. In summary, the exciton diffusion efficiency η_{ED} is almost 100% but the charge dissociation efficiency η_{CD} is limited by the geminate recombination in amorphous polymer solar cells as will be discussed in Sect. 6.6.

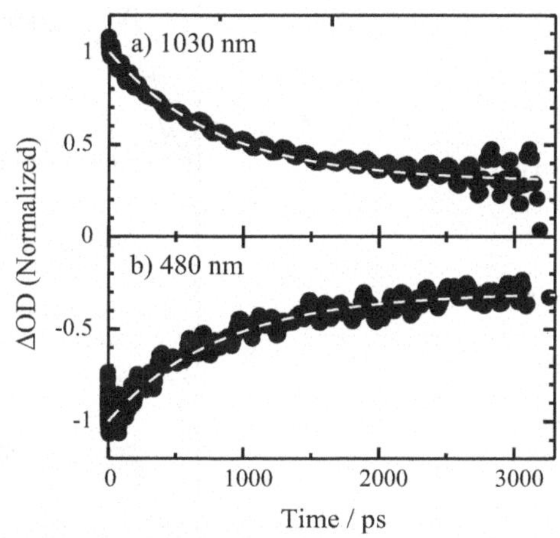

Fig. 6.7 Transient absorption decays of RRa-P3HT/PCBM blend films excited at 400 nm with a fluence of ~12 $\mu J\ cm^{-2}$, which were measured at **a** 1030 nm and **b** 480 nm. The decays were fitted with an exponential function and a constant: $\Delta OD(t) = A_D \exp(-t/\tau_D) + B$. The broken lines represent the best-fitting curves. Reprinted with the permission from [27]. Copyright 2010 American Chemical Society

6.4.2 Crystalline Polymer Solar Cells

Figure 6.8 shows the transient absorption spectra of RR-P3HT pristine and RR-P3HT/PCBM blend films excited at 400 or 600 nm [27]. For the RR-P3HT pristine film, a large absorption band was observed at around 1200 nm immediately after the laser excitation. This band is ascribed to polymer singlet excitons, which decay with a time constant of 330 ps under a low excitation intensity, as reported previously [26]. The negative signals at around 500 nm are ascribed to the GSB. For the RR-P3HT/PCBM blend films, as shown in the figure, the singlet exciton band is reduced to ~30–60% of that observed for the pristine film at 0 ps depending on the weight

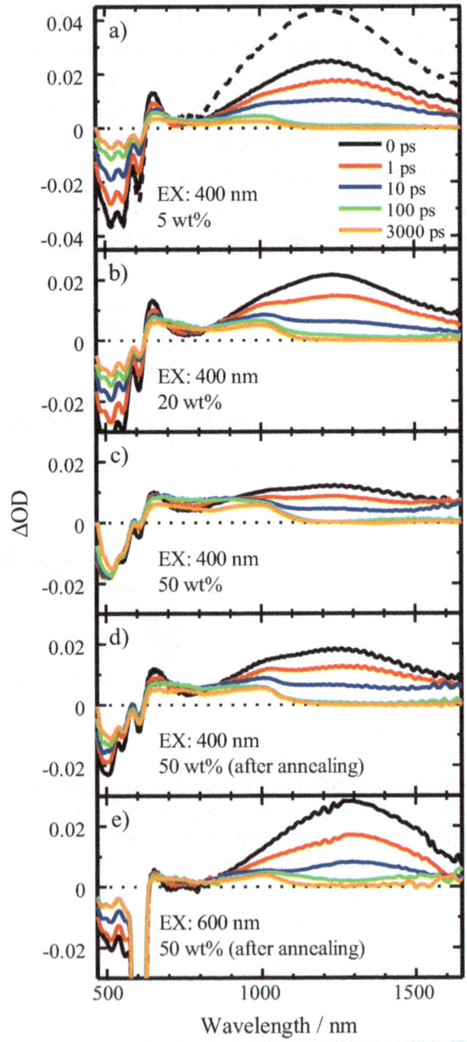

Fig. 6.8 Transient absorption spectra of RR-P3HT pristine (broken line in the panel a) and RR-P3HT/PCBM blend (solid lines) films measured at 0 (black), 1 (red), 10, (blue), 100 (light green), and 3000 (orange) ps after the laser excitation. The excitation wavelengths were **a**–**d** 400 nm (~15 μJ cm^{-2}) and **e** 600 nm (~10 μJ cm^{-2}). The PCBM concentrations were **a** 5, **b** 20, and **c**–**e** 50 wt%. The spectra in the panels d and e were measured after thermal annealing at 140 °C for 30 min. The ΔOD was corrected for variations in the absorption at the excitation wavelength. Reprinted with the permission from [27]. Copyright 2010 American Chemical Society

fraction of PCBM: It is more quenched with increasing PCBM fraction. Instead, new absorption bands are observed over 700 and 1100 nm, which are ascribed to polymer polarons. In other words, polymer polarons are promptly generated even at 0 ps in the blend films. This is similar to that observed for RRa-P3HT/PCBM blends. Thus, it would be ascribed to the prompt charge generation in amorphous P3HT domains mixed with PCBM. Subsequently, the broadband of singlet excitons decays and completely disappears at 100 ps. As described in [27], the singlet exciton signals decay with different time constants, which shorten with increasing PCBM fraction and extend after thermal annealing at 140 °C for 30 min. In other words, the decay time is dependent upon P3HT domain size in the blend films. Indeed, the decay time of singlet excitons is dependent upon polymer crystallinity: It was reported to be as short as 0.8 ps for less crystalline PSBTBT and PCBM and as long as 100 ps for highly crystalline PNOz4T and $PC_{71}BM$ blend films. As such, it is ascribed to the delayed charge generation following the exciton diffusion in large crystalline domains. In summary, there are two pathways for the charge generation in crystalline polymer solar cells: one is prompt charge generation at amorphous polymer domains mixed with PCBM or at an interface of crystalline polymer domains and PCBM where no exciton diffusion is needed to generate charge carriers and the other is delayed charge generation following exciton diffusion in large crystalline polymer domains.

At a later time stage after several tens of picoseconds, as shown in Fig. 6.8, singlet excitons disappear and instead polymer polarons are observed from 700 to 1100 nm. As shown in Fig. 6.9, the decay dynamics of these bands is dependent upon the excitation intensity. For the transient absorption signals at 700 and 1000 nm, the absorption decays slower with decreasing excitation intensities, suggesting bimolecular

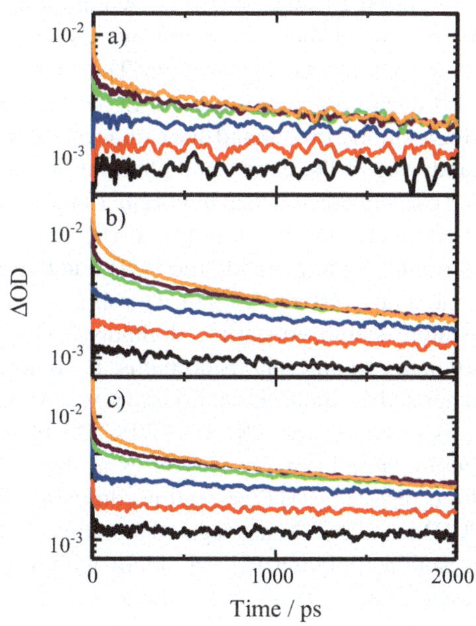

Fig. 6.9 Transient absorption signals measured at **a** 700, **b** 850, and **c** 1000 nm of RR-P3HT/PCBM (50 wt%) blend films excited at 400 nm with a fluence of 6, 12, 24, 48, 72, and 120 μJ cm^{-2} from bottom to top in each panel. Reprinted with the permission from [27]. Copyright 2010 American Chemical Society

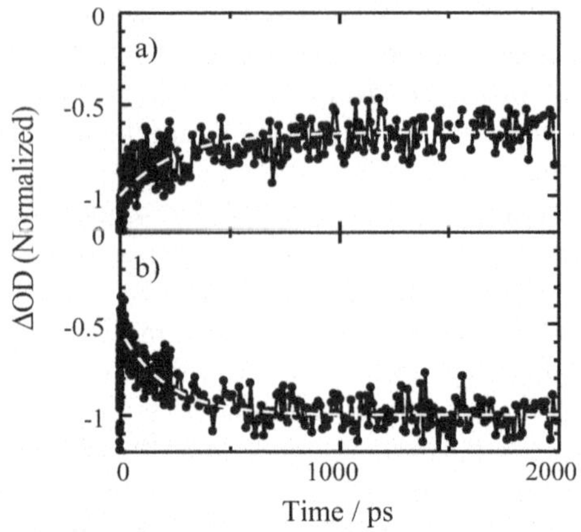

Fig. 6.10 Transient absorption signals of RR-P3HT/PCBM (50 wt%) blend films excited at 400 nm (~12 μJ cm^{-2}), measured at **a** 480 and **b** 610 nm, which were fitted by $\Delta OD(t) = A_R[1 - \exp(-t/\tau_R)] + B$ and $\Delta OD(t) = A_D \exp(-t/\tau_D) + B$, respectively. The broken lines represent the best-fitting curves Adapted with the permission from [27]. Copyright 2010 American Chemical Society

recombination. Interestingly, no decay is observed under lower excitation intensities, suggesting that no geminate recombination is involved: The dissociation efficiency η_{CD} is almost 100%. For the transient absorption signals at 850 nm, on the other hand, the absorption decays slower with decreasing excitation intensities and finally decays with the same time constant under lower excitation intensities, suggesting geminate recombination. This is similar to the decay observed for RRa-P3HT/PCBM blend films. Therefore, as described in [27], the transient absorption signals are ascribed to delocalized polarons in crystalline P3HT domains for 700 nm, localized polarons loosely bound to PCBM at the interface in disordered amorphous P3HT domains for 850 nm, and localized polarons in disordered P3HT domains for 1000 nm.

Figure 6.10 shows the time evolution of the GSB measured at 400 and 610 nm. As mentioned above, transient species observed are not singlet excitons but polymer polarons at this later time stage. Thus, these negative signals are also ascribed to the GSB due to polymer polarons. In the steady-state absorption, RR-P3HT crystalline films exhibit a broadband due to amorphous phase at shorter wavelengths and vibronic bands due to the crystalline phase at around 600 nm. Thus, the GSB bands at 400 and 610 nm are ascribed to polymer polarons in amorphous and crystalline domains, respectively. As shown in Fig. 6.10, the GSB band at 400 nm recovers with a time constant of 250 ps while the GSB band at 610 nm is negatively increased with the same time constant (250 ps). This agreement suggests hole transfer from amorphous to more stable crystalline domains. This would improve the charge dissociation efficiency in amorphous domains. If no hole transfer is involved, polymer polarons generated in amorphous domains would suffer from the geminate recombination as is observed for RRa-P3HT/PCBM blend films. This hole transfer is probably due to cascade energy structures at the P3HT/PCBM interface, which consists of three phases: P3HT crystalline domains, P3HT amorphous domains mixed with PCBM, and PCBM aggregated domains. A similar hole transfer has been reported for crystalline polymer/PCBM blends with three-phase structures [39]. In other words, cascade energy structures at the interface would be beneficial for the efficient charge dissociation in crystalline polymer solar cells like RR-P3HT/PCBM. The charge dissociation efficiency in other crystalline polymer solar cells will be discussed in Sect. 6.6.

6.4.3 Polymer Solar Cells with Small Photon Energy Loss

Figure 6.11 shows spectroscopic data of PNTz4T and PNOz4T pristine and PNTz4T/PC_{71}BM and PNOz4T/PC_{71}BM blend films [36]. As shown in Fig. 6.11a and d, the photoluminescence is significantly quenched to a few % for PNTz4T/PC_{71}BM but modestly quenched to ~35% for PNOz4T/PC_{71}BM blend films. These findings suggest that the exciton diffusion efficiency η_{ED} is almost 100% for PNTz4T/PC_{71}BM but as low as ~65% for PNOz4T/PC_{71}BM blend films. As shown in Fig. 6.11 b and e, transient absorption spectra of PNTz4T/PC_{71}BM and PNOz4T/PC_{71}BM blend films exhibit slightly complicated spectral changes

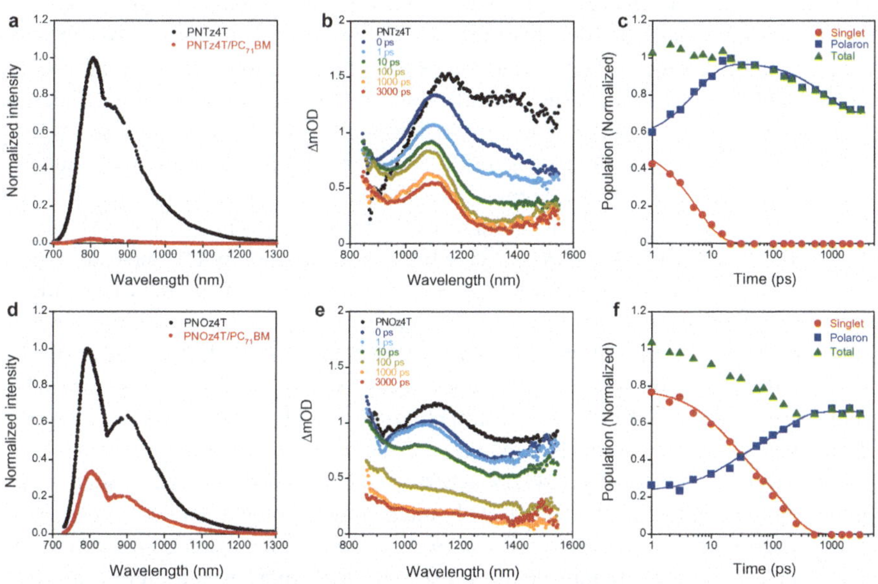

Fig. 6.11 Spectroscopic data of PNTz4T and PNOz4T pristine and PNTz4T/PC_{71}BM and PNOz4T/PC_{71}BM blend films. Photoluminescence spectra of **a** PNTz4T and **d** PNOz4T pristine films (red dots) and **a** PNTz4T/PC_{71}BM and **d** PNOz4T/PC_{71}BM blend films (black dots). Transient absorption spectra of **b** PNTz4T and e) PNOz4T pristine films at 0 ps (black dots) and **b**) PNTz4T/PC_{71}BM and **e** PNOz4T/PC_{71}BM blend films measured at 0 (blue), 1 (light blue), 10 (green), 100 (light brown), 1000 (orange), and 3000 ps (red dots). Time evolution of singlet excitons (red circles), polarons (blue squares), and the total of singlet excitons and polarons (green triangles) in **c** PNTz4T/PC_{71}BM and **f** PNOz4T/PC_{71}BM blend films. Reprinted by permission from Macmillan Publishers Ltd [36]. Copyright 2015

with time but can be resolved to two components of polymer singlet excitons and polarons as described in Ref. 36. Figure 6.11c and f show the time evolution of polymer singlet excitons, polarons, and the total of them. As shown in the figure, polymer polarons are generated at 0 ps more efficiently in PNTz4T/PC_{71}BM than in PNOz4T/PC_{71}BM blend films, suggesting that PNTz4T/PC_{71}BM blends have a larger fraction of mixed regions and or small domains. Subsequently, polymer singlet excitons decay and polymer polarons rise with the same time constant: 5.5 ps for PNTz4T/PC_{71}BM and 100 ps for PNOz4T/PC_{71}BM blend films. These time constants are consistent with the photoluminescence quenching efficiency mentioned above because the singlet exciton lifetime is as long as a few hundreds of picoseconds. For PNTz4T/PC_{71}BM, the polaron signals decay after 100 ps, which is ascribed to the geminate recombination because the decay is independent of the excitation intensities. For PNOz4T/PC_{71}BM, on the other hand, the polaron signals do not decay at all on this time scale, suggesting efficient charge dissociation. In summary, polymer polarons are generated from singlet excitons with >95% but ~30% of them recombine geminately in PNTz4T/PC_{71}BM blend films. In other words, the exciton diffusion efficiency η_{ED} is as high as >95% but the charge dissociation efficient η_{CD} is limited to ~70% in PNTz4T/PC_{71}BM blend films. On the other hand, ~35% of singlet excitons radiatively deactivate to the ground state and hence only ~65% of them are converted into polymer polarons with a dissociation efficiency of ~100% in PNOz4T/PC_{71}BM blend films. In other words, the exciton diffusion efficiency η_{ED} is as low as ~65% but the charge dissociation efficient η_{CD} is as high as ~100%. The difference in η_{ED} is due to the difference in crystalline domain size, which is revealed by TEM and AFM images of the blend films. As described in [36], the LUMO–LUMO energy offset is as large as 0.31 eV for PNTz4T/PC_{71}BM and as small as 0.12 eV for PNOz4T/PC_{71}BM, which are estimated by CV measurements. The small energy offset in PNOz4T/PC_{71}BM is consistent with an effective bandgap energy evaluated from temperature dependence of V_{OC} and electroluminescence being similar to the photoluminescence of PNOz4T pristine films. These results show that the charge dissociation efficiency η_{CD} is as high as ~100% even for such a small energy offset of ~0.1 eV. As a result, photon energy loss, which is defined by the difference between the optical bandgap E_g and the open-circuit voltage energy eV_{OC}, is as small as 0.56 eV for PNOz4T/PC_{71}BM solar cells. As described in Sect. 6.2, it has been believed that the energy offset should be large enough to break the Coulomb binding energy of electron–hole pairs in excitons, which has been considered to be more than 0.3 eV for efficient devices [40], although there was a study reporting that the minimum driving force is as small as 0.1 eV for the charge transfer from singlet excitons to the CT state [41]. In other words, our finding clearly contradicts the conventional photovoltaic conversion mechanism. After this study, similar small photon energy losses have been reported for many systems [42]. Thus, further studies are needed for revising the conventional photovoltaic conversion mechanism.

6.4.4 Ternary Blend Polymer Solar Cells

Figure 6.12 shows transient absorption spectra of RR-P3HT/PCBM binary and RR-P3HT/PCBM/SiPc6 ternary blend films upon the laser excitation of the P3HT absorption at 400 nm. As described before, the large absorption band at around 1250 nm is ascribed to polymer singlet excitons. The negative signals at around 500 nm are ascribed to the GSB of P3HT. A broad depression is observed from 650 to 850 nm, which is ascribed to stimulated emission from P3HT. At 100 ps, broad absorption bands are observed from 650 to 1000 nm, which is ascribed to polymer polarons. For the RR-P3HT/PCBM binary blends, polymer singlet excitons decay with a time constant of 25 ps (85%) and 330 ps (15%). Polymer polarons are generated with the same time constant of 25 ps. Thus, this is ascribed to the charge generation following exciton diffusion. The longer and minor time constant is ascribed to the lifetime of polymer singlet excitons. For the RR-P3HT/PCBM/SiPc6 ternary blends, on the other hand, polymer singlet excitons decay with a time constant of 4 ps (50%) and the same time constants observed for the binary blends. Polymer polarons are generated with a time constant of 6 ps. This disagreement in the two time constants suggests that polymer polarons are not generated directly from polymer singlet excitons. There should be an intermediate state. As shown in Fig. 6.12c, a small but sharp

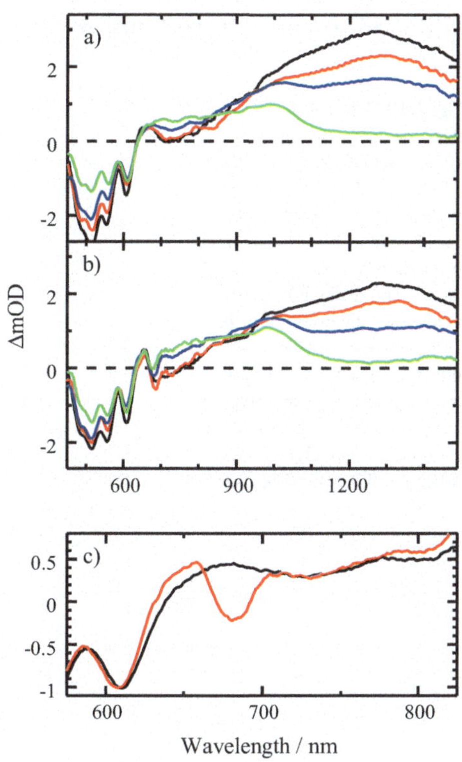

Fig. 6.12 Transient absorption spectra of **a** RR-P3HT/PCBM and **b** RR-P3HT/PCBM/SiPc6 blend films measured at 0 (black), 1 (red), 10, (blue), and 100 (light green) ps after the laser excitation. The excitation wavelength and intensity were 400 nm and 6 μJ cm^{-2}, respectively. **c** Transient absorption spectra of RR-P3HT/PCBM (black line) and RR-P3HT/PCBM/SiPc6 (red line) blend films measured at 10 ps Adapted with the permission from [13]. Copyright 2010 American Chemical Society

depression is observed at around 680 nm only for the RR-P3HT/PCBM/SiPc6 ternary blends. This depression band is ascribed to the GSB of SiPc6 dye molecules because the wavelength is the same as that of the dye absorption. As described in [13], the depression fraction increases with a time constant of 4 ps, which is in good agreement with the lifetime of polymer singlet excitons observed for the ternary blends. Note that SiPc6 has negligible absorption at 400 nm and hence cannot be generated directly by the laser excitation at 400 nm. Therefore, the rapid decay of polymer singlet excitons (4 ps) is ascribed to an efficient energy transfer from polymer singlet excitons to SiPc6 dye molecules. This is consistent with an effective spectral overlap between P3HT emission and SiPc6 absorption bands. In summary, polymer singlet excitons migrate to a donor/acceptor interface with a time constant of 25 ps followed by charge generation in RR-P3HT/PCBM binary blend films upon the photoexcitation of the P3HT absorption band at 400 nm. On the other hand, polymer singlet excitons are efficiently transferred to SiPc6 dye molecules by long-range energy transfer with a time constant of 4 ps, and then polymer polarons are generated from the dye singlet excitons with a time constant of 6 ps, which includes the energy transfer time.

Figure 6.13 shows transient absorption spectra of RR-P3HT/SiPc6 binary, PCBM/SiPc6 binary, and RR-P3HT/PCBM/SiPc6 ternary blend films upon the laser excitation of the SiPc6 absorption at 680 nm. The dye concentration was fixed to 3.4 wt%. In other words, SiPc6 molecules are isolated in RR-P3HT/SiPc6 or PCBM/SiPc6 binary blend films. As shown in Fig. 6.13a, a flat absorption band

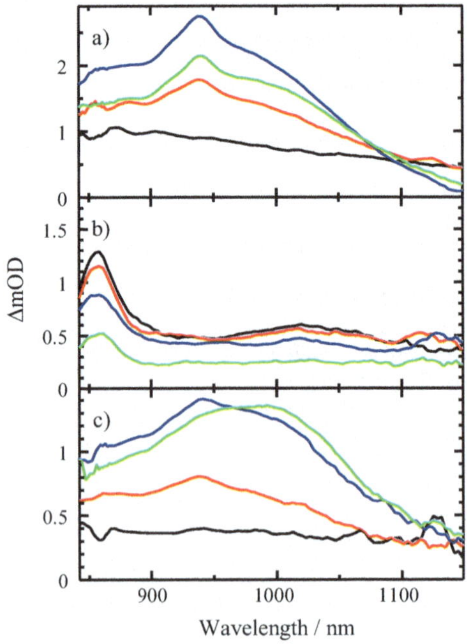

Fig. 6.13 Transient absorption spectra of **a** RR-P3HT/SiPc6, **b** PCBM/SiPc6 binary, and **c** RR-P3HT/PCBM/SiPc6 ternary blend films measured at 0 (black), 1 (red), 10, (blue), and 200 (light green) ps after the laser excitation. The excitation wavelength and intensity were 680 nm and 7 μJ cm^{-2}, respectively. Adapted with the permission from [13]. Copyright 2010 American Chemical Society

is observed first, which is ascribed to SiPc6 singlet excitons because of the selective excitation at 680 nm. Subsequently, a broad absorption and a small but sharp absorption bands are observed at around 850–1050 nm and at 940 nm, respectively. As described in [13], the broad and sharp absorption bands are ascribed to polymer polarons and dye anions, respectively. These two bands increase with a time constant of 2 ps and then decay with a time constant of 2 ns. In summary, polymer polarons are generated with a time constant of 2 ps in RR-P3HT/SiPc6 binary blends. As shown in Fig. 6.13b, a large absorption and a small and broad absorption are observed at 855 and 1030 nm, respectively. As described in [13], the large and small absorption bands are ascribed to SiPc6 anion and PCBM anion, respectively. These two bands are observed even at 0 ps and then decay with time constants of 35 ps (40%) and 2 ns (60%). In summary, polymer polarons are promptly generated within a laser pulse width of 100 fs in PCBM/SiPc6 binary blends. In other words, charge generation dynamics is dependent on the location of dye molecules in blend films.

For RR-P3HT/PCBM/SiPc6 ternary blend films excited at the dye absorption, as shown in Fig. 6.13c, a flat absorption band due to SiPc6 singlet excitons is observed first immediately after the laser excitation, and then a broad absorption band with a small and sharp absorption peak, which is due to P3HT polarons and SiPc anions, respectively, increase with a time constant of 2 ps. This charge generation dynamics is the same as that observed for RR-P3HT/SiPc6 binary blends, suggesting that SiPc6 dye molecules are basically located in P3HT domains. If SiPc6 dye molecules were isolated in P3HT domains as is the case for RR-P3HT/SiPc6 binary blends, polymer polarons and SiPc6 anions recombine geminately with a time constant of 2 ns. However, this is not the case. As shown in Fig. 6.13c, the absorption at 940 nm does not decay at all. Instead, the absorption peak is shifted from 940 to 1000 nm with time. The absorption at 1000 nm is ascribed to P3HT polarons and PCBM anions observed for RR-P3HT/PCBM blends as described in Sect. 6.4.2. In other words, this peak shift is indicative of charge shift (electron transfer) from SiPc6 anions to PCBM.

The time evolution of these transient species can be obtained from the spectral simulation. As shown in Fig. 6.14, SiPc6 anions rapidly decay with a time constant of

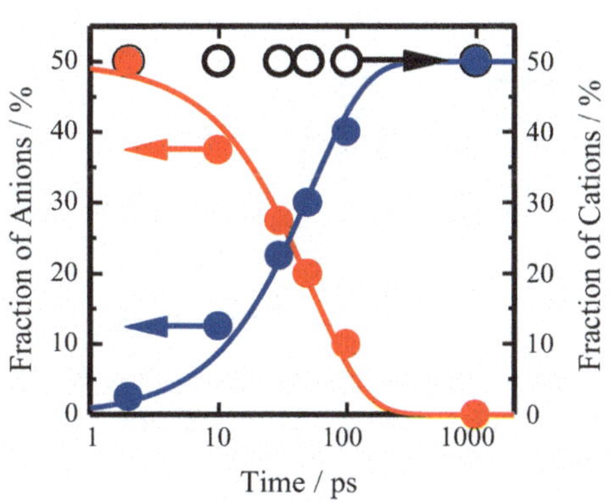

Fig. 6.14 The fraction of SiPc6 anions (red circles), PCBM anions (blue circles), and P3HT polarons (open circles) in RR-P3HT/PCBM/SiPc6 ternary blend films after the laser excitation of SiPc6 at 680 nm with a fluence of 7 μJ cm^{-2}. Adapted with the permission from [13]. Copyright 2010 American Chemical Society

50 ps, which is much faster than the geminate recombination of polymer polarons and dye anions observed for RR-P3HT/SiPc6 blend films. Instead, PCBM anions increase with the same time constant, suggesting a charge shift from SiPc6 to PCBM. This finding suggests that SiPc6 dye molecules contact with PCBM as well as P3HT in RR-P3HT/PCBM/SiPc6 ternary blend films. In other words, SiPc6 dye molecules are located at the P3HT/PCBM interface in the ternary blend films. As described in [13], only a small part of polymer polarons decay during the charge shift, suggesting that a majority of dye molecules are located at the interface and a small minor part of them are isolated in P3HT domains. Interestingly, no decay of polymer polarons is observed at all during the charge shift for RR-P3HT/PCBM/SiPc6 ternary blend films fabricated by solvent annealing. In this case, almost all the dye molecules are spontaneously located at the P3HT/PCBM interface in the ternary blend films. As discussed in [11], this is partly because dye molecules are expelled from polymer crystalline or PCBM aggregated domains to disordered mixed domains, and partly because dye molecules have an intermediate surface energy between RR-P3HT and PCBM. In other words, dye locations can be controlled by careful design of dye molecules as demonstrated in [43].

6.5 Charge Recombination Dynamics

This section describes the bimolecular recombination dynamics of dissociated charge carriers on a timescale of microseconds in polymer/fullerene blend films. Here, our attention is focused on bimolecular recombination dynamics in RR-P3HT/PCBM blend films. Figure 6.15 shows photovoltaic materials studied in this chapter [38,

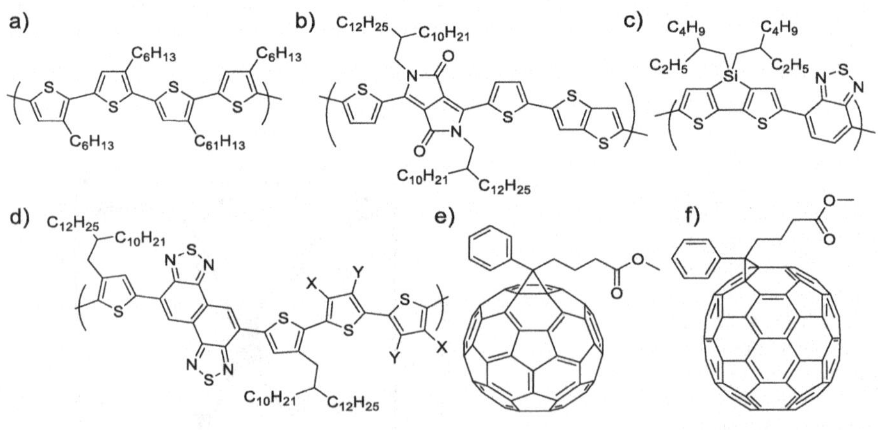

Fig. 6.15 Chemical structures of photovoltaic materials studied in this chapter: **a** RR-P3HT, **b** DT-PDPP2T-TT, **c** PSBTBT, **d** PNTz4T (X = Y = H), PNTz4TF2 (X = H, Y = F), PNTz4TF4 (X = Y = F), **e** PCBM, and **f** $PC_{71}BM$

44–46]. It will be discussed later how bimolecular recombination is reduced in these polymer/fullerene blend films in Sect. 6.6.

6.5.1 Bimolecular Recombination Dynamics

Figure 6.16 shows the transient absorption spectra of RR-P3HT/PCBM blend films excited at 400 nm measured on a time scale of microseconds [44]. On this time stage, singlet excitons completely disappear but triplet excitons may be observed. However, this is not the case because these bands are not quenched at all under an oxygen atmosphere. As shown in the figure, two large absorption bands are observed at around 700 and 1000 nm. The absorption band at 700 nm is larger than that at 1000 nm at an early time stage of <10 μs but smaller than that at 1000 nm at a later time stage of >20 μs. In other words, the absorption band at 700 nm decays faster than that at 1000 nm. This finding suggests that there are at least two different polymer polarons in the RR-P3HT/PCBM blend film. As described in Sect. 6.4.2, the band at 700 nm is ascribed to delocalized polarons in crystalline domains and the band at 1000 nm is ascribed to localized polarons in disordered domains.

As shown in Fig. 6.17, these two bands decay faster at an early time stage under higher excitation intensities, suggesting bimolecular recombination. The decay dynamics can be well-fitted with an empirical power-law equation

$$n(t) = \frac{n(0)}{(1 + at)^{\alpha}} \tag{6.4}$$

where $n(t)$ is the charge carrier density at time t, $n(0)$ is the initial carrier density at $t = 0$, and a and α are kinetic parameters. This power-law decay is characteristic of

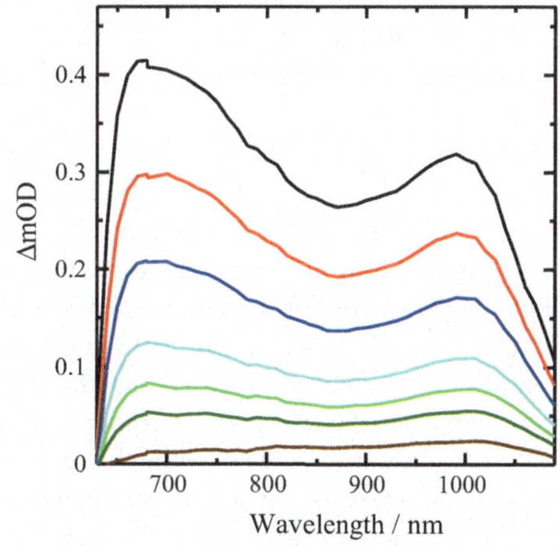

Fig. 6.16 Transient absorption spectra of RR-P3HT/PCBM blend films excited at 400 nm with a fluence of 30 μJ cm^{-2}. The spectra were measured at 0.5 (black), 1 (red), 2 (blue), 5, (light blue), 10 (light green), 20 (green), and 100 μs (brown). Adapted with the permission from [44]. Copyright 2010 American Chemical Society

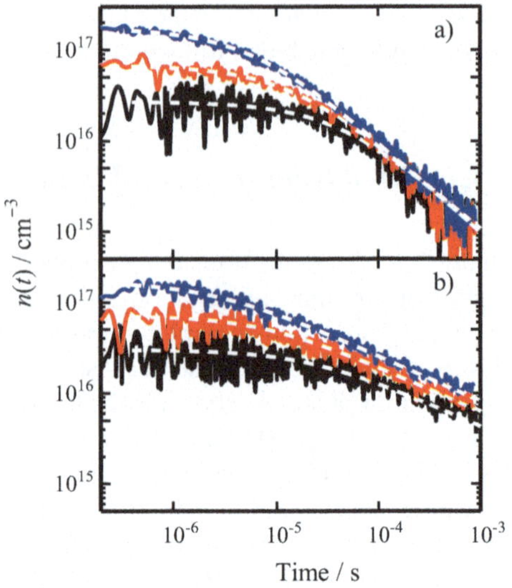

Fig. 6.17 Charge density decays of RR-P3HT/PCBM blend films excited at 400 nm with a fluence of 0.8 (blue lines), 1.8 (red lines), and 4.7 μJ cm^{-2} (black lines). The spectra were measured at **a** 700 nm and **b** 1000 nm. The broken lines represent fitting curves with an empirical equation: $n(t) = n(0)/(1 + at)^{\alpha}$. Reprinted with the permission from [44]. Copyright 2010 American Chemical Society

trap-limited charge recombination in a medium with an energetic disorder [47, 48]. For diffusion-limited bimolecular recombination, the rate equation is given by

$$\frac{dn(t)}{dt} = -\gamma(t)n^2(t) \tag{6.5}$$

By substituting Eq. (6.4) into Eq. (6.5), the time-dependent bimolecular recombination rate $\gamma(t)$ is obtained as

$$\gamma(t) = -\frac{1}{n^2(t)}\frac{dn(t)}{dt} = \frac{a\alpha}{n(0)}(1 + at)^{\alpha-1} \tag{6.6}$$

By substituting Eq. (6.4) into Eq. (6.6) again, the carrier density-dependent bimolecular recombination rate $\gamma(n)$ is obtained as

$$\gamma(n) = \frac{a\alpha}{n}\left(\frac{n}{n(0)}\right)^{\frac{1}{\alpha}} \tag{6.7}$$

From the fitting curves in Fig. 6.17, the exponent α is evaluated to be unity for the delocalized polaron observed at 700 nm and ~0.5 for the localized polaron observed at 1000 nm. As shown in Fig. 6.18, $\gamma(t)$ and $\gamma(n)$ are constant for the delocalized polaron. This finding suggests that the decay kinetics is ascribed to bimolecular recombination of free carriers. In other words, delocalized polarons can diffuse freely as free charge carriers. On the other hand, $\gamma(t)$ and $\gamma(n)$ obey a power-law with a slope of ~0.5 and ~1, respectively, for the localized polaron. This finding suggests that the decay kinetics is ascribed to trap-limited bimolecular recombination. In other words,

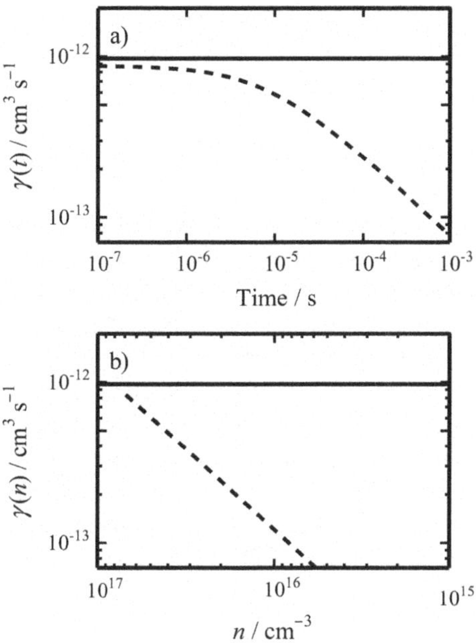

Fig. 6.18 Log–log plots of the bimolecular recombination rate as a function of **a** time t (Eq. (6.5)) and **b** carrier density n (Eq. (6.6)). The solid and broken lines represent the bimolecular recombination rate for the charge carriers observed at 700 and 1000 nm, respectively. Adapted with the permission from ref. [44]. Copyright 2010 American Chemical Society

localized polarons are trapped in disordered domains with an energetic disorder. At an early time stage, an effective trap depth is shallower because most trap sites are filled with more charge carriers, and hence, the recombination rate is faster. At a later time stage, an effective trap depth is deeper because trap sites are partly filled with less charge carriers, and hence, the recombination rate is slower. These assignments are consistent with the temperature dependence of the decay kinetics as described in [44]. The activation energy for delocalized polarons is as small as ~0.078 eV, which is independent of the carrier density. On the other hand, the activation energy for localized polarons decreases exponentially from 0.178 to 0.097 eV with increasing carrier density.

From Eq. (6.7), the carrier lifetime τ_n is given by

$$\tau_n = \frac{1}{\gamma(n)n} = \frac{n(0)^{\frac{1}{\alpha}}}{a\alpha} n^{-\frac{1}{\alpha}} \tag{6.8}$$

By comparing Eq. (6.8) with Eq. (6.3) for TPV/TPC measurements, the following relationship is obtained.

$$\alpha = \frac{1}{\lambda} \tag{6.9}$$

$$a = \frac{\lambda n(0)^{\lambda}}{(\lambda + 1)\tau_{\Delta n_0} n_0^{\lambda}} \tag{6.10}$$

Thus, the empirical equation Eq. (6.4) can be expressed by using kinetic parameters for TPV/TPC measurements.

$$n(t) = n(0)\left[1 + \frac{\lambda n(0)^{\lambda}}{(\lambda + 1)\tau_{\Delta n_0} n_0^{\lambda}} t\right]^{-\frac{1}{\lambda}} \quad (6.11)$$

Durrant and his co-workers demonstrated that the decay dynamics observed by transient absorption spectroscopy is consistent with that observed by TPV/TPC measurements [23]. At a charge density of ~10^{17} cm^{-3}, which corresponds to 1 sun illumination condition, both charge carriers exhibit a similar bimolecular recombination rate of ~10^{-12} cm^3 s^{-1} at an early time stage of <1 μs as shown in Fig. 6.18. This is because localized polarons also contribute to charge transport owing to trap filling. More interestingly, the recombination rate is two orders of magnitude smaller than the Langevin recombination rate ($\gamma_L \approx 10^{-10}$ cm^3 s^{-1}). Here, γ_L is estimated by $\gamma_L = e\mu_h/\varepsilon_r\varepsilon_0$ with a slower charge mobility [49] of $\mu_h \approx 10^{-4}$ cm^2 V^{-1} s^{-1} [50]. Such a slow recombination rate would be the key for highly efficient charge collection in RR-P3HT/PCBM solar cells as described in the next section.

6.5.2 Charge Carrier Lifetime

For highly efficient charge collection, the charge carrier lifetime should be longer than the charge collection time. The charge carrier lifetime can be estimated by transient absorption spectroscopy as mentioned above. The carrier density has been reported to be ~10^{17} cm^{-3} in RR-P3HT/PCBM blend films under 1 sun illumination. Thus, the carrier lifetime τ_n is estimated to be $\tau_n = 14$ μs from Eq. (6.8) with kinetic parameters a and α, which are obtained by fitting analysis for the decay curve in Fig. 6.17. On the other hand, the charge collection time τ_{CC} is roughly estimated by Eq. (6.12)

$$\tau_{CC} = \frac{L^2}{2\mu V} \quad (6.12)$$

where L is the thickness of the active layer, μ is the charge mobility, and V is the applied voltage, which is typically assumed to be V_{OC} at the short circuit. For RR-P3HT/PCBM blend films, τ_{CC} is estimated to be $\tau_{CC} = 0.8$ μs for a thickness of 100 nm. In other words, the charge carrier lifetime τ_n is two orders of magnitude longer than the charge collection time τ_{CC}. This is consistent with highly efficient charge collection in RR-P3HT/PCBM blends. This would be true for RR-P3HT/PCBM solar cells under device operation conditions. We note that the charge carrier dynamics observed by transient absorption with pulsed laser excitations might be different from the charge carrier dynamics under steady-state solar illumination.

TPV/TPC measurements are a useful method for estimating charge carrier lifetime under device operation conditions.

Figure 6.19 shows TPV and TPC decays of RR-P3HT/PCBM solar cells under 1 sun illumination [45]. Under steady-state illumination, the photovoltage is constant at V_{OC}. Upon a small perturbation laser pulse excitation, minority excess charges are generated, resulting in a small increase in the photovoltage ΔV_0 at a decay time of $t = 0$. As shown in Fig. 6.19a, TPV signals decay exponentially because the decay of the minority excess charges follows pseudo-first-order kinetics as mentioned in Sect. 6.3. From the slope in the figure, the lifetime of minority excess charges $\tau_{\Delta n}$ can be estimated to be $\tau_{\Delta n} = 4.8$ μs. With increasing V_{OC}, $\tau_{\Delta n}$ decreases exponentially as shown in Eq. (6.1). From the slope in Logarithmic plots of $\tau_{\Delta n}$ against V_{OC}, ν is estimated to be $\nu = 1.5$. As shown in Fig. 6.19b, TPC signals decay with a time constant of 0.5 μs, which is almost one order of magnitude faster than TPV signals due to bimolecular recombination in the blend film. The decay time constant is rather consistent with the charge collection time $\tau_{CC} = 0.8$ μs. Thus, the rapid TPC decay is ascribed to the charge collection to the electrode because of the shirt-circuit condition. Consequently, the amount of the minority excess charges Δq can be evaluated from the integral of the TPC decay signals over time. As described in [23, 24], the differential capacitance dC is defined as follows:

$$\mathrm{d}C(V_{OC}) = \frac{\Delta q}{\Delta V_0(V_{OC})} \tag{6.13}$$

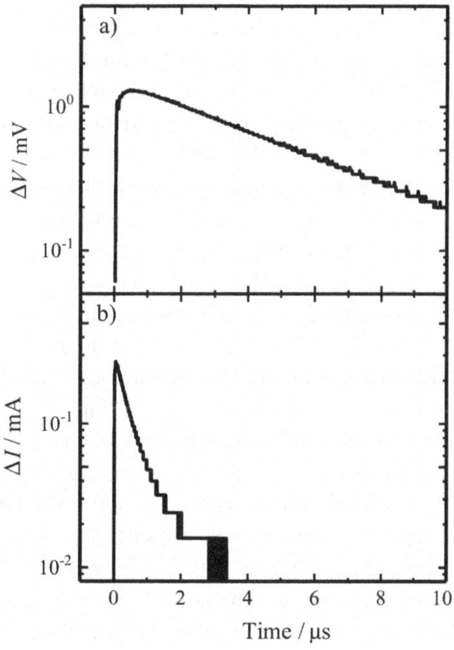

Fig. 6.19 **a** TPV and **b** TPC decays of RR-P3HT/PCBM solar cells under 1 sun illumination. Adapted with the permission from ref. [45]. Copyright 2016 The Society of Photopolymer Science and Technology

where $\Delta V_0(V_{OC})$ is the initial increment in the TPV signals observed under the same condition as the TPV measurement. Consequently, the charge carrier density n is given by Eq. (6.14)

$$n = \frac{1}{eAL}\int_0^{V_{OC}} dC(V_{OC})dV \tag{6.14}$$

where A and L are the area and the thickness of the active layer, respectively. With increasing V_{OC}, n increases exponentially as shown in Eq. (6.2). Under 1 sun condition, n and m are evaluated to be $n = 5.6 \times 10^{16}$ cm^{-3} and $m = 3.6$, respectively. Thus, λ is estimated to be $\lambda = m/\nu = 2.4$. Finally, the charge carrier lifetime τ_n is estimated to be $\tau_n = (1 + \lambda)\tau_{\Delta n} = 16$ μs, which is in good agreement with $\tau_n =$ 14 μs estimated from the transient absorption spectroscopy described above. On the other hand, bimolecular recombination rate γ is estimated to be $\gamma = 1/\tau_n n = 1.1 \times 10^{12}$ cm^3 s^{-1}, which is also consistent with that evaluated from the transient absorption spectroscopy. Thus, the reduction factor ζ is estimated to be $\zeta = 8.7 \times 10^{-3}$ for RR-P3HT/PCBM solar cells. This good agreement indicates that charge carriers generated by pulsed laser light are in thermal equilibrium in RR-P3HT/PCBM blends on a timescale of microseconds.

6.6 Challenging Issues and Concluding Remarks

This section describes challenging issues to be solved for further improvements in photovoltaic performance of polymer solar cells. As described in Sect. 6.4, polymer morphology has impact on photovoltaic conversion efficiencies. Table 6.1. summarizes η_{ED} and η_{CD} in polymer/fullerene blends with different polymer crystallinities. As shown in the table, η_{ED} is almost unity for amorphous polymer/fullerene blends. This is probably because fullerene small molecules are likely to be distributed relatively homogeneously in amorphous matrices. For crystalline polymers, η_{ED} decreases with increasing polymer crystallinity. This is because a part of excitons cannot arrive at a donor/acceptor interface because of large crystalline domains. On the other hand, η_{CD} increases with increasing polymer crystallinity. One possible explanation is that Coulomb binding energy would be reduced for polymer polarons delocalized in crystalline domains. Assuming an effective dielectric constant of 3.5, the Coulomb binding energy is as large as 0.41 eV for an electron–hole pair at a separation distance of 1 nm, but decreases to 0.14 eV at a separation distance of 3 nm. This is still much larger than the thermal energy at room temperature ($k_B T = 26$ meV). Durrant and his co-workers have suggested that the effective Coulomb binding energy would be comparable to the entropy term by considering the number of states for the charge separation [56, 57]. Gregg has demonstrated that such an entropy effect would be larger for three-dimensional aggregates such as fullerene molecules [58]. As shown in Fig. 6.20, the entropy term $-T\Delta S$ would be dominant at longer separation distances. As a result, the Gibbs free energy ΔG exhibits a maximum at a

Table 6.1 Photovoltaic conversion efficiency in polymer solar cells with different crystallinities

Polymers	Morphology	η_{ED}	η_{CD}	L_C/nm[c]	d_π/nm[d]	ΔE_{LL}/eV	Refs.
PNOz4T	Highly crystalline	~0.6	~1			~0.1	36
RR-P3HT (TA)[a]	Highly crystalline	~0.9	>0.9	~12	0.38	~1.1	27,51,52
RR-P3HT	Less crystalline	~0.95	~0.8	5.7	0.38	~1.1	27,53
PNTz4T	Less crystalline	>0.95	~0.75	2.7	0.35	~0.3	36,38
PSBTBT	Less crystalline	~1	~0.75	4.6	0.35	~0.4	34,54,55
PCPDTBT (DIO)[b]	Less crystalline	~1	~0.7			~0.3	29
PCPDTBT	Slightly ordered	~1	~0.5			~0.3	29
N-P7	Amorphous	~1	~0.65			~0.4	30
RRa-P3HT	Amorphous	~1	~0.3			~1.2	27

[a]TA represents thermal annealing at 140 °C for 30 min. [b]DIO represents 1,8-diiodeoctane employed as a solvent additive. [c]L_C represents a coherence length in the π-stacking direction (010). [d]d_π represents a π–π stacking distance of crystallites

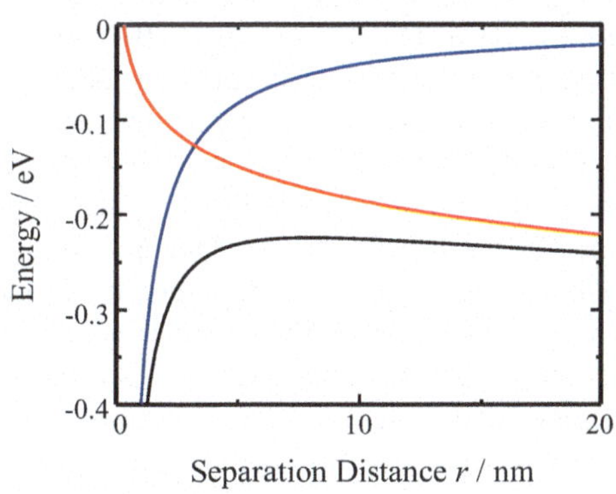

Fig. 6.20 The Coulomb potential energy ΔH (blue line), the entropy contribution $-T\Delta S$ (red line), and the Gibbs free energy ΔG (black line) for an electron–hole pair at a separation distance r. The dielectric constant is assumed to be 3.5. The entropy term is calculated for 3D aggregates with a lattice constant of 1 nm

separation distance of around a few nanometers. In other words, electron–hole pairs can dissociate into free charge carriers at the threshold distance. Deibel and his co-workers have demonstrated by kinetic Monte Carlo simulation that polaron pairs can dissociate efficiently by considering delocalized charge carriers along conjugated polymer chain segments [59]. On the other hand, it has been suggested that fullerene aggregates also have impact on efficient charge dissociation [60, 61]. As such, charge delocalization in polymer crystalline domains or fullerene aggregates would be one of the key factors for highly efficient charge dissociation in polymer solar cells. As described in Sect. 6.4.3, recent studies have shown that singlet excitons can efficiently dissociate into free charge carriers even for an energy offset as small as <0.1 eV. We should understand why such efficient charge dissociation is possible for small energy offset and also the minimum energy offset for efficient charge generation in polymer solar cells.

Table 6.2 Kinetic and device parameters for bimolecular recombination in polymer solar cells

Blends	γ/cm^3 s^{-1}	γ_L/cm^3 s^{-1}	ζ	FF	L/nm	Ref.
RR-P3HT/PCBM	1.1×10^{-12}	1.3×10^{-10}	0.009	0.67	210	45,62
DT-PDPP2T-TT/$PC_{71}BM$	1.7×10^{-11}	4.0×10^{-9}	0.004	0.61	300	45,63
PSBTBT/$PC_{71}BM$	6.1×10^{-12}	1.7×10^{-10}	0.04	0.46	180	46
PNTz4T/$PC_{71}BM$	7.5×10^{-12}	5.8×10^{-10}	0.01	0.73	290	38
PNTz4TF2/$PC_{71}BM$	2.1×10^{-11}	7.9×10^{-10}	0.03	0.63	295	38
PNTz4TF4/$PC_{71}BM$	9.3×10^{-12}	9.4×10^{-11}	0.1	0.50	285	38

Another challenging issue to be solved is to understand the origin of long-lived charge carriers due to reduced bimolecular recombination observed for several highly efficient polymer solar cells. Table 6.2 summarizes kinetic parameters for bimolecular recombination and device parameters in several polymer/fullerene solar cells. As shown in the table, the reduction factor is seemingly not closely related to FF of the devices. This is partly because FF depends not only on the reduction factor but also other photovoltaic and diode parameters. Among them, the most effectively suppressed reduction factors are found in RR-P3HT/PCBM, DT-PDPP2T-TT/$PC_{71}BM$, and PNTz4T/$PC_{71}BM$ solar cells. For these solar cells, FF remains high even for thick devices. For the other devices, FF monotonically decreases with increasing thickness. As such, the reduced bimolecular recombination would be one of the key factors for realizing high FF even for thick devices. These polymers are all highly crystalline polymers and hence would be likely to form cascade energy structures at an interface due to three phases: polymer crystalline domains, polymer/fullerene mixed domains, and fullerene aggregated domains. Such cascaded energy structures might suppress bimolecular recombination effectively. Indeed, recent studies have shown that the lifetime of charge carriers is longer in ternary devices with cascade energy structures than in binary devices [12, 64]. We need to clarify further how bimolecular recombination is suppressed in polymer solar cells.

Acknowledgements Most of the studies in this chapter were conducted with Professor Emeritus Ito at Kyoto University and Associate Professor Benten at Nara Institute of Science and Technology. They have been done mainly by Dr Guo, Dr Yamamoto, Dr Honda, and Dr Tamai. They were partly supported by the JST PRESTO program (Photoenergy and Conversion Systems and Materials for the Next-Generation Solar Cells) and the JST ALCA program (Solar Cell and Solar Energy Systems: JPMJAL1404).

References

1. Yu, G., Gao, J., Hummelen, J.C., Wudl, F., Heeger, A.J.: Polymer photovoltaic cells: Enhanced efficiencies via a network of internal donor–acceptor heterojunctions. Science **270**, 1789–1791 (1995)
2. Hall, J.J.M., Walsh, C.A., Greenham, N.C., Marseglia, E.A., Friend, R.H., Moratti, S.C., Holmes, A.B.: Efficient photodiodes from interpenetrating polymer networks. Nature **376**, 498–500 (1995)
3. Meng, L., Zhang, Y. Wan, X., Li, C., Zhang, X., Wang, Y., Ke, X., Xiao, Z., Ding, L., Xia, R., Yip, H.L., Cao, Y., Chen, Y.: Organic and solution-processed tandem solar cells with 17.3% efficiency. Science **361**, 1094–1098 (2018)
4. Liu, Q., Jiang, Y., Jin, K., Qin, J., Xu, J., Li, W., Xiong, J., Liu, J., Xiao, Z., Sun, K., Yang, S., Zhang, X., Ding, L.: 18% Efficiency organic solar cells. Bull. Sci. **65**, 272–275 (2020)
5. Shaheen, S.E., Brabec, C.J., Sariciftci, N.S., Padinger, F., Fromherz, T., Hummelen, J.C.: 2.5% Efficient organic plastic solar cells. Appl. Phys. Lett. **78**, 841–843 (2001)
6. Padinger, F., Rittberger, R.S., Sariciftci, N.S.: Effects of postproduction treatment on plastic solar cells. Adv. Funct. Mater. **13**, 85–88 (2003)
7. Peet, J., Kim, J.Y., Coates, N.E., Ma, W.L., Moses, D., Heeger, A.J., Bazan, G.C.: Efficiency enhancement in low-bandgap polymer solar cells by processing with alkane dithiols. Nat. Mater. **6**, 497–500 (2007)
8. Lin, Y., Wang, J., Zhang, Z.-G., Bai, H., Li, Y., Zhu, D., Zhan, X.: An electron acceptor challenging fullerenes for efficient polymer solar cells. Adv. Mater. **27**, 1170–1174 (2015)
9. Honda, S., Nogami, T., Ohkita, H., Benten, H., Ito, S.: Improvement of the light-harvesting efficiency in polymer/fullerene bulk heterojunction solar cells by interfacial dye modification. ACS Appl. Mater. Interf. **1**, 804–810 (2009)
10. Honda, S., Ohkita, H., Benten, H., Ito, S.: Multi-colored dye sensitization of polymer/fullerene bulk heterojunction solar cells. Chem. Commun. **46**, 6596–6598 (2010)
11. Honda, S., Ohkita, H., Benten, H., Ito, S.: Selective dye loading at the heterojunction in polymer/fullerene solar cells. Adv. Energy Mater. **1**, 588–598 (2011)
12. Xu, H., Ohkita, H., Tamai, Y., Benten, H., Ito, S.: Interface engineering for ternary blend polymer solar cells with a heterostructured near-IR dye. Adv. Mater. **27**, 5868–5874 (2015)
13. Honda, S., Yokoya, S., Ohkita, H., Benten, H., Ito, S.: Light-harvesting mechanism in polymer/fullerene/dye ternary blend studied by transient absorption spectroscopy. J. Phys. Chem. C **115**, 11306–11317 (2011)
14. Ohkita, H., Ito, S.: Transient absorption spectroscopy of polymer-based thin-film solar cells. Polymer **52**, 4397–4417 (2011)
15. Ohkita, H., Ito, S.: Exciton and charge dynamics in polymer solar cells studied by transient absorption spectroscopy. In: Choy, W.C.H. (ed.) Organic Solar Cells, pp. 103–137. Springer-Verlag, London (2013)
16. Ohkita, H., Tamai, Y., Benten, H., Ito, S.: Transient absorption spectroscopy for polymer solar cells. IEEE J. Sel. Top Quantum Electron. **22**, 100–111 (2016)
17. Norrish, R.G.W., Porter, G.: Chemical reactions produced by very high light intensities. Nature **164**, 658 (1949)
18. Porter, G.: Flash photolysis and spectroscopy. A new method for the study of free radical reactions. Proc. R. Soc. London, Ser. A **200**, 284–300 (1950)
19. Norrish, R.G.W.: Some fast reactions in gases studied by flash photolysis and kinetic spectroscopy. The official website of the Nobel Prize. https://www.nobelprize.org/prizes/chemistry/1967/norrish/lecture/ (1967). Accessed 8 November 2019
20. Porter, G.: Flash photolysis and some of its applications. The official website of the Nobel Prize. https://www.nobelprize.org/prizes/chemistry/1967/porter/lecture/ (1967). Accessed 8 November 2019
21. Zewail, A.: Femtochemistry: Atomic-scale dynamics of the chemical bond using ultrafast lasers. The official website of the Nobel Prize. https://www.nobelprize.org/prizes/chemistry/1999/zewail/lecture/ (1999). Accessed 8 November 2019

22. Shuttle, C.G., O'Regan, B., Ballantyne, A.M., Nelson, J., Bradley, D.D.C., de Mello, J., Durrant, J.R.: Experimental determination of the rate law for charge carrier decay in a polythiophene:fullerene solar cell. Appl. Phys. Lett. **92**, 093311 (2008)
23. Maurano, A., Shuttle, C.G., Hamilton, R., Ballantyne, A.M., Nelson, J., Zhang, W., Heeney, M., Durrant, J.R.: Transient optoelectronics analysis of charge carrier losses in a selenophene/fullerene blend solar cell. J. Phys. Chem. C **115**, 5947–5957 (2011)
24. Credgington, D., Durrant, J.R.: Insights from transient optoelectronic analyses on the open-circuit voltage of organic solar cells. J. Phys. Chem. Lett. **3**, 1465–1478 (2012)
25. Kirchartz, T., Nelson, J.: Meaning of reaction orders in polymer: fullerene solar cells. Phys. Rev. B **86**, 165201 (2012)
26. Guo, J., Ohkita, H., Benten, H., Ito, S.: Near-IR femtosecond transient absorption spectroscopy of ultrafast polarons and triplet formation in polythiophene films with different regioregularities. J. Am. Chem. Soc. **131**, 16860–16880 (2009)
27. Guo, J., Ohkita, H., Benten, H., Ito, S.: Charge generation and recombination dynamics in poly(3-hexylthiophene)/fullerene blend films with different regioregularites and morphologies. J. Am. Chem. Soc. **132**, 6154–6164 (2010)
28. Ohkita, H., Kosaka, J., Guo, J., Benten, H., Ito, S.: Charge generation dynamics in polymer/polymer solar cells studied by transient absorption spectroscopy. J. Photon. Energy **1**, 011118 (2011)
29. Yamamoto, S., Ohkita, H., Benten, H., Ito, S.: Role of interfacial charge transfer state in charge generation and recombination in low-bandgap polymer solar cells. J. Phys. Chem. C **116**, 14804–14810 (2012)
30. Yamamoto, S., Ohkita, H., Benten, H., Ito, S., Yamamoto, S., Kitazawa, D., Tsukamoto, J.: Efficient charge generation and collection in amorphous polymer-based solar cells. J. Phys. Chem. C **117**, 10277–10284 (2013)
31. Ide, M., Koizumi, Y., Saeki, A., Izumiya, Y., Ohkita, H., Ito, S., Seki, S.: Near-infrared absorbing thienoisoindigo-based copolymers for organic photovoltaics. J. Phys. Chem. C **117**, 26859–26870 (2013)
32. Ie, Y., Karakawa, M., Jinnai, S., Yoshida, H., Saeki, A., Seki, S., Yamamoto, S., Ohkita, H., Aso, Y.: Electron-donor function of methanofullerenes in donor–acceptor bulk heterojunction systems. Chem. Commun. **50**, 4123–4125 (2014)
33. Yamamoto, S., Yasuda, H., Ohkita, H., Benten, H., Ito, S., Miyanishi, S., Tajima, K., Hashimoto, K.: Charge generation and recombination in fullerene-attached poly(3-hexylthiophene)-based diblock copolymer films. J. Phys. Chem. C **118**, 10584–10589 (2014)
34. Tamai, Y., Tsuda, K., Ohkita, H., Benten, H., Ito, S.: Charge-carrier generation in organic solar cells using crystalline donor polymers. Phys. Chem. Chem. Phys. **16**, 20338–20346 (2014)
35. Mori, D., Benten, H., Ohkita, H., Ito, S.: Morphology-limited free carrier generation in donor/acceptor polymer blend solar cells composed of poly(3-hexylthiophene) and fluorene-based copolymer. Adv. Energy Mater. **5**, 1500304 (2015)
36. Kawashima, K., Tamai, Y., Ohkita, H., Osaka, I., Takimiya, K.: High-efficiency polymer solar cells with small photon energy loss. Nat. Commun. **6**, 10085 (2015)
37. Tamai, Y., Ohkita, H., Namatame, M., Marumoto, K., Shimomura, S., Yamanari, T., Ito, S.: Light-induced degradation mechanism in poly(3-hexylthiophene)/fullerene blend solar cells. Adv. Energy Mater. **6**, 1600171 (2016)
38. Kawashima, K., Fukuhara, T., Suda, Y., Suzuki, Y., Koganezawa, T., Yoshida, H., Ohkita, H., Osaka, I., Takimiya, K.: Implication of fluorine atom on electronic properties, ordering structures, and photovoltaic performance in naphthobisthiadiazole-based semiconducting polymers. J. Am. Chem. Soc. **138**, 10265–10275 (2016)
39. Gallaher, J.K., Prasad, S.K.K., Uddin, M.A., Kim, T., Kim, J.Y., Woo, H.Y., Hodgkiss, J.M.: Spectroscopically tracking charge separation in polymer: fullerene blends with a three-phase morphology. Energy Environ. Sci. **8**, 2713–2724 (2015)
40. Dennler, G., Scharber, M.C., Brabec, C.J.: Polymer-fullerene bulk-heterojunction solar cells. Adv. Mater. **21**, 1323–1338 (2009)

41. Veldman, D., Meskers, S.C.J., Janssen, R.A.: The energy of charge-transfer states in electron donor–acceptor blends: Insight into the energy losses in organic solar cells. Adv. Funct. Mater. **19**, 1939–1948 (2009)
42. Wang, Y., Qian, D., Cui, Y., Zhang, H., Hou, J., Vandewal, K., Kirchartz, T., Gao, F.: Optical gaps of organic solar cells as a reference for comparing voltage losses. Adv. Energy Mater. **8**, 1801352 (2018)
43. Xu, H., Wada, T., Ohkita, H., Benten, H., Ito, S.: Molecular design of near-IR dyes with different surface energy for selective loading to the heterojunction in blend films. Sci. Rep. **5**, 9321 (2015)
44. Guo, J., Ohkita, H., Yokoya, S., Benten, H., Ito, S.: Bimodal polarons and hole transport in poly(3-hexylthiophene):fullerene blend films. J. Am. Chem. Soc. **132**, 9631–9637 (2010)
45. Fukuhara, T., Osaka, M., Tamai, Y., Ohkita, H., Benten, H., Ito, S.: Reduced bimolecular recombination in polymer solar cells. J. Photopolym. Sci. Technol. **29**, 575–580 (2016)
46. Fukuhara, T., Tamai, Y., Osaka, I., Ohkita, H.: Bimolecular recombination and fill factor in crystalline polymer solar cells. Jpn. J. Appl. Phys. **57**, 08RE01 (2018)
47. Nelson, J.: Diffusion-limited recombination in polymer–fullerene blends and its influence on photocurrent collection. Phys. Rev. B **67**, 155209 (2003)
48. Tachiya, M., Seki, K.: Theory of bulk electron–hole recombination in a medium with energetic disorder. Phys. Rev. B **82**, 085201 (2010)
49. Koster, L.J.A., Mihailetchi, V.D., Blom, P.W.M.: Bimolecular recombination in polymer/fullerene bulk heterojunction solar cells. Appl. Phys. Lett. **88**, 052104 (2006)
50. Pivrikas, A., Juška, G., Mozer, A., Scharber, M., Arlauskas, K., Sariciftci, N.S., Stubb, H., Österbacka, R.: Bimolecular recombination coefficient as a sensitive testing parameter for low-mobility solar-cell materials. Phys. Rev. Lett. **94**, 176806 (2005)
51. Chen, D., Nakahara, A., Wei, D., Nordlund, D., Russell, T.P.: P3HT/PCBM bulk heterojunction organic photovoltaics: correlating efficiency and morphology. Nano Lett. **11**, 561–567 (2011)
52. Salleo, A., Kline, R.J., DeLongchamp, D.M., Chabinyc, M.L.: Microstructural characterization and charge transport in thin films of conjugated polymers. Adv. Mater. **22**, 3812–3838 (2010)
53. Motaung, D.E., Malgas, G.F., Arendse, C.J.: Correlation between the morphology and photophysical properties of P3HT:fullerene blends. J. Mater. Sci. **45**, 3276–3283 (2010)
54. Chen, H.-Y., Hou, J., Hayden, A.E., Yang, H., Houk, K.N., Yang, Y.: Silicon atom substitution enhances interchain packing in a thiophene-based polymer system. Adv. Mater. **22**, 371–375 (2010)
55. Guilbert, A.A., Frost, J.M., Agostinelli, T., Pires, E., Lilliu, S., Macdonald, J.E., Nelson, J.: Influence of bridging atom and side chains on the structure and crystallinity of cyclopentadithiophene-bennzothiadiazole polymers. Chem. Mater. **26**, 1226–1233 (2014)
56. Ohkita, H., Cook, S., Astuti, Y., Duffy, W., Tierney, S., Zhang, W., Heeney, M., McCulloch, I., Nelson, J., Bradley, D.D.C., Durrant, J.R.: Charge carrier formation in polythiophene/fullerene blend films studied by transient absorption spectroscopy. J. Am. Chem. Soc. **130**, 3030–3042 (2008)
57. Clarke, T.M., Durrant, J.R.: Charge photogeneration in organic solar cells. Chem. Rev. **110**, 6736–6767 (2010)
58. Gregg, B.A.: Entropy of charge separation in organic photovoltaic cells: The benefit of higher dimensionality. J. Phys. Chem. Lett. **2**, 3013–3015 (2011)
59. Deibel, C., Strobel, T., Dyakonov, V.: Origin of the efficient polaron-pair dissociation in polymer-fullerene blends. Phys. Rev. Lett. **103**, 036402 (2009)
60. Savoie, B.M., Rao, A., Bakulin, A.A., Gelinas, S., Movaghar, B., Friend, R.H., Marks, T.J., Ratner, M.A.: Unequal partnership: asymmetric roles of polymeric donor and fullerene acceptor in generating free charge. J. Am. Chem. Soc. **136**, 2876–2884 (2014)
61. Gélinas, S., Rao, A., Kumar, A., Smith, S.L., Chin, A.W., Clark, J., van der Poll, T.S., Bazan, G.C., Friend, R.H.: Ultrafast long-range charge separation in organic semiconductor photovoltaic diodes. Science **343**, 512–516 (2014)
62. Li, G., Shrotriya, V., Huang, J., Yao, Y., Moriarty, T., Emery, K., Yang, Y.: High-efficiency solution processable polymer photovoltaic cells by self-organization of polymer blends. Nat. Mater. **4**, 864–868 (2005)

63. Li, W., Hendriks, K.H., Roelofs, W.S.C., Kim, Y., Wienk, M.M., Janssen, R.A.J.: Efficient small bandgap polymer solar cells with high fill factors for 300 nm thick films. Adv. Mater. **25**, 3182–3186 (2013)
64. Izawa, S., Nakano, K., Suzuki, K., Hashimoto, K., Tajima, K.: Dominant effects of first monolayer energetics at donor/acceptor interfaces on organic photovoltaics. Adv. Mater. **27**, 3025–3031 (2015)

Chapter 7
First-Principles Investigations of Electronically Excited States in Organic Semiconductors

Takatoshi Fujita

7.1 Introduction

Understanding the charge photogeneration process is essential for the rational design of novel materials for organic solar cells. The charge photogeneration process includes several elementary steps [16]: (i) light absorption by an organic molecule or polymer and the formation of an electronically excited state, which is frequently refereed as an exciton state, (ii) the exciton diffusion to a donor/acceptor (D/A) interface, (iii) a charge-transfer reaction at the interface and the subsequent formation of an interfacial charge-transfer state, comprising an electron in the acceptor material and a hole in the donor material, and (iv) a charge-separation reaction, which gives rise to free charge carriers. Although the organic solar cell with the internal quantum efficiency near 100% has been reported [85], the underlying mechanism enabling efficient charge separation remains unclear [32, 46]. In addition, charge recombination to the ground state has recently become a focus of research because the non-radiative charge recombination results in a considerable energy loss in an open-circuit voltage [9, 122]. A thorough understanding of electronically excited states is necessary for elucidating the charge photogeneration process. The representative excited states in the charge photogeneration process are an initial excited-state absorbing photon, an interfacial CT states, and the charge-separated (CS) states; it is thus essential to predict those energy levels and characterize the excited states.

An ab initio method solving the electronic Schrödinger equation has become a powerful tool to explore the electronic states of organic semiconductors [4, 64, 94]. In addition to the energy, the wave function properties of excited states, such as the electron-hole (e–h) separation and their spatial extent, can be directly obtained within a given approximation. Because of increases in computational capabilities and development of program packages [34, 65], an electronic structure calculation for an

T. Fujita (✉)
Institute for Molecular Science, 38 Nishigo-Naka, Aichi 444-8585, Japan
e-mail: tfujita@ims.ac.jp

M. Hiramoto and S. Izawa (eds.), *Organic Solar Cells*,
https://doi.org/10.1007/978-981-15-9113-6_7

isolated molecule or a molecular crystal has now become a routine task. However, an organic solar cell consists of a combination of donor and acceptor materials; the packing structures in the vicinity of the D/A interface can be significantly disordered and may differ from those in the bulk molecular crystal. Localized (e.g., molecular orbital (MO)) and delocalized (e.g., Bloch orbital) electronic states can coexist in such disordered molecular aggregates. In addition, materials of different crystallinity, ranging from planar heterojunctions to bulk heterojunctions, have been used in organic solar cells. Unified descriptions of localized and delocalized electronic states are necessary to explore organic materials of different crystallinity.

Predicting the electronic states of disordered molecular aggregates requires large-scale electronic structure calculations. However, the computation time rapidly increases with increasing system size; thus, such calculations are difficult to perform for disordered molecular aggregates, including D/A interfaces. We have developed large-scale methods that are suitable for large molecular assemblies [38, 39, 42–45]. In particular, one recent development [38] has enabled an excited-state calculation for a system containing over 2,000 atoms [44, 45]. In this chapter, we present our recent efforts to explore the excited states of organic semiconductors on the basis of the large-scale calculations. After briefly discussing the effects of intermolecular interactions and molecular aggregation on the electronic states, we present applications of our methods to molecular clusters and a D/A interface.

The remainder of this chapter is organized as follows: In Sect. 7.2, we discuss the effects of molecular assemblies on electronic states and introduce the polarization and delocalization effects. In Sect. 7.3, we overview our ab initio approaches based on the many-body Green's function method and the fragment molecular orbital method. In Sect. 7.4, we present numerical results for electronic states of pentacene (PEN) clusters and illustrate the roles of polarization and delocalization effects. In Sect. 7.5, we investigate the interfacial CT states in a D/A interface and discuss the role of induced polarization in the e–h pair separation. In Sect. 7.6, we summarize this chapter and briefly discuss future challenges for theoretical approaches.

7.2 Effects of Molecular Aggregation on Electronic States: Polarization and Delocalization

In this section, we describe the key concepts regarding the electronic states of organic materials. Because organic materials consist of molecules bound by relatively weak non-covalent interactions, an organic molecule in the condensed phase exhibits electronic properties similar to those of an isolated molecule. Therefore, we consider the electronic states of organic aggregates by exploring how intermolecular interactions and molecular aggregations influence molecular electronic states. Here, we introduce the polarization and delocalization effects. The polarization effect indicates the change in electronic states between a molecule in the gas phase and that in the condensed phase, caused by the interaction between the considered molecule and surrounding molecules. In some manuscript, the polarization energy may include reorganization energy arising from rearrangement of molecular structure between the neutral and charged electronic states. However, in this manuscript, we use the polarization energy to refer to the energy change arising from the electronic interactions. The latter indicates the delocalization of an MO or an electronic excitation over multiple molecules. The interactions among localized states (e.g., MOs) can result in the formation of a delocalized state (e.g., a Bloch orbital) that is a coherent superposition of multiple localized states.

The present section is not intended to be a comprehensive review on this topic. Readers may refer to existing reviews [2, 6, 21, 23, 25, 33, 54, 89, 110] for this purpose. Our aim here is to focus on the intermolecular interaction on the electronic states, which will be illustrated by numerical results for realistic structures of organic semiconductors in the latter sections.

First, we present the polarization and delocalization effects, taking the energy levels of highest occupied molecular orbital (HOMO) and lowest unoccupied molecular orbital (LUMO) as examples. The evolution of energy levels from a gas to a solid phase illustrates the effects of polarization and delocalization. Next, we briefly consider the polarization energies of electronically excited states. Finally, we classify electronically excited states formed in molecular aggregates. We introduce a model Hamiltonian for the excited states and discuss how exciton states can be derived as an approximation of this model.

7.2.1 Ionization Potential and Electron Affinity from the Gas to Solid Phase

We describe the polarization and delocalization, considering the energy levels of the HOMO and LUMO. The evolution of the HOMO and LUMO levels from the gas phase to crystalline phase is represented in Fig. 7.1. The solid-state effects on the HOMO or LUMO level can be decomposed into the polarization and dispersion effects. The interaction between a considered molecule and surrounding molecules results in energy shifts for the HOMO and LUMO. Here, the HOMO and LUMO

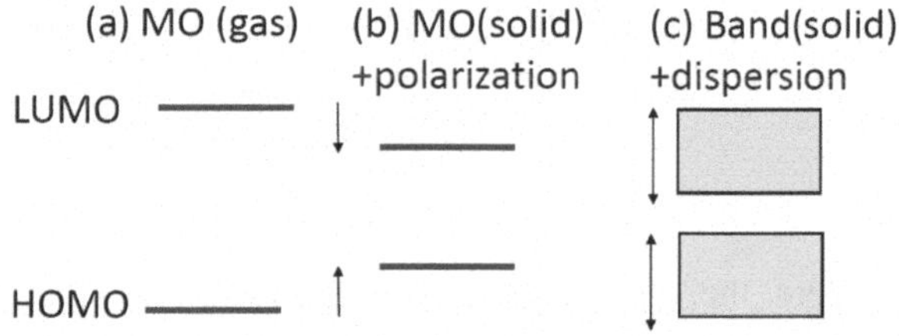

Fig. 7.1 Evolution of the HOMO and LUMO levels from the gas phase to the solid phase

levels tend to increase and decrease, respectively, resulting in a reduced HOMO–LUMO gap. The HOMOs and LUMOs can form delocalized orbitals via the intermolecular electronic couplings; these delocalized orbitals can be measured as the band dispersion by angle-resolved photoelectron spectroscopy [63, 78].

Here, the polarization energy of the HOMO and LUMO is defined as the energy difference between the orbital energy in a gas phase $\epsilon^{(g)}$ and that in the solid phase $\epsilon^{(s)}$:

$$\epsilon_H^{(s)} = \epsilon_H^{(g)} + P^+, \tag{7.1}$$

$$\epsilon_L^{(s)} = \epsilon_L^{(g)} - P^-, \tag{7.2}$$

where superscripts (s) and (g) refer to the solid and gas phases, respectively, and subscripts H and L refer to the HOMO and LUMO, respectively. Here, P^+ and P^- indicate the polarization energy for the cation (LUMO) and anion (HOMO) states. The definition of these polarization energies is consistent with those reported in recent photoelectron spectroscopy study [118]. Recent studies have shown that the polarization energy can be further divided into the electrostatic (ES) and induced polarization (IP) contributions (see recent reviews [2, 25] for a more comprehensive discussion). On the one hand, the ES contribution represents the interaction between the charge carrier and the permanent electrostatic moments of surrounding molecules. The ES effects are predominantly responsible for the orientation dependence of the ionization potential [123]; moreover, they also control the morphology dependence of the energy levels of the D/A heterojunctions [44, 98]. On the other hand, the IP contribution represents the charge carrier and induced dipole moments of surrounding molecules. The latter is also referred to as electronic polarization [118] or induction contribution [25, 88]. In contrast to the long-range nature of the ES effect, the IP effect has a relatively short range. Furthermore, the IP effects are less sensitive to the orientations, morphology, or materials, as compared to ES effects.

The MOs in molecular crystals can form a delocalized state because of intermolecular electronic couplings. Here, we introduce a model Hamiltonian for one-electron orbitals in an aggregate. Model Hamiltonians for the HOMO-derived states (HDSs) and the LUMO-derived states (LDSs) are given by

$$H_{HDS} = \sum_I \epsilon_H^I |\psi_H^I\rangle \langle\psi_H^I| + \sum_{I \neq J} t_{HH}^{IJ} |\psi_H^I\rangle \langle\psi_H^I|, \tag{7.3}$$

$$H_{LDS} = \sum_I \epsilon_L^I |\psi_L^I\rangle \langle\psi_L^I| + \sum_{I \neq J} t_{LL}^{IJ} |\psi_L^I\rangle \langle\psi_L^I|. \tag{7.4}$$

Here, the superscript (s) was dropped, and ϵ_H^I and ϵ_L^I denote the HOMO and LUMO energies of an Ith molecule, respectively; t_{HH}^{IJ} and t_{LL}^{IJ} are HOMO–HOMO or LUMO–LUMO transfer integrals, respectively, between Ith and Jth molecules.

Diagonalization of a single-electron Hamiltonian provides delocalized orbitals, $|\psi_{HB}^{M}\rangle = \sum_{I} C_{MI}|\psi_{H}^{I}\rangle$, where M indexes a HOMO-derived state.

In a periodic system comprising molecules of equivalent MO energies, an eigenstate from the Hamiltonian is equally delocalized over all the constituent molecules. In this limit, the one-electron orbital is described as a Bloch orbital labeled by a wavevector in the first Brillouin zone. By contrast, in a realistic molecular aggregate with energy variations among MO energies, delocalized and localized orbitals can coexist. A convenient measure for quantifying the extent of delocalization is the inverse participation ratio (IPR) [57]. For example, the IPR of the Mth HOMO-derived state is given by

$$IPR(M) = \frac{1}{\sum_{I} |C_{MI}|^4} \tag{7.5}$$

This molecular IPR quantifies the number of molecules over which the wave function is delocalized.

As compared to the polarization effect, an experimental characterization of the spatial extent of electronic states is more challenging. The observation of band-like transport [105] possibly indicates the existence of delocalized Bloch orbitals, although the extent of the delocalization still remains to be determined. Matsui et al. have used the electron spin resonant technique to obtain the spatial extent of charge carriers [74], characterizing the distribution of trapped states in a PEN thin-film transistor. They found that at 20 K the major trap states comprise localized states spanning around 1.5 and 5 molecules, with a broad distribution of states extending over 6–16 molecules. Motivated by their work, Hoshi et al. [57] have presented the large-scale electronic structure calculations for the disordered PEN thin film and analyzed the IPRs of the HOMO-derived states. The roles of disorder in the electronic states, as well as its influence on the charge-transport mechanism, have been intensively discussed [33, 110].

7.2.2 Polarization Energies of Excited States

Here, we briefly discuss the polarization energy of an excited state. The HOMO or LUMO energy corresponds to the energy level of a single charge (electron or hole) carrier. By contrast, an electronic excitation simultaneously creates an electron and a hole; thus, the excitation energy corresponds to the energy of an e–h pair. Therefore, we consider the polarization energy of an excited state as the interaction between the e–h pair and the environment.

Similar to the polarization energies of an anion or cation state, the polarization energy of an excited state, P^{ex}, is introduced as

$$E^{(s)} = E^{(g)} - P^{ex}, \tag{7.6}$$

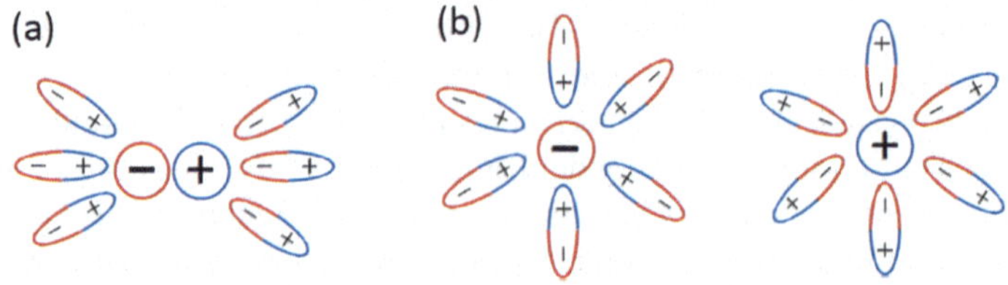

Fig. 7.2 Schematics of polarization of **a** short-range and **b** long-range electron-hole pairs

where $E^{(s)}$and $E^{(g)}$refer to the excitation energies in solid and gas phases, respectively. P^{ex} is the polarization energy of the excited state, corresponding to the stabilization energy of the surrounding molecules that is induced by the creation of the e–h pair. The plus sign on P^{ex} indicates that the excitation energy in a solid phase is lower than that in a gas phase. The polarization energy of the excited state can be decomposed into one-body and two-body terms:

$$P^{ex} = P^{+} + P^{-} - P^{\pm}. \tag{7.7}$$

$P^{+} + P^{-}$ is the sum of the polarization energies of the electron and hole, while $P^{\pm}$ denotes the e–h correlation term of the polarization energy which depends on the e–h separation. $P^{\pm}$ is canceled out $P^{+} + P^{-}$ and reduces the polarization energy at shorter e–h separation, representing the situation in which an e–h pair with a small e–h separation effectively acts as a neutral composite particle. In other words, the polarization energy of an excited state increases with increasing the e–h separation, as schematically shown in Fig. 7.2.

If an excited state is a pure HOMO → LUMO transition, P^{+} or P^{-} in Eq. 7.7 is the same as that introduced in Eq. 7.1 or Eq. 7.2. However, in general, an excited state can involve MOs other than HOMO and LUMO and is a superposition of multiple electronic transitions between occupied MOs (e.g., HOMO, HOMO-1, …) and unoccupied MOs (e.g., LUMO, LUMO+1, …). In such cases, an electron or a hole wave function which constitutes the excite state may differ from that for the cation or the anion state.

The polarization energy of an excited state can be compared to the solvation energy of donor-acceptor complex within classical electrostatics form,

$$P(r_{DA}) = \left(\frac{1}{\epsilon_{\infty}} - \frac{1}{\epsilon_{0}}\right)\left(\frac{1}{2r_A} + \frac{1}{2r_D} - \frac{1}{r_{DA}}\right), \tag{7.8}$$

where ϵ_{∞} and ϵ_0 are the optical and zero frequency relative dielectric constants of the media, and r_A and r_D are the Born radius of the donor and acceptor, respectively. $r_A(r_D)$ qualitatively corresponds to the spatial extent of the electron (hole) wave function. The third term is equivalent to $P^{\pm}$ and cancels the one-body polarization energies. In the classical electrostatic model, the two-body polarization energy is inversely proportional to the donor-acceptor (or e–h) distance, r_{DA}. However, the dependence of $P^{\pm}$ on the e–h separation in an organic material has yet to be elucidated.

7.2.3 Excited States in Molecular Aggregates

In this section, we classify excited states in molecular aggregates. Figure 7.3 schematically presents the four limiting cases of excited states in molecular aggregates, categorizing excited states with respect to the e–h separation, and the spatial extent of an excited state. The e–h separation denotes the center-of-mass distance between the electron and hole wave functions that constitute an excited state. The extent of the delocalization of an excited state can be characterized by the number of molecules involved in the excited state. An exciton state was originally introduced to refer to a delocalized excited state; thus, we also use an exciton state to refer to a delocalized excited state in organic aggregates. It should be noted that in some manuscripts an "exciton state" may refer to a state with small e–h separation, including an intramolecular excited state (Fig. 7.3a) and a Frenkel exciton state (Fig. 7.3c). An excited state with small e–h separation is also denoted as a "local exciton state".

Localized excited states include intramolecular excited states (Fig. 7.3a) and intermolecular CT states (Fig. 7.3b). Here, an intramolecular excited state corresponds to

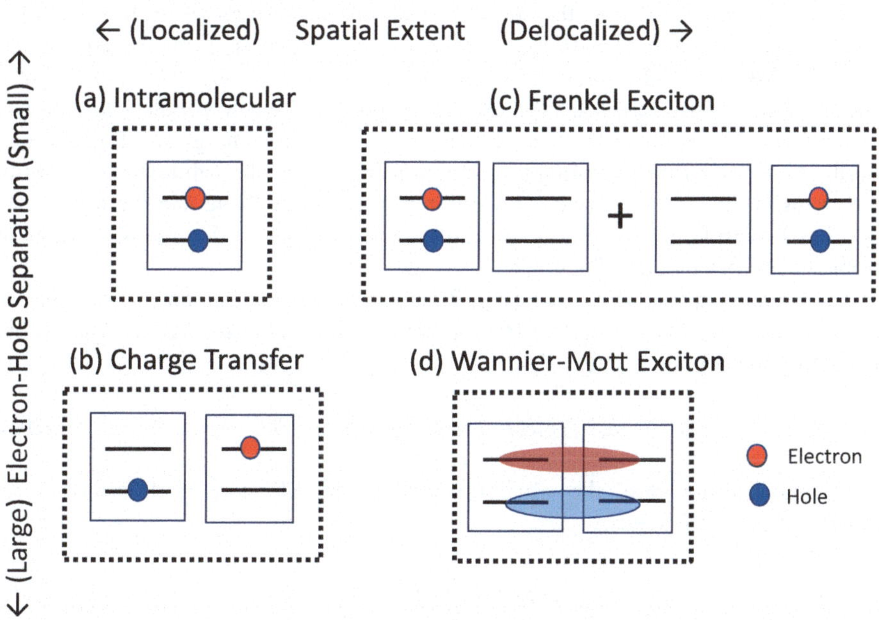

Fig. 7.3 Classification of excited states in molecular aggregates with respect to an electron-hole separation and excited-state delocalization

an electronic transition within a single organic molecule, such as a $\pi \rightarrow \pi^*$ transition in most π conjugated molecules. Hereafter, we denote a localized intramolecular excited state as an LE state. The intermolecular CT state consists of an electron localized within a molecule and a hole localized within another molecule. In the following, we use the term "CT state" to refer to a localized CT state comprising a localized electron and a localized hole. In this classification, the CT states include CS states in an OSC, although the distance at which a CT state can be regarded as a free electron and a free hole is unclear. The CT state should be distinguished from the Wannier-Mott exciton state, as explained later. The number of molecules involved in a localized CT state is two. By contrast, the Wannier-Mott exciton state is composed of delocalized electron and hole wave functions, in which the excitation is shared by multiple molecules via these delocalized orbitals.

We introduce a model Hamiltonian which includes the four representative states as limiting cases. For simplicity, a tight-binding model consisting of the HOMO and LUMO for one molecule is considered. The matrix elements of the excited-state Hamiltonian can be represented by an e–h configuration, $|e_I h_J\rangle = |\psi_L^I\rangle \otimes |\psi_H^I\rangle$. $|e_I h_J\rangle$ denotes an excited state comprising an electron in Ith molecule and a hole in Jth molecule, corresponding to an LE state for $I = J$ and a CT state for $I \neq J$. The singlet excited-state Hamiltonian represented in the two-body basis is written as

$$\langle e_I h_J | H | e_K h_L \rangle = \delta_{JL} H_{IK}^e - \delta_{IK} H_{JL}^h + 2V_{IJ,KL} - U_{IK,JL}. \tag{7.9}$$

Here, H_{IK}^e and H_{JL}^e denote the one-body Hamiltonians for an electron and a hole, respectively. For a model with two MOs per molecule, these terms are equivalent to the HOMO- or LUMO-derived Hamiltonian, i.e., $H_{IK}^e = \delta_{IK}\epsilon_L^I + (1 - \delta_{IK})t_{LL}^{IK}$. The third and fourth terms are the exchange and direct electron-hole interactions, respectively. The exchange term describes the singlet-triple energy splitting and the dipole-dipole electronic coupling for excitation energy transfer, while the direct term is responsible for the attractive e–h Coulomb interactions. Note that the third term vanishes for triplet excited states. Ab initio computations of these e–h interaction terms will be discussed in Sect. 7.3.

The Frenkel exciton (FE) state (Fig. 7.3c) can be written as a coherent superposition of multiple LE states. The Hamiltonian for the FE states can be derived by neglecting the CT states in Eq. 7.9,

$$\langle e_I h_I | H_{FE} | e_J h_J \rangle = E_I \delta_{IJ} + V_{IJ}^C - V_{IJ}^{Ex}. \tag{7.10}$$

In this limiting case, the wave function of the FE state is described as a superposition of LE states,

$$|\Psi_{FE}\rangle = \sum_I C_I |e_I h_I\rangle. \tag{7.11}$$

Here, the excitation energy of the Ith molecule E_I is given as, $E_I = \epsilon_L^I - \epsilon_H^I + 2V_{HLHL} - U_{HHLL}$, where V_{HLHL} and U_{HHLL} are the exchange and attractive e–h interactions, corresponding the third and fourth terms in Eq. 7.9, respectively. An off-diagonal element of the FE Hamiltonian, $\langle e_I h_I | H | e_J h_J \rangle$, is known as an excitonic coupling and is responsible for excitation energy transfer and dispersion of the FE state. The excitonic coupling can be divided into the long-range Coulomb and short-range exchange interactions as $\langle e_I h_I | H | e_J h_J \rangle = V^C + V^{Ex}$ [36, 58]. For example, the long-range Coulomb coupling is given by

$$V^C = \int d\mathbf{r}_1 \int d\mathbf{r}_2 \frac{\rho_I^{(eg)*}(\mathbf{r}_1, \mathbf{r}_1)\rho_J^{(eg)}(\mathbf{r}_2, \mathbf{r}_2)}{|\mathbf{r}_1 - \mathbf{r}_2|}, \tag{7.12}$$

Here, $\rho_I^{(eg)}$ denotes the transition density between the ground and excited states, $\rho_I^{(eg)} = \psi_H^I(r)\psi_L^{I*}(r)$. V^C can be well approximated as a dipole-dipole interaction, $V^C = (\boldsymbol{\mu}_I \cdot \boldsymbol{\mu}_J - 3(\boldsymbol{\mu}_I \cdot \mathbf{e}_R)(\boldsymbol{\mu}_J \cdot \mathbf{e}_R))/R_{IJ}^3$, in long molecular distances, where μ_{I} is the transition dipole moment between the ground and excited states, $\boldsymbol{\mu}_I = -\int \mathbf{r}\rho_I^{(eg)}(\mathbf{r}, \mathbf{r})$. By contrast, V^{Ex} describes the short-range exchange interaction. In most organic systems, the long-range interaction is much stronger than the exchange interaction. V^C vanishes for triplet excited states, with only V^{Ex} determining the triplet exciton diffusion.

Although the FE state can be regarded as being delocalized over multiple molecules, the mechanism of the delocalization should be distinguished from that of Bloch orbitals. The FE state is formed by the long-range Coulomb interactions between transition densities of organic molecules, without the wave function overlaps. The FE state can also be denoted as a collective excited state, in which multiple e–h configurations are involved in the electronic excitation. In J-aggregates [117], for example, FE states emerge as the J-band characterized by red-shifted absorption spectrum, larger oscillator strength, narrow bandwidth, and small stokes shifts [54, 102]. An FE state can also show super radiant decay, a coherence enhancement of the radiative decay rate, which was reported by Dicke [27].

Another limit of delocalized excited states is the Wannier-Mott (WM) exciton state (Fig. 7.3d). The WM exciton states can be derived when the e–h interaction terms are much smaller than the transfer integrals. If the e–h interaction terms can be treated perturbatively, the excited-state wave function is written as a direct product of the eigenfunctions of the HOMO-derived state and the LUMO-derived state as

$$\Psi_{WM} = |\psi_{LDS}\rangle \otimes |\psi_{HDS}\rangle, \tag{7.13}$$

where $|\psi_{LDS}\rangle$ or $|\psi_{HDS}\rangle$ is an eigenfunction of H^e or H^h. In a periodic system, the WM Hamiltonian can be rewritten in the same form as that of a hydrogen atom. The diagonalization of the one-body Hamiltonian under periodic boundary conditions provides the energy-band dispersion, e.g., $H^h|\psi_{\mathbf{k}_h}\rangle = E(\mathbf{k}_h)|\psi_{\mathbf{k}_h}\rangle$. Using the one-body energies (e.g., $E(\mathbf{k}_h) \simeq \mathbf{k}_h^2/2m_h$) and neglecting the e–h exchange term, we can reduce the Hamiltonian to:

$$H_{WM} = \frac{\mathbf{k}_e^2}{2m_e} + \frac{\mathbf{k}_h^2}{2m_h} - \frac{1}{\mathbf{r}_e - \mathbf{r}_h}. \quad (7.14)$$

Here, $m_e(m_h)$ is the effective mass of an electron (hole), $\mathbf{k}_e$ ($\mathbf{k}_h$) is the wavenumber vector, and $\mathbf{r}_e(\mathbf{r}_h)$ is the position of the electron (hole). Because of its similarity to the Hamiltonian of hydrogen atom, the optical spectra of the WM exciton states resemble the Rydberg transitions.

The term "CT exction" can be found in recent studies (e.g., [6, 21, 101]). It appears that in some manuscripts CT excitons and CT states are used interchangeably. Herein, we follow the discussion given by Cudazzo et al. [21]. The excited-state Hamiltonian is divided into the LE-LE and CT-CT blocks as follows,

$$H = \begin{pmatrix} \langle LE|H|LE\rangle & \langle LE|H|CT\rangle \\ \langle CT|H|LE\rangle & \langle CT|H|CT\rangle \end{pmatrix}. \quad (7.15)$$

The LE-LE block is identical to Eq. 7.11; thus, the diagonalization of it yields the wave functions of FE states as in Eq. 7.10. The LE-CT block, $\langle LE|H|CT\rangle$, describes the interaction between LE and CT states, which is dominated by a transfer integral, e.g., $\langle e_I h_I|H|e_J h_I\rangle \simeq t_{LL}^{IJ}$. Eigenstates of the CT-CT block, $\langle CT|H|CT\rangle$, may be considered as CT exciton (CTX) states as

$$\Psi_{CTX} = \sum_{I \neq J} C_{IJ}\,|e_I h_J\rangle. \quad (7.16)$$

A CT exciton state described by Eq. 7.16 may be regarded as a collective state comprising multiple e–h configurations of a localized electron and a localized hole. Apparently, this is in contrast to a WM exciton state consisting of single e–h configuration of a delocalized electron and a delocalized hole. However, the distinction between a CT exciton state and a WM exciton state is non-trivial. Regarding that a delocalized electron or hole state is a superposition of LUMOs or HOMOs, a WM state can be rewritten as $\Psi_{WM} = \sum_I C_I|\psi_L^I\rangle \otimes \sum_J C_I|\psi_L^I\rangle = \sum_{I,J} C_I C_J\,|e_I h_J\rangle$. Therefore, a measure quantifying the collectivity [87] of excited states must be introduced to distinguish CT exciton states from WM exciton states. We leave it to a future study as to how CT exciton states are distinguished from WM exciton states and are characterized in organic semiconductor materials.

Before closing this section, we consider an interfacial CT states [69, 71, 114] formed at a D/A interface. An interfacial CT state is composed of an electron in an acceptor material and a hole in a donor material and plays a central role in the charge photogeneration process. The energy offset between the interfacial CT state and a LE (or an FE) state provides the driving energy for the charge separation but constitutes the energy loss in the open-circuit voltage. The extent and the roles of the delocalization of interfacial CT states have attracted broad attention [5, 26, 44, 48, 106, 125]. It is argued that the delocalization of an electron or a hole wave function forming the interfacial CT state reduces the exciton binding energy, allowing for the efficient charge separation. If interfacial CT states are similar to a WM exciton state, in the sense that it consists of a delocalized electron and a delocalized hole, the efficient charge separation would take place as a result of the reduction of exciton binding energy. However, as we have seen, a WM exciton state can be formed when e–h interactions are sufficiently smaller than transfer integrals. This is often not the case, particularly for organic bulk heterojunction solar cells, in which the transfer integrals would be suppressed by the inherent structural disorder. Characterizing and engineering the interfacial CT states are key for designing efficient and low-energy-loss organic solar cells.

7.3 Electronic Structure Calculations for Optoelectronic Properties

In this section, we present an overview of our theoretical developments for large-scale electronic state calculations, aiming toward applications for organic materials. Accurate estimations of e–h interactions are important for predicting the energy levels of excited states. To this end, we have employed the many-body Green's function theory within the GW approximation. The effects of molecular aggregation on the electronic states can be treated in combination with the fragment molecular orbital method, a large-scale method suitable for organic materials.

First, the many-body Green's function methods are briefly summarized. Next, the advantages of the Green's function methods will be discussed in comparison to standard density functional theories, with benchmark numerical results for the exciton binding energy of an isolated PEN molecule. Finally, the fragment molecular orbital method is presented as to how the polarization and delocalization effects can be incorporated.

7.3.1 Many-Body Green's Function Method Within GW Approximation

Here, we briefly summarize the many-body Green's function methods. In general, quantitative determinations of the energy levels of electronic states require a highly accurate method, which goes beyond the standard density functional theory (DFT). However, a highly accurate method requires a considerable computational time, particularly for large molecular systems, and is not practical for organic materials containing a large number of organic molecules or polymers. The many-body Green's function methods within the GW approximation can offer practical schemes for calculating electronic states with reasonable accuracy. Here, we briefly describe the GW and GW/Bethe-Salpeter equation (GW/BSE) methods. More comprehensive descriptions on this approach can be found elsewhere [3, 11, 84].

The GW is based on the one-body Green's function [53]. The one-body Green's function describes an electron addition or an electron removal process; the poles of the Green's function correspond to the energy differences between the neutral and charged (anionic or cationic) electronic states, including the ionization potential and electron affinity. Within the perturbative approach of the GW method, the charged excitation energy is obtained as the energy of a quasiparticle state for an electron ($p \geq$ LUMO) or a hole ($p \leq$ HOMO):

$$\epsilon_{GW,p} = \epsilon_p + \Sigma_{GW,p}(\epsilon_{GW,p}), \tag{7.17}$$

where ϵ_p is an orbital energy from the Hartree-Fock (HF) or a Kohn-Sham (KS) method, and $\epsilon_{GW,p}$ is the quasiparticle energy. Here, $\Sigma_{GW,p}(\epsilon_{GW,p})$ is the energy-dependent potential called self-energy, including the many-body effects between the added particle (electron or hole) and the electrons in the system.

In the GW method, the self-energy is approximated as the product of the one-body Green's function and a screened Coulomb potential W,

$$\Sigma_{GW}(\mathbf{r}_1, \mathbf{r}_2, E) = \frac{i}{2\pi} \int d\omega G(\mathbf{r}_1, \mathbf{r}_2, E + \omega) W(\mathbf{r}_1, \mathbf{r}_2, \omega) e^{i\omega\eta}, \tag{7.18}$$

where η is a positive infinitesimal. Here, the screened Coulomb potential is defined as the convolution of the dielectric function, $\epsilon\ (\mathbf{r}_1, \mathbf{r}_3,)$, and the bare Coulomb potential:

$$W(\mathbf{r}_1, \mathbf{r}_2, \omega) = \int d\mathbf{r}_3 \epsilon(\mathbf{r}_1, \mathbf{r}_3, \omega)^{-1} \frac{1}{|\mathbf{r}_3 - \mathbf{r}_2|}. \tag{7.19}$$

The GW method can accurately describe the induced polarization effects, via the screened Coulomb potential in the self-energy. In the HF or KS method with Koopman's theorem [103, 112], an ionization potential or an electron affinity is approximated as the negative value of the HOMO or LUMO that is determined for a neutral state. In this case, the orbital relaxation and electronic polarization effects, which are induced by the electron removal or addition, are not taken into account. By contrast, the GW self-energy includes the response of the system to the electron removal or addition via the spatially- and energy-dependent dielectric function; therefore, the induced polarization effects can be accurately incorporated in the GW self-energy. Previous studies have confirmed that the HOMO–LUMO gap renormalization by a polarizable environment (e.g., Figure 7.1a→ b) can be successfully reproduced by the GW [68, 79], while it cannot be described using standard DFT functionals.

Electronically excited states can be computed in combination with the BSE [93], which is based on the particle-hole Green's function. The particle-hole Green's function describes the creation of an e–h pair which is equivalent to an electronic excitation from an occupied orbital to an unoccupied orbital. The pole of the particle-hole Green's function corresponds to excitation energy, i.e., the energy difference between the ground and an excited electronic state. The BSE can be transformed into a non-Hermitian eigenvalue problem for excited states, which is similar to linear-response time-dependent DFT (TDDFT).

Instead of providing a detailed theoretical formulation, we briefly consider the excitation energy from GW/BSE. Within the Tamm-Dancoff approximation (TDA) [56, 93] of the GW/BSE, the HOMO–LUMO excitation energy of a spin-singlet state is given by:

$$E = \epsilon_{GW,\,L} - \epsilon_{GW,H} + 2(HL|HL) - (HH|W|LL). \tag{7.20}$$

The HOMO–LUMO exchange integral, $(HL|HL)$, is

$$(HL|HL) = \int d\mathbf{r}_1 \int d\mathbf{r}_2 \psi_H(\mathbf{r}_1)\psi_L^*(\mathbf{r}_1)\frac{1}{|\mathbf{r}_1 - \mathbf{r}_2|}\psi_L(\mathbf{r}_2)\psi_H^*(\mathbf{r}_2). \tag{7.21}$$

Here, this exchange integral corresponds to Eq. 7.9. The fourth term, $(HH|W|LL)$, describes the e–h attractive interaction. Within the GW approximation of the BSE, the e–h attraction is given by the screened Coulomb potential:

$$(HH|W|LL) = \int d\mathbf{r}_1 \int d\mathbf{r}_2 |\psi_H(\mathbf{r}_1)|^2 W(\mathbf{r}_1, \mathbf{r}_2)|\psi_L(\mathbf{r}_2)|^2. \tag{7.22}$$

Because the exchange integral decays rapidly with increasing the e–h separation, the long-range asymptotic behavior of the CT states, e.g., the energy difference between the CT and CS states, is governed by the attractive e–h interaction. In the GW/BSE, the environmental polarization effects can be incorporated into the e–h interaction via the screened Coulomb potential. Therefore, the GW/BSE approach is useful for calculating CT excited states in a polarizable media, including CS states in an OSC.

7.3.2 Benchmark Results for Exciton Binding Energy

Here, we contrast the GW/BSE method with other single-reference methods [28] such as configuration interaction single (CIS) and TDDFT. For further information, we refer interested readers to the review article by Blase et al. [11], which provides detailed comparisons between the GW/BSE and TDDFT. We discuss the difference among the CIS, TDDFT, and GW/BSE in terms of the exciton binding energy.

Within the CIS, TDA-TDDFT, and GW/TDA-BSE, the Hamiltonian for singlet excited state can be written as

$$H_{ia,jb} = \delta_{ij}\delta_{ab}(\epsilon_a - \epsilon_b) + 2(ia|jb) - (ij|U|ab) \tag{7.23}$$

Here, *i* and *j* refer to occupied orbitals, and and *b* refer to unoccupied orbitals. In the TDDFT or GW/BSE, matrix elements among resonant transitions correspond to

Table 7.1 Comparison of different excited-state theories

Theory	Orbital	$U(\mathbf{r}_1, \mathbf{r}_2)$	Asymptotic U
CIS	Hartree-Fock	$1/\lvert\mathbf{r}_1 - \mathbf{r}_2\rvert$	$1/r$
TDDFT(GGA)	Kohn-Sham	$V_{GGA}(\mathbf{r}_1)\delta(\mathbf{r}_1, \mathbf{r}_2)$	0
GW/BSE	Dyson	$W(\mathbf{r}_1, \mathbf{r}_2)$	$1/(\varepsilon_M r)$

those of excited-state Hamiltonian in the CIS method. $(ia|jb)$ is the exchange integral as in 1.21, and $(ij|U|ab)$ describes the e–h attractive interaction by a potential U $(\mathbf{r}_1, \mathbf{r}_2)$. Note that the exchange integral is same for the CIS, TDA-TDDFT, and GW/TDA-BSE, except that different orbitals are used. The CIS, TDA-TDDFT, and GW/TDA-BSE differ in the single-electron orbitals and e–h attractions, as summarized in Table 7.1. As explained in the previous section, the GW/BSE writes the e–h attraction as the screened Coulomb potential, $U_{GW/BSE}(\mathbf{r}_1, \mathbf{r}_2) = W(\mathbf{r}_1, \mathbf{r}_2)$, which takes account of the induced polarization effects through the dielectric function. By contrast, the CIS describes the e–h attraction as the bare Coulomb potential, $U_{CIS}(\mathbf{r}_1, \mathbf{r}_2) = V(\mathbf{r}_1, \mathbf{r}_2) = 1/|\mathbf{r}_1 - \mathbf{r}_2|$. The linear-response TDDFT uses a semi-empirical exchange-correlation potential to describe the e–h attraction. In combination with the local-density approximation (LDA) or generalized gradient approximation (GGA), the e–h attraction is treated by a (semi)local potential:

$$U_{GGA}(\mathbf{r}_1, \mathbf{r}_2) = V_{GGA}(\mathbf{r}_1)\delta(\mathbf{r}_1, \mathbf{r}_2). \tag{7.24}$$

The qualitative difference between these excited-state theories can be readily seen by considering the long-range asymptotic behavior of the e–h attractions. Because the exchange interaction has a short range, the long-range behavior of the CT states, particularly the dependence of excitation energy on the e–h separation, is governed by the e–h attraction. The CIS uses bare Coulomb interaction, and the long-range CT states behave as $1/r$, where r denotes the distance between the HOMO and LUMO. By contrast, in the TDDFT with the LDA or GGA, the e–h Coulomb attraction is described by the local potential which decays for long-range e–h separation. This result is well known as the incorrect asymptotic behavior of the long-range CT excitation in the TDDFT [29], and thus, the CT excitation cannot be appropriately described using (semi)local functionals. The GW/BSE incorporates the dielectric function into the e–h attraction with its long-range behavior being given by $1/(\epsilon_M r)$, where $1/\epsilon_M$ is the macroscopic dielectric constant.

In the KS-DFT, a variety of hybrid functionals, in which the HF exchange and DFT exchange functionals are mixed using empirical parameters, have been proposed. The TDDFT with a hybrid functional combines the bare Coulomb potential and (semi)local potential for describing e–h attraction. The e–h attraction in the TDDFT based on a global hybrid (GH), such as the B3LYP [8] and PBE0 [1], is given by:

$$U_{GH}(\mathbf{r}_1, \mathbf{r}_2) = c_{HF}V(\mathbf{r}_1, \mathbf{r}_2) + (1 - c_{HF})V_{GGA}(\mathbf{r}_1)\delta(\mathbf{r}_1, \mathbf{r}_2), \tag{7.25}$$

Table 7.2 Asymptotic behavior of U for the TDDFT with various hybrid functionals, global hybrid (GH), range-separated hybrid (RSH), and CAM-type range-separated hybrids, with representative functionals

Type	Asymptotic U	Functionals
GH	c_{HF}/r	B3LYP, PBE0
RSH	$\mathrm{erf}(\mu r)/r \sim 1/r$	LC-BLYP, LC-ωPBE
CAM	$(\alpha + \beta\mathrm{erf}\,(\mu r))/r \sim (\alpha + \beta)/r$	CAM-B3LYP

where its long-range asymptotic behavior is determined by the fraction of the HF exchange, c_{HF}/r. For example, $c_{HF} = 0.2$ and $c_{HF} = 0.25$ are used in the representative B3LYP [8] and PBE0 [1] functionals, respectively.

More sophisticated hybrid functionals are based on the long-range correction (LC) scheme. The LC-DFT has become a standard tool for exploring organic semiconductor molecules because of its excellent balance between accuracy and computational efficiency [64]. In the LC scheme [60, 116], the Coulomb operator is divided into short- and long-range potentials using the error function and a range-separation parameter (μ), $1/r = \mathrm{erf}(\mu r)/r - (1 - \mathrm{erf}(\mu r))/r$. The DFT functional and HF exchange are used for describing the short- and long-range potentials, respectively. For the LC-TDDFT with a single range-separation parameter, the asymptotic e–h Coulomb attraction is given by $\mathrm{erf}(\mu r)/r$. Because typical values for the range-separation parameter are 0.2–0.4 bohr^{-1}, the long-range CT states asymptotically behave as $1/r$. The more general range-separation scheme refereed as the Coulomb-attenuating method (CAM) [120] have been proposed, in which the Coulomb operator is divided using three parameters, $1/r = (1 - \alpha - \beta\mathrm{erf}(\mu r))/r + (\alpha + \beta\mathrm{erf}(\mu r))/r$. For the TDDFT with this range-separation scheme, the asymptotic e–h attraction becomes $(\alpha + \beta\mathrm{erf}(\mu r))/r \sim (\alpha + \beta)/r$. The asymptotic behaviors of U for various hybrid functionals are summarized in Table 7.2.

To demonstrate those differences in the e–h attraction, we compare the exciton binding energies of an isolated PEN molecule from CIS, TDDFT, and GW/BSE. Experimentally, the exciton binding energy can be estimated as the difference between the absorption peak and the fundamental gap. To estimate the exciton binding energy, the ground-state and subsequent excited-state calculations were performed for the isolated PEN molecule. The electron affinity and ionization potential were estimated as the negative values of HOMO and LUMO energies, respectively. The ionization potential and electron affinity of the single PEN molecule were determined as 6.59 eV [17] and 1.39 eV [19], respectively, resulting in the experimentally-estimated HOMO-LUMO gap as 5.20 eV. The S_1 excitation energy of a single PEN molecule was obtained as 2.29 eV [52].

The DFT functionals studied here are the GGA functionals without the HF exchange (PBE [86] and BLYP [7, 67]), the global hybrid functionals (PBE0 [1] and B3LYP [8]), and the several range-separated hybrid functionals (LC-BLYP [60], CAM-B3LYP [120], LC-ωPBE [116], and ωB97XD [14]). To compare CIS, TDDFT, and GW/BSE, the TDA was employed for all the TDDFT and GW/BSE calculation.

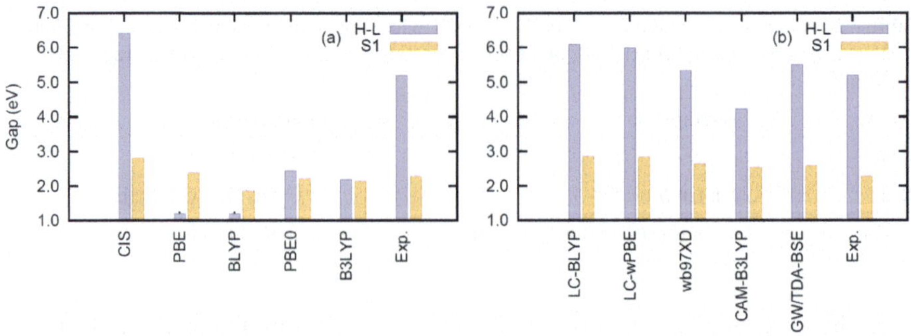

Fig. 7.4 HOMO-LUMO (H-L) gap and S_1 excitation energies: **a** CIS and TDDFT with GGA and global hybrid functionals, **b** TDDFT with range-separated hybrid functionals and GW/BSE. Here, TDDFT and BSE calculations were performed with the Tamm-Dancoff approximation

The structure optimized by B3LYP/6-31G* was used as the PEN structure. The CIS and TDDFT calculations were performed using Gaussian 16 [34], whereas the GW/BSE calculation was performed using the ABINIT-MP [76, 77, 107]. The one-shot GW calculation was performed using HF starting point, and the BSE excitation energy was obtained with the statically screened Coulomb potential. 6-311G** basis set was used in all calculations.

The HOMO–LUMO gap and the first singlet (S_1) excitation energy from the CIS, GGA, and global hybrid functionals are shown in Fig. 7.4a. The CIS tends to overestimate both the HOMO–LUMO gap and the S_1 excitation energy, with the exciton binding energy being moderately overestimated. For the GGA and global hybrid functionals, the excitation energies are well reproduced. However, the HOMO–LUMO gap is significantly underestimated relative to the experimental value. In particular, the PBE and BLYP provide the spurious result that the HOMO–LUMO gap is lower than the S_1 excitation energy. Although the results are improved to a certain extent by including the HF exchange (PBE0 and B3LYP), the exciton binding energies are still considerably underestimated. Overall, the CIS method may be useful for estimating the exciton binding energy, but both the HOMO–LUMO gap and the S_1 excitation energy are overestimated. In contrast, the exciton binding energy cannot be reproduced by the pure GGA or global hybrid functionals, including the commonly used B3LYP.

The results from the several range-separated hybrid functionals and the GW/BSE are shown in Fig. 7.4b. The severe underestimations by the GGA or global hybrid functionals can be drastically improved by applying range-separated hybrid functionals. Overall, the long-range corrected functionals offer balanced descriptions for the both HOMO–LUMO gap and the S_1 excitation energy and reasonably reproduce the exciton binding energy, with the exception of CAM-B3LYP. The moderate underestimation of exciton binding energy from the CAM-B3LYP is attributed to the range-separation parameters, where the long-range asymptotic e–h attraction is described as $(\alpha + \beta)/r$, $\alpha + \beta$=0.65 at CAM-B3LYP. The GW/BSE shows a performance similar to that of the range-separated functionals.

Finally, we add a couple of comments to the benchmark results. First, the TDA was employed in all the TDDFT calculations. As explained, we used the TDA to contrast the CIS, TDDFT, and GW/BSE. For the S_1 excitation energy of the PEN molecule, the full diagonalization mixing resonant and anti-resonant transitions decreases excitation energies by 0.2–0.3 eV than those with the TDA. However, the inclusion of anti-resonant transitions has minor effects on the long-range CT states, and thus, the discussion regarding the long-range asymptotic behavior is valid without the TDA. Second, although the range-separated functionals provide the reasonable results for the isolated PEN molecule, they do not necessarily provide accurate results for a crystal or a molecule in a condensed media. This is because parameters in range-separated functionals are generally determined so as to reproduce the properties of gas-phase molecules. A recent development of the optimally tuned range-separation scheme [90] allows for the accurate prediction of exciton binding energies for solid-state systems. By contrast, the GW/BSE method can yield reasonable exciton binding energies for both gas- and condensed-phase systems.

7.3.3 Fragment Molecular Orbital Method

In this section, we summarize our recent attempt to extend the applicability of the GW to disordered molecular aggregates. The GW and GW/BSE implementation have been established for isolated molecular systems [10, 92, 99] and periodic systems [20, 100, 109]. Although recent developments of the efficient algorithms and implementations have enabled GW calculations for systems containing more than 100 atoms [47, 51, 81, 82], it is still computationally formidable to apply the GW to disordered molecular aggregates of more than 1,000 atoms. We briefly summarize our recent developments of a large-scale GW based on the fragment molecular orbital method [39, 45].

Our implementation adopts a fragment-based electronic structure method [50]. In a fragment-based method, an entire system is first divided into numerous small subsystems, and the physical properties and the wave function of the entire system are approximated from the multiple quantum-chemistry calculations for the subsystems. In the FMO method proposed by Kitaura and coworkers [61], an entire system is divided into many small parts refereed as fragments. The self-consistent field (SCF) calculations for fragment monomers, dimers, and optionally trimers are then performed to approximate physical properties of the entire system. Theoretical formulations of the FMO method and its applications have been widely reviewed [30, 104, 107].

If a molecule is assigned as an independent fragment in an FMO calculation, the SCF calculation for a fragment monomer provides the one-electron orbitals localized within the molecule. The orbital energies ϵ_p^I and MOs ($\left|\psi_p^I\right\rangle$) of an Ith fragment (molecule) are given by

$$\left(F^I + V_{EESP}^I\right)\left|\psi_p^I\right\rangle = \epsilon_p^I\left|\psi_p^I\right\rangle, \tag{7.26}$$

where F^I denotes the Hartree-Fock or Kohn-Sham operator, and V_{EESP}^I is the environmental electrostatic potential. V_{EESP}^I describes the electrostatic Coulomb interaction between the electrons in the fragment I and electrostatic potential from all other fragments. In the FMO method, SCF calculations for fragment monomers are repeated until the charge density of an entire system in the electronic ground state is converged. Therefore, the FMO method can effectively describe the ES contributions of the polarization energy discussed in Sect. 7.2.1, the interaction between the charge and permanent electrostatic moments of all the other fragments.

The induced polarization energy can be also treated in combination with the GW method. The quasiparticle energy for the fragment is given by [39, 45],

$$\epsilon_{FMO-GW,\,p}^I = \epsilon_p^I + \Sigma_{FMO-GW,p}^I, \tag{7.27}$$

where $\epsilon_{GW,p}^I$ indicates a quasiparticle energy, and $\Sigma_{FMO-GW,p}^I$ is the GW self-energy a for fragment I. As explained, ϵ_p^I in the FMO method includes the ES contribution which represents the interaction between the electron localized within Ith fragment and the permanent multipole moments of all the other fragments. By contrast, the FMO-GW self-energy ($\Sigma_{FMO-GW,p}^I$) includes the IP contribution of polarization energy, which describes the interaction between the electron and the induced dipole moments of all the fragments. Because the dielectric function of the entire system is computed in the FMO-GW calculations, the electronic polarization effect induced by the electron addition or removal in the Ith fragment can be treated.

The MOs obtained in the FMO method are localized within a fragment in an entire system, and intermolecular electronic coupling must be considered to describe orbital delocalization over multiple fragments. Within the FMO method, such delocalization effects can be incorporated by calculating intermolecular orbital interactions and constructing the total Fock matrix [31, 113]. The intermolecular orbital interaction can be obtained as:

$$t_{pq}^{IJ} = \langle\psi_p^I|F^{IJ}|\psi_q^J\rangle, \tag{7.28}$$

where F^{IJ} is a one-electron Hamiltonian for the fragment dimer. t_{pq}^{IJ} can be regarded as a transfer integral [43, 62]. For example, if p = HOMO and q = HOMO, we obtain the HOMO–HOMO transfer integral, which is responsible for hole transfer rate and valence-band dispersion. If the HOMOs are collected from all the fragments, the monomer MOs energies and their transfer integrals can be combined to define the Hamiltonian for the HOMO-derived state in the form of Eq. 7.3. Furthermore, it is straightforward to include MOs other than HOMOs and LUMO in the model Hamiltonian.

Delocalized excited states can be also treated in the FMO method. We have developed the excited-state methods for large systems based on the FMO and exciton model [38]. In this method, an excited-state wave function for an entire system is written as a superposition of LE and CT states as,

$$|\Psi\rangle = \sum c_{LE}|LE\rangle + \sum c_{CT}|CT\rangle . \tag{7.29}$$

Here, |*LE*> denotes an excited state localized within a fragment [15, 76], whereas |*CT*> is an interfragment CT excited state. These states correspond to the LE and CT states in Fig. 7.3a, b, respectively. Based on the wave function ansatz of Eq. 7.29, the total excited-state Hamiltonian is calculated in a same form as Eq. 7.15. Matrix elements of the excited-state Hamiltonian can be found elsewhere [38].

7.4 Pentacene Clusters

In this section, we present an application of our method to PEN clusters. We investigate the change of electronic states with respect to the cluster size, illustrating the roles of polarization and delocalization. We note that numerous first-principles studies on PEN systems, including single molecule [10, 66], clusters [18], and crystals [20, 68, 100, 101, 109]. Although the two limiting cases of isolated and periodic systems have been well studied, the changes in electronic states from the gas to solid phase remain unclear. After we consider the effect of polarization on the localized electronic states, we characterize the spatial extent of the one-electron orbitals and excited states.

Here, the computational details are briefly summarized. A thin-film structure of a PEN crystal [97] (CCDC number: 665900) was used. From the crystallographic information file, the PEN cluster structures ($(\mathrm{PEN})_N$, (N = 3, 14, and 33)) were extracted, as shown in Fig. 7.5. The FMO-GW/BSE calculations were performed with the B3LYP starting point and the 6-31G* basis set. In the FMO calculations, the Mulliken-point charge approximation [77] was adopted for environmental electrostatic potentials; the electrostatic approximation for separated fragment pairs was also used with threshold value of 2.0 in van der Waals unit [77]. The calculations were performed using the ABINIT-MP software [76, 77, 107].

7.4.1 Polarization Energies of Charged and Excited States

We first present the MO energies as a function of cluster size, as shown in Fig. 7.6. For the PEN clusters, the MO energy is obtained as the average of the three central PEN molecules (the molecules depicted in purple color in Fig. 7.5a–c) in the cluster structures. The HOMO and LUMO energy levels increase and decrease, respectively, as the cluster size increases. The energy shifts result in the HOMO–LUMO gap reduction, from 4.83 eV (N = 1, i.e., an isolated PEN) to 3.85 eV (N = 33). This

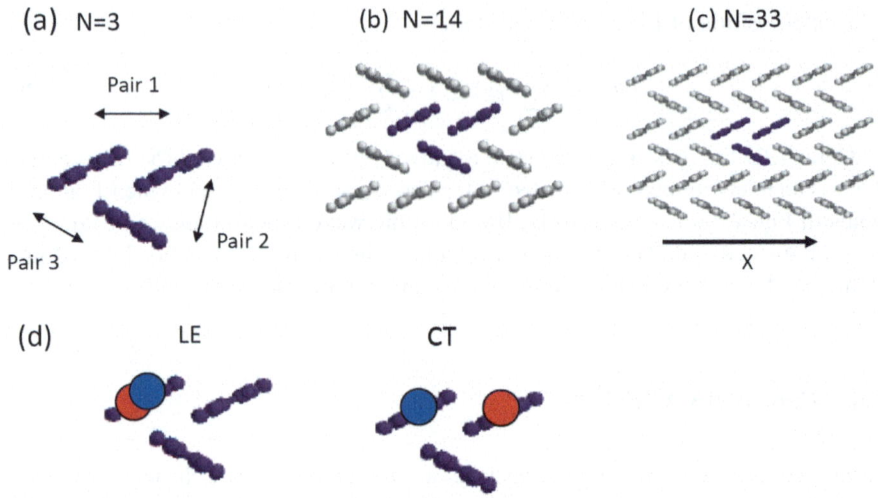

Fig. 7.5 **a–c** PEN cluster structures ($N = 3$, 14, and 33). **d** Schematic of the LE and CT states, where the red and blue circles represent electrons and holes, respectively

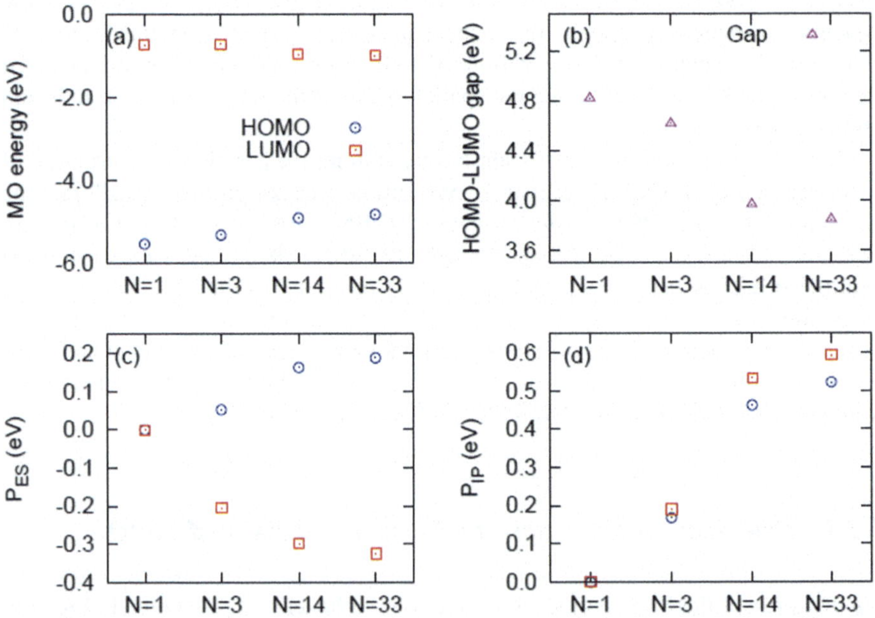

Fig. 7.6 Average **a** MO energy, **b** HOMO–LUMO gap, and **c** ES and **d** IP polarization energies as a function of cluster size. In **a**, **c** and **d**, the red squares and blue circles indicate the LUMO (electron) and HOMO (hole) results, respectively

large HOMO-LUMO gap reduction from the gas to the condensed phases is consistent with the previous findings [68].

As discussed in Sect. 7.2.1, the MO energy change can be understood in terms of the polarization energy. Here, we introduce the polarization energies as:

$$\epsilon_{LUMO}^{(N)} = \epsilon_{LUMO}^{(N=1)} - P^{-}, \tag{7.30}$$

$$\epsilon_{HOMO}^{(N)} = \epsilon_{HOMO}^{(N=1)} + P^{+}. \tag{7.31}$$

where ϵ^{N} denotes the average HOMO or LUMO energy of the three PEN molecules in the (PEN)$_N$ cluster. As explained in Sect. 1.2.1 the ES and IP effects contribute to the total polarization energy. Thus, the calculated polarization energies were decomposed into the ES and IP contributions as follows:

$$P^{+/-} = P_{ES}^{+/-} + P_{IP}^{+/-}. \tag{7.32}$$

Here, P_{ES} represents the ES contribution of the polarization energies and is approximately described as the interaction between positive (or negative) charge and the quadrupole moments of surrounding molecules. On the other hand, P_{IP} indicates the interaction between the charge and the induced dipole moments of surrounding molecules. The calculated ES and IP polarization energies are shown Fig. 7.6c, d, respectively. Here, the plus signs on P^{+}/P^{-} indicate that the HOMO/LUMO energies are increased/decreased compared to the gas-phase values. The ES contributions tend to increase both the HOMO and LUMO levels. By contrast, the IP contribution increases and decreases the HOMO and LUMO levels, respectively. Thus, the IP effect is predominantly responsible for the HOMO–LUMO gap reduction in the PEN clusters. This finding is consistent with the earlier experimental suggestions that the IP terms are primarily responsible for the HOMO–LUMO gap reduction, whereas the ES terms play a decisive role in the orientation dependence [118, 123].

We found a difference between the anion and cation polarization energies. Yamada et al. [118] attempted to decompose experimental polarization energies into ES and IP terms by assuming that magnitudes of the cation and anion energies are identical, $P_{ES}^{+} = -P_{ES}^{-}$ and $P_{IP}^{-} = P_{IP}^{+}$. By contrast, according to our calculations, the polarization energy of the anion state is larger than that of the cation state. The absolute values of P_{ES}^{-} and P_{IP}^{-} are larger than P_{ES}^{+} and P_{IP}^{+} by 137 and 72 meV, respectively, in the $N = 33$ cluster. This discrepancy may be attributed to the larger spatial extent of the LUMO compared to the HOMO.

Next, we turn to the polarization energies of localized excited states. We consider the S_1 excited state of a PEN molecule in the cluster as an LE state. The CT states were defined as the intermolecular HOMO → LUMO transition. The computed average energies of the LE and CT states as a function of cluster size are shown in Fig. 7.7. Here, the LE energy was obtained as the average of the three PEN molecules, whereas the CT energies were obtained as the average of six intermolecular CT states within the three PEN molecules. The excitation energies of the LE states increase

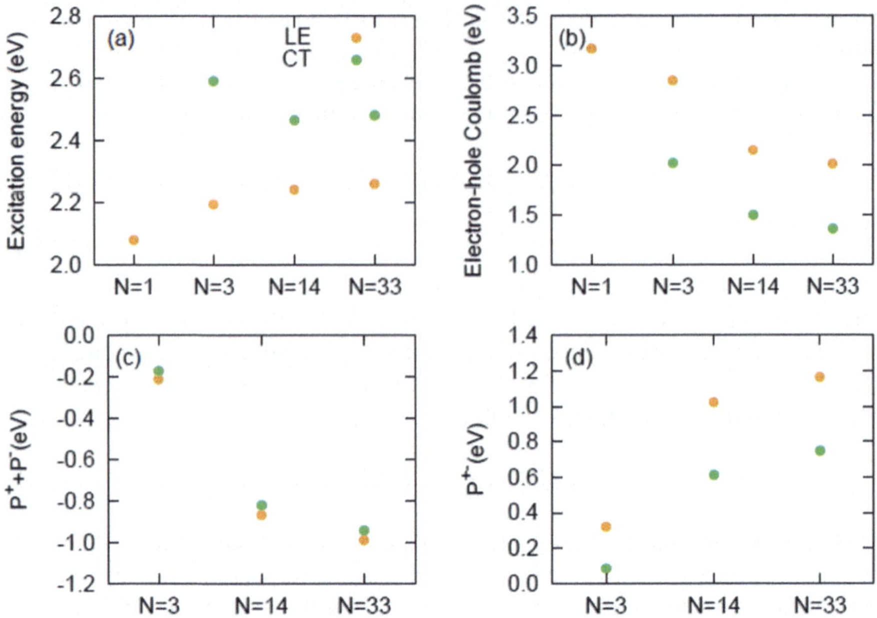

Fig. 7.7 Average **a** excitation energy, **b**, e–h Coulomb interaction, **c** one-body polarization energy, and **d** two-body polarization energy for the LE (orange) and CT (green) states in the PEN clusters

with increasing cluster size from 2.08 eV ($N = 1$) to 2.26 eV ($N = 33$). By contrast, the CT energy decreases from 2.59 eV ($N = 3$) to 2.48 eV ($N = 33$); thus, the energy difference among the LE and CT states decrease with increasing cluster size. Because the CT states have a large electron density difference between the ground and excited states, they are more stabilized by the surrounding molecules than the LE states.

As discussed in Sect. 7.2.3, the polarization energy of an excited state can be decomposed into one-body and two-body polarization energies. The calculated one-body and two-body polarization energies are shown in Figs. 7.7c, d, respectively. The polarization energies of the CT states were defined with respect to the PEN dimer, and the results for $N = 3$ correspond to the energy difference between the trimer and the dimer. As expected from the considerable HOMO–LUMO gap reduction, the one-body polarization terms reduce the excitation energies relative to the gas-phase values. In contrast, the two-body polarization term largely cancels the decrease in the one-body polarization energy. The two-body polarization should decrease with increasing the e–h separation; thus, the $P^{\pm}$ values of the CT states are smaller than those of the LE states, as confirmed by Fig. 7.7d. For the LE states, the decrease in S_1 excitation energies caused by the one-body polarization is canceled out by the two-body term, such that the excitation energy is increased. Conversely, the decrease in the CT excitation energies from the one-body polarization is not fully compensated for by the two-body term, resulting in the lower energies relative to the gas-phase values ($N = 2$).

In summary, we have investigated the polarization energies of the HOMO, LUMO, and localized excited states. Our calculations confirmed the previous suggestion that the IP effects are predominantly responsible for the HOMO–LUMO gap reduction. In contrast to the previous assumption that the anion and cation polarization energies are same, we have found the unequal polarization between the cation and anion states. We also examined the polarization energies of excited states, highlighting the cancelation between the one-body and two-body polarization energies in the LE and CT states.

7.4.2 Delocalized Electronic States

Having investigated the polarization energies of the localized states, we next turn to the delocalized electronic states in the PEN clusters. The HOMO-derived and LUMO-derived states were described as the superposition of HOMOs and LUMOs of all the constituent PEN molecules in the cluster structures, as given by Eqs. 7.3 and 7.4. As well as the single-electron states, the delocalized excited states were calculated as the superposition of the LE and CT states (Eq. 7.29).

Here, we briefly discuss factors governing the spatial extent of the electronic states. One factor contributing the spatial extent of the electronic states is an electronic coupling, i.e., an electronic interaction between localized states. Table 7.3 presents the electronic couplings among the central PEN molecules in the $N = 33$ cluster as a representative case. The transfer integrals determine the formation of the HOMO- and LUMO-derived states and hybridization between the LE and CT states. The excitonic couplings lead to the formation of the FE states. We note that the electronic couplings are not sensitive to the molecular environment. For example, the electronic coupling values in the $N = 33$ cluster are almost identical to corresponding values in the $N = 3$ cluster, with a difference of less than 1 meV, indicating the minor environmental effect.

Another influential factor is the energy variations among the constituent localized states. The energy variations of the HOMO energies, LUMO energies in the $N =$ 33 cluster are presented in Fig. 7.8. Note that the PEN molecular geometries were taken from the crystal structures without any structure optimization, and the energy variations observed in 1.8 are solely result from their molecular environments. The

Table 7.3 The LUMO–LUMO transfer integrals, HOMO–HOMO transfer integrals, and excitonic couplings in units of meV among the PEN trimer in the cluster structure of $N = 33$ (See Fig. 7.5a for pair labeling)

	t_e	t_h	V_F
Pair 1	49	52	27
Pair 2	76	74	8
Pair 3	69	70	8

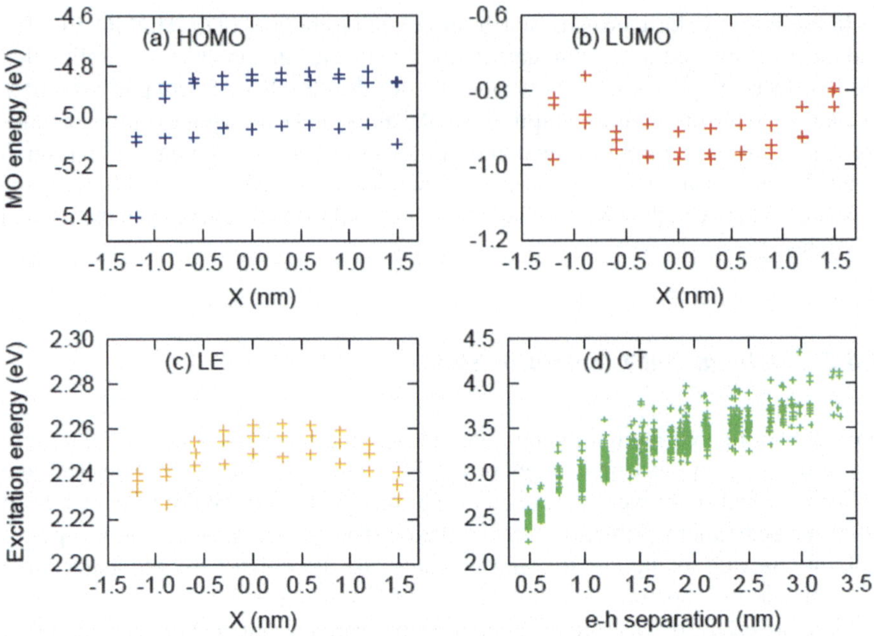

Fig. 7.8 **a** HOMO energy, **b** LUMO energy, and **c** LE excitation energy for PEN molecules in the $N = 33$ cluster with respect to X axis (see Fig. 7.5c). **d** CT excitation energy with respect to the e–h separation in the $N = 33$ cluster

PEN molecules in this system are embedded in heterogeneous polarizable environments: The electronic states of molecules in the edge regions are similar to those of an isolated molecule, whereas those in the central region are more similar to electronic states in the solid phase. As expected form the considerable HOMO and LUMO energy changes from $N = 1$ to $N = 33$ cluster, considerable energy variations are present among the PEN molecules in the cluster structures.

The energy variations among the localized excited states are shown in Figs. 7.8c, d, where the LE states are shown with respect to the X axis given in Fig. 7.5c, and the CT states are depicted with respect to their e–h separation. In contrast to the HOMO and LUMO energies, almost no energy variation is observed for the LE states. Because of the cancelation between the one-body and two-body polarization terms, the LE states are not sensitive to their molecular environment. By contrast, the CT states have stronger interaction with surrounding molecular environment, and the energetic disorder is observed for CT states with a same e–h separation. For example, the energy variation of around 0.2 eV was observed for CT states with an e–h separation of 0.5 nm.

Now, we examine the spatial extent of the HOMO-derived and LUMO-derived states. In Fig. 7.9, the IPRs for the HOMO-derived states and LUMO-derived states in the $N = 33$ cluster are shown. The relatively localized electronic states appear near the highest HOMO-derived state and lowest LUMO-derived states, corresponding

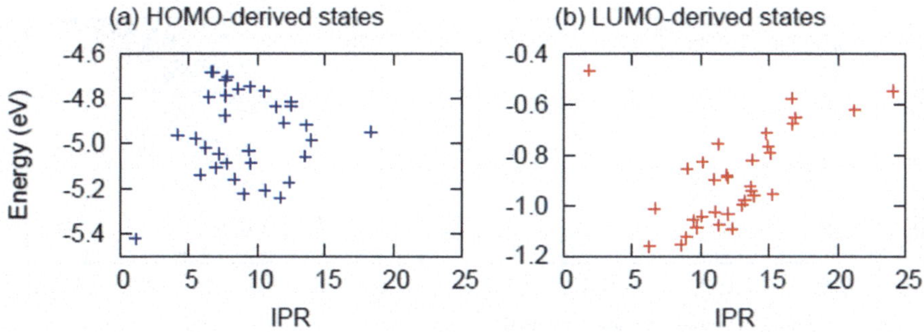

Fig. 7.9 IPRs for **a** HOMO-derived and **b** LUMO-derived states with respect to energy in the $N = 33$ cluster

Table 7.4 Energy (E) in units of eV and IPRs for the highest and lowest states of the HOMO-derived and LUMO-derived states in the $N = 3$, 14, and 33 clusters. The gap between the highest HOMO-derived state and lowest LUMO-derived state is also shown

		HDS		LDS		
N		Low.	High.	Low.	High.	Gap
3	E	−5.52	−5.15	−0.96	−0.45	4.20
	IPR	1.27	1.59	1.13	1.16	
14	E	−5.45	−4.80	−1.10	−0.48	3.70
	IPR	1.16	3.08	3.29	1.91	
33	E	−5.42	−4.68	−1.16	−0.46	3.52
	IPR	1.14	6.66	6.23	1.86	

to the band edges. Meanwhile, more delocalized states arise as the energy moves from the band edge. The IPRs for the highest HOMO-derived state and the lowest LUMO-derived state slightly increases and decreases, respectively, with increasing the cluster sizes, as summarized in Table 7.4. The electronic states near the band edge in the cluster structures are localized by the energy variations arising from the heterogeneous polarizable environment.

We now focus on the delocalized excited states. Here, we attempt to characterize the excited states in terms of the e–h separation and the spatial extent of the excited states. It is not straightforward to quantify the spatial extent of excited states, and various measures have been proposed [87]. Because the excited-state wave function can be described from the electron and hole wave functions, we separately calculated the IPRs for the electron and hole wave functions [26, 44]. In Fig. 7.10, the e–h separation, and the electron and hole IPRs are shown for the low-energy excited states. At energies below 2.5 eV, the excited states are characterized by the small e–h separation and IPRs > 10. Therefore, according to the categorization shown in Fig. 7.3, the low-energy excited states in the PEN clusters are regarded as FE states.

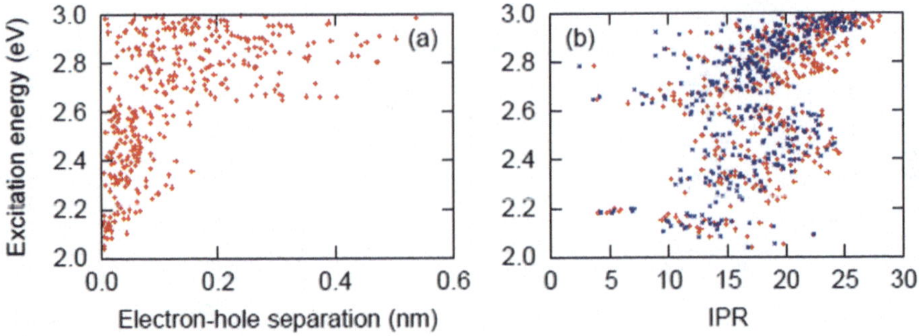

Fig. 7.10 **a** Excitation energies with respect to the e–h separation and **b** IPRs for the electron (red) and hole (blue) wave functions which constitute the excited state

Table 7.5 Excitation energies (E) in units of eV, CT characters (P_{CT}), electron and hole IPRs (IPRe and IPRh), and cooperative factor (CF) for the lowest (Low) and highest (High) LE-dominant states in the $N = 3$, 14, and 33 clusters

N		E	P_{CT}	IPRe	IPRh	CF
3	Low	2.12	0.18	2.81	2.61	1.88
	High	2.19	0.20	1.99	2.47	0.03
14	Low	2.06	0.25	7.93	7.24	6.17
	High	2.19	0.32	4.23	4.47	0.17
33	Low	2.04	0.27	17.5	17.2	13.3
	High	2.20	0.34	5.33	6.78	0.01

In the presence of an energy gap between the LE and CT states, LE-dominant and CT-dominant states are energetically separated. Our calculations for excited states also provided N LE-dominant states with partial admixture of CT states in the low-energy regions. The excited-state characters of the lowest and highest of the LE-dominant states in the PEN clusters are given in Table 7.5. Here, CT character quantifies the participation of the CT states in the excited state. The CT character of the FE states slightly increases with increasing the cluster sizes, which is consistent with the reduction of energy difference between the LE and CT states. In the $N = 33$ cluster, the CT character of approximately 30% is present in the LE-dominant states.

The states in the energy range of 2.2–2.6 eV primarily consist of the CT state, while their e–h separation is smaller than that of localized CT states. Those states may be regarded as charge-resonant states: For the dimer system, the FE state can be written as $\Psi_{EF} = |e_1h_1\rangle \pm |e_2h_2\rangle$. By contrast, the charge-resonant (CR) state is $\Psi_{CR} = |e_1h_2\rangle \pm |e_2h_1\rangle$. Therefore, the charge-resonant state does not have net dipole moment, with the small electron-hole separation. However, it is not clear as to how the charge-resonant states can be distinguished from FE states.

The spatial extent characterized by the IPRs should be carefully interpreted because the large electron (hole) IPR does not indicate the existence of the delocalized LUMO (HOMO)-derived state. In other words, an electron or hole IPR cannot distinguish the WM state, which is direct product of delocalized orbitals, from the FE state comprising many e–h configurations. In our calculations, the FE states were confirmed by the small e–h separation and minor participation of the CT states. The distinction between the FE and WM exciton states may be clarified by introducing the natural transition orbital analysis [73, 87], which is warranted in future studies.

It is interesting to compare the IPRs to other measures for characterizing the excited states. In particular, several measures have been introduced to quantify the delocalization length of the FE state [22, 102, 108]. Here, we introduce the exciton cooperativity factor (CF) as [59]:

$$\mathrm{CF} = \frac{|\mu_M|^2}{|\mu_{LE}|^2}. \tag{7.33}$$

CF quantifies the enhancement of the transition dipole moment of an Mth delocalized excited state (μ_N) relative to that of the monomer LE state (μ_{LE}); it provides the upper bound of the exciton delocalization and vanishes for dark states. The PEN molecules have transition dipole moment in their short molecular axis; thus, in the herringbond structure, the PEN cluster acts as a J-aggregate along the short molecular axis [55]. Consistent with the earlier suggestion, the lowest LE-dominant states are characterized by large CFs and thus have the enhanced transition dipole moments. For the lowest LE-dominant state, the CFs well correlate with the electron and hole IPRs. By contrast, the highest LE-dominant states are optically dark, as shown by the vanishing CFs.

Finally, we compare our results with previous theoretical studies. GW/BSE studies on PEN crystals reported by different groups [21, 101, 109] have concluded that the low-lying excited states of PEN crystals have significant CT characters. By contrast, our calculations for the PEN clusters suggest that the low-energy excited states of the PEN cluster can be regarded as FE states with the approximately 30% CT character. This discrepancy may arise from the treatment of polarizable environments. Increasing the number of molecules may further decrease the energies of CT states relative to those of LE states. The different CT characters in the cluster and crystals indicate that the polarization effect has a critical impact on the nature of excited states.

In summary, we have investigated the spatial extent of the HOMO- and LUMO-derived states and the excited states. Consistent with the energy changes from a single molecule to clusters, the large energy variations are present for the HOMOs and LUMOs because of the heterogeneous polarizable environments. These energy variations result in the coexistence of the localized and delocalized electronic states. We have characterized excited states in the PEN clusters in terms of the e–h separation, electron and hole IPRs, CT character, and CFs. Our calculations suggest that the low-energy excited states in the PEN cluster can be regarded as the FE states.

7.5 Pentacene/C_{60} Planar Heterojunction

In this section, we present preliminary results for a PEN/C_{60} interface structure as a model for organic solar cells. We investigate interfacial CT states comprising an electron on a C_{60} molecule and a hole on a PEN molecule, with a focus on the effect of induced polarization on the e–h energies.

Because PEN and C_{60} molecules are typical p-type and n-type organic semiconductor molecules, their blends have been studied from several aspects. The electronic levels at the PEN/C_{60} interfaces have been determined by the photoelectron spectroscopy [80, 95, 119]. The CT absorption for the standing-up orientation of the PEN/C_{60} planar heterojunction was revealed by Rand and coworkers [13] using the polarized external quantum efficiency measurements. Later, they investigated the interfacial CT states of the PEN/C_{60} interfaces under different conditions of blend morphologies and D/A ratios [70], observing the CT energy variations as large as 0.3 eV. In the computational approaches, excited-state calculations for PEN/C_{60} have been performed for model D/A systems [12, 72, 75, 121, 124, 125]. In particular, Bredas and coworkers performed TDDFT excited-state calculations for PEN/C_{60} clusters of up to ten molecules [125]. Although interfacial CT states can be modeled by using a small D/A cluster, CS states cannot be defined in such a small system. We note that in some first-principles studies for D/A systems, interfacial CT states are not distinguished from CS states. Recently, we performed the excited-state calculations for the local interface structures of the PEN/C_{60} planar heterojunction based on the FMO method [44]. However, the e–h interactions were described as the bare Coulomb interactions, and the screening of e–h interactions by the excited-state polarization was not included. This is seen in the relative overestimation of the CS state energies relative to the short-range CT states [44]. The importance of the polarizable environments has been emphasized by Beljonne and coworkers [24, 115], who treated the state-specific polarization using microelectrostatic calculations.

The computational details are briefly provided here. According to the X-ray diffraction measurement [13], we prepared the edge-on orientation of the PEN/C_{60} planar heterojunctions, as shown in Fig. 7.11. The details of the modeling and subsequent molecular dynamics simulations are presented elsewhere [44]. Although the molecular dynamics simulations performed by Fu et al. [35] have indicated the coexistence of the edge-on and face-on orientations of PEN-C_{60} pairs, we used the edge-on orientation as a limiting case. The FMO-GW calculations were performed for the local interface structure Fig. 7.11b, which includes 36 PEN and 15 C_{60} molecules. The remaining molecules in the total structure in Fig. 7.11a were incorporated as the external point charges. The B3LYP was used as a starting point for the one-shot FMO-GW calculation with 6-31G* basis set. The electronic structure calculations were performed using the ABINIT-MP software [76, 77, 107]. Other details for the FMO-GW calculations can be found in [45].

The HOMO and LUMO energies along the direction perpendicular to the PEN/C_{60} interface are shown in Fig. 7.12. Near the interface, the averaged HOMO/LUMO energies are −4.26/− 0.86 eV for PEN molecules and −5.56/− 0.74 eV for C_{60}

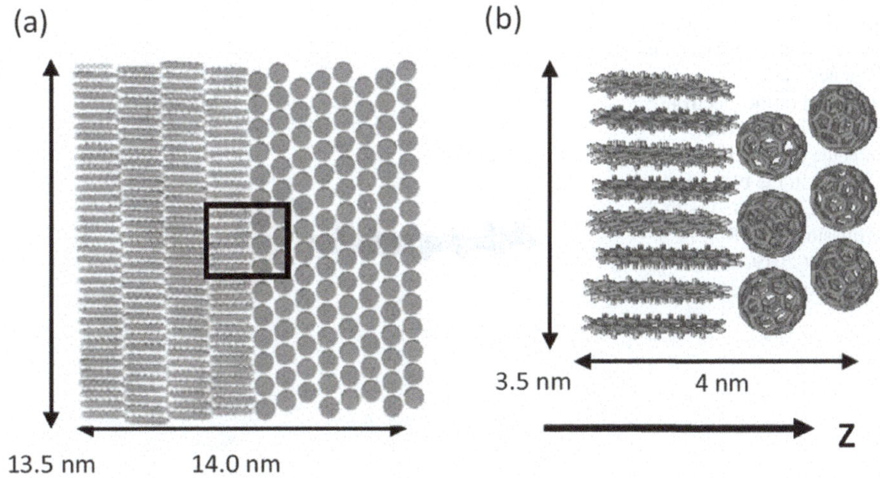

Fig. 7.11 **a** Atomistic structures for the edge-on orientation of the PEN/C_{60} interface. **b** The local interface structure treated quantum-mechanically by the excited-state method. Adapted from ref [44] with permission from the PCCP owner societies

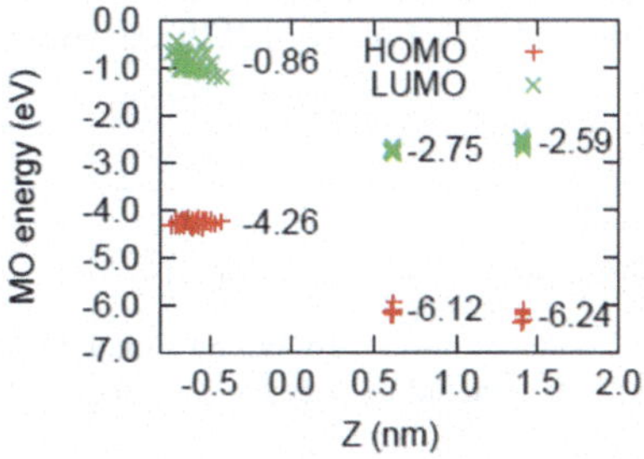

Fig. 7.12 FMO-GW quasiparticle energies for the HOMO and LUMO of PEN or C_{60} molecule in the local interface structure. Reprinted with permission from Ref. [45] Copyright 2019, American Institute of Physics

molecules. For comparison, the HOMO/LUMO energies of isolated molecules are −5.56/− 0.74 eV for PEN and −6.74/− 1.84 eV for C_{60}. Therefore, the HOMO and LUMO energies in the interface structure are significantly higher and lower, respectively, than those of isolated molecules. In particular, the PEN HOMO–C_{60} LUMO gap was estimated as 1.51 eV, which is significantly lower than the corresponding gas-phase value of 3.72 eV.

Herein, we investigate the environmental effects on the CT state, using the representative state comprising a nearest-neighbor PEN–C_{60} pair depicted in Fig. 7.13 as an example. The CT energy was estimated from the PEN HOMO–C_{60} LUMO gap and the e–h Coulomb interactions, $E_{CT} = \epsilon_L^{C_{60}} - \epsilon_H^{PEN} - U_{HH,LL}$. The results for the representative CT state in the different environmental treatments are presented in Table 7.6, where we compare the CT energies of the corresponding isolated PEN–C_{60} pair ($E_{CT}(Gas)$), the CT energy including only the ES effect in the interface structure ($E_{CT}(ES)$), and the CT energy including both ES and IP effects ($E_{CT}(ES + IP)$). The

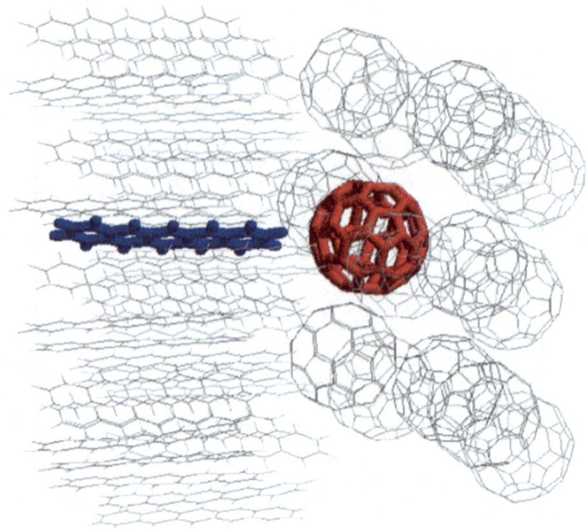

Fig. 7.13 Visualization of the representative interfacial CT states, which is composed of the electron localized in the red C_{60} molecule and the hole localized in the blue PEN molecule. Reprinted with permission from Ref. [45] Copyright 2019, American Institute of Physics

Table 7.6 Excitation energy (E_{CT}), orbital-energy difference ($\Delta\epsilon$), and e–h interaction (U_{eh}) in units of eV for the representative interfacial CT states shown in Fig. 7.13. Reprinted with permission from Ref. [45]. Copyright 2019, American Institute of Physics

	E_{CT}	$\Delta\epsilon$	U_{eh}
Gas[a]	1.91	3.18	1.27
ES[b]	1.70	2.96	1.26
ES + IP	0.91	1.58	0.68

[a]CT state of the corresponding isolated PEN-C_{60} pair
[b]CT state including only the ES effect in the interface structure

bare e–h Coulomb interaction for this CT state is -1.19 eV. The orbital-energy difference, $\Delta\epsilon = \epsilon_L^{C_{60}} - \epsilon_H^{PEN}$, for gas-phase value (3.18 eV) is lower than that estimated from isolated PEN and C_{60} molecules (3.52 eV); the deformation of their molecular structure causes the reduction in $\Delta\varepsilon$. A comparison between $E_{CT}(Gas)$ and $E_{CT}(ES)$ indicates that the ES effect marginally decreases the orbital-energy difference by 0.22 eV, but does not affect the e–h attraction. By contrast, the IP effect significantly decreases the orbital-energy difference by 1.38 eV and weakens the e–h interaction by 0.58 eV, resulting in the decrease in the CT excitation energy by 0.89 eV.

We turn to the interfacial CT energies with respect to the e–h separation. A CT energy diagram was obtained by calculating the interfacial CT states for all PEN–C_{60} pairs shown in Fig. 7.11b, where the three degenerate LUMOs of the C_{60} molecules were included to define the CT states. Figure 7.14a shows the CT energy as a function of the e–h separation. The energies of the nearest-neighbor CT states were obtained in the energy range of 0.8–1.1 eV, which is consistent with the external quantum

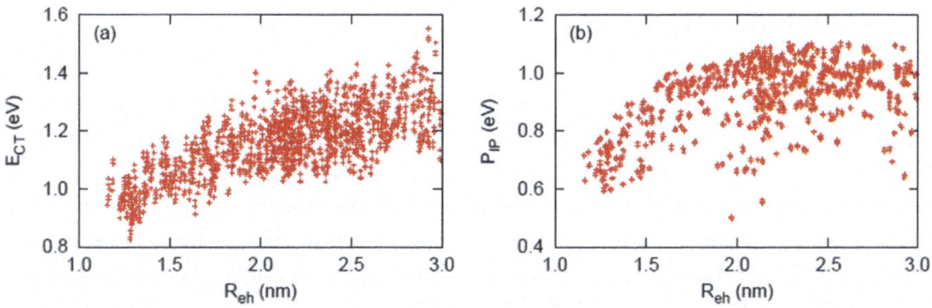

Fig. 7.14 **a** Excitation energies (E_{CT}) and **b** induced polarization energies (P_{IP}) for CT states with respect to the **e**– separation (R_{eh}) in the PEN/C_{60} interface. Reproduced with permission from Ref. [45] Copyright 2019, American Institute of Physics

efficiency measurement reported by Rand and coworkers [13, 70]. The energy landscape for the e–h separation can be characterized by the bound e–h pairs at short distance (R_{eh}< 2.0 nm) and a relatively flat energy profile (R_{eh}> 2.0 nm) compared with that reported in the previous study [44].

We now clarify the impact of the IP effect on the charge separation. The IP contribution of polarization energy (P^{ex}_{IP}) of a CT state was introduced as follows:

$$P^{ex}_{IP} = E_{CT}(ES) - E_{CT}(ES + IP). \tag{7.34}$$

Qualitatively, this IP energy includes the one-body and two-body contributions, $P^{ex}_{IP} = P^{+}_{IP} + P^{-}_{IP} - P^{\pm}$, discussed in Sect. 7.2.2. Figure 7.3b presents the IP energies for the CT states, indicating that P^{ex}_{IP} increases with increasing R_{eh}. This behavior demonstrates that two separated charges are more strongly stabilized by their respective environments than a bound e–h pair, as schematically shown in Fig. 7.2. As shown in Fig. 7.3b, E_{IP} increases from 0.7 eV at 1.3 nm to 1.1 eV at 2.5 nm, favoring separated charges by ~0.4 eV. P^{ex}_{IP} may converge near 2.5 nm; thus, the CT states with an e–h separation is greater than 2.5 nm may be regarded as CS states, in the sense that the e–h correlation of the IP effects vanishes.

In summary, we have examined the interfacial CT states in the PEN/C_{60} interface by performing the large-scale GW calculations. Our calculation indicates that the IP effect has a strong influence on the energy diagram of the CT states. Although we focused on the localized electronic states in this section, the effects of charge delocalization in the charge-separation energetics have also been discussed [26, 44, 106]. A comprehensive discussion on the combined effects of polarization and delocalization, as well as a comparison with experiments, will be presented in future works.

7.6 Concluding Remarks

In this chapter, we have presented our first-principle studies on the electronically excited states of organic semiconductor materials. In particular, we have focused on the effects of intermolecular interactions and molecular aggregation on the electronic states. The development of the fragment-based GW method has enabled the accurate determinations of energy levels of charged and electronically excited states and the elucidation of the effects of polarization and delocalization effects. By characterizing the electronic states of the PEN clusters in detail, we have illustrated the roles of polarization and delocalization. Moreover, we have investigated the essential role of the IP effect in the CT states in the PEN/C_{60} interface structure. Further studies on other D/A interface systems are expected to provide a deeper understanding of excited states in an OSC.

We have presented the computation results within the Born-Oppenheimer approximation, i.e., electronic states obtained from the electronic Schrödinger equation with a fixed nuclear geometry. However, interactions between electronic states and a nuclear degree of freedom strongly influence optical and transport properties. In particular, organic semiconductor materials are characterized by a relatively strong electron–vibration (phonon) interaction compared with inorganic materials. For example, nuclear motion has a strong effect on charge-transport mechanics [110, 111]. The role of nuclear vibration in optical spectra has been intensively studied within the Frenkel–Holstein model [102]. In the time-resolved spectroscopy studies on OSCs [83], charge-separation dynamics have been detected as the formation of polarons. However, the microscopic mechanism of the polaron formation, i.e., how an electron or a hole is dressed by molecular vibration (phonon) with increasing the e–h separation, remains unclear. Electron–vibration interactions result in considerable non-radiative decay to the ground state [9], limiting the open-circuit voltage. From a computational standpoint, ab initio computations of electron-vibration [91] or electron-phonon [49] interactions in molecular aggregates are more challenging than single-point calculations, in which the electronic Schrödinger equation is solved with a fixed nuclear position. Future theoretical developments in this direction are necessary to elucidate the roles of molecular vibration and phonons in the excited-state properties.

Finally, the dynamical aspects of charge photogeneration should also be addressed. Accordingly, an FMO calculation for an aggregate provides a model excited-state Hamiltonian, which can be used to simulate real-time dynamics in combination with quantum dynamic theories [37, 40–43, 59, 96]. We have already applied this strategy to excited-state dynamics in the organic semiconductor thin film [42] and the D/A interface [37]. Future developments of electronic structure methods and quantum dynamic theories would provide further microscopic insights into the photophysical processes in organic optoelectronic materials.

Acknowledgements T.F. thanks Prof. Yuji Mochizuki, Dr. Tatsuya Nakano, and Dr. Yoshio Okiyama for their collaborations on the FMO-based excited-state calculations and their implementation in the ABINIT-MP program. T.F. also thanks Prof. Yoshifumi Noguchi at Shizuoka

University for collaboration on the developments of FMO-GW/BSE methods, Prof. Takeo Hoshi at Tottori University for collaboration on organic materials, and Prof. Seiichiro Izawa at Institute for Molecular Science for discussion on the organic solar cells. T.F. acknowledges the financial supports from the Ministry of Education, Culture, Sports, Science and Technology (MEXT) as a Building of Consortia for the Development of Human Resources in Science and Technology and from a Grant-in-Aid for Scientific Research (C) (19K05255) from the Japan Society for the Promotion of Science (JSPS). The majority of the FMO calculations in this work were performed using the facilities at the Supercomputer Center, the Institute for Solid State Physics, the University of Tokyo. The Gaussian calculations in this work were conducted at the Research Center for Computational Science, Okazaki, Japan.

References

1. Adamo, C., Barone, V.: Toward reliable density functional methods without adjustable parameters: The pbe0 model. J. Chem. Phys. **110**(13), 6158–6170 (1999)
2. Akaike K.: Advanced understanding on electronic structure of molecular semiconductors and their interfaces. Jpn. J. Appl. Phys. 57(3S2):03EA03 (2018)
3. Aryasetiawan, F., Gunnarsson, O.: The gw method. Rep. Prog. Phys. **61**(3), 237 (1998)
4. Atahan-Evrenk, S., Aspuru-Guzik, A.: Prediction and theoretical characterization of p-type organic semiconductor crystals for field-effect transistor applications. In: Atahan-Evrenk, S., Aspuru-Guzik, A. (eds) Prediction and Calculation of Crystal Structures, Topics in Current Chemistry, vol 345, Springer International Publishing (2014), pp 95–138
5. Bakulin, A., Rao, A., Pavelyev, V.G., van Loosdrecht, P.H.M., Pshenichnikov, M.S., Niedzialek, D., Cornil, J., Beljonne, D., Friend, R.H.: The role of driving energy and delocalized States for charge separation in organic semiconductors. Science **335**(6074), 1340 (2012)
6. Bardeen, C.J.: The structure and dynamics of molecular excitons. Annu. Rev. Phys. Chem. **65**(1), 127–148 (2014)
7. Becke, A.D.: Density-functional exchange-energy approximation with correct asymptotic behavior. Phys. Rev. A **38**(6), 3098 (1988)
8. Becke, A.D.: Density-functional thermochemistry. iv. a new dynamical correlation functional and implications for exact-exchange mixing. J. Chem. Phys. **104**(3), 1040–1046 (1996)
9. Benduhn, J., Tvingstedt, K., Piersimoni, F., Ullbrich, S., Fan, Y., Tropiano, M., McGarry, K.A., Zeika, O., Riede, M.K., Douglas, C.J., Barlow, S., Marder, S.R., NeherD, Spoltore D., Vandewal, K.: Intrinsic non-radiative voltage losses in fullerene-based organic solar cells. Nat. Energy **2**(April), 17053 (2017)
10. Blase, X., Attaccalite, C., Olevano, V.: First-principles GW calculations for fullerenes, porphyrins, phtalocyanine, and other molecules of interest for organic photovoltaic applications. Phys. Rev. B **83**, 115103 (2011)
11. Blase, X., Duchemin, I., Jacquemin, D.: The bethe–salpeterequation in chemistry: relations with td-dft, applications and challenges. Chem. Soc. Rev. **47**(3), 1022–1043 (2018)
12. Brédas, J.L., Norton, J.E., Cornil, J., Coropceanu, V.: Molecular understanding of organic solar cells: The challenges. Acc. Chem. Res. **42**(11), 1691–1699 (2009)
13. Brigeman, A.N., Fusella, M.A., Yan, Y., Purdum, G.E., Loo, Y.L., Rand, B.P., Giebink, N.C.: Revealing the full charge transfer state absorption spectrum oforganic solar cells. Adv. Energy Mater. **6**(21), 1601001 (2016)
14. Chai, J.D., Head-Gordon, M.: Long-range corrected hybrid density functionals with damped atom–atom dispersion corrections. Phys. Chem. Chem. Phys. **10**(44), 6615–6620 (2008)
15. Chiba, M., Fedorov, D.G., Kitaura, K.: Time-dependent density functional theory with the multilayer fragment molecular orbital method. Chem. Phys. Lett. **444**(4–6), 346–350 (2007)

16. Clarke, T.M., Durrant, J.R.: Charge photogeneration in organic solar cells. Chem. Rev. **110**(11), 6736–6767 (2010)
17. Coropceanu, V., Malagoli, M., da Silva Filho, D.A., Gruhn, N.E., Bill, T.G., Brédas, J.L.: Hole- and electron-vibrational couplings in oligoacene crystals: Intramolecular contributions. Phys. Rev. Lett. **89**, 275503 (2002)
18. Coto, P.B., Sharifzadeh, S., Neaton, J.B., Thoss, M.: Low-lying electronic excited states of pentacene oligomers: A comparative electronic structure study in the context of singlet fission. J. Chem. Theory Comput. **11**(1), 147–156 (2014)
19. Crocker, L., Wang, T., Kebarle, P.: Electron affinities of some polycyclic aromatic hydrocarbons, obtained from electron-transfer equilibria. J. Am. Chem. Soc. **115**(17), 7818–7822 (1993)
20. Cudazzo, P., Gatti, M., Rubio, A.: Excitons in molecular crystals from first-principles many-body perturbation theory: Picene versus pentacene. Phys. Rev. B **86**, 195307 (2012). https://doi.org/10.1103/physrevb.86.195307, https://link.aps.org/doi/10.1103/PhysRevB.86.195307
21. Cudazzo, P., Sottile, F., Rubio, A., Gatti, M.: Exciton dispersion in molecular solids. J. Phys.: Condens. Matter **27**(11), 113204 (2015)
22. Dahlbom, M., Pullerits, T., Mukamel, S., Sundström, V.: Exciton delocalization in the b850 light-harvesting complex: Comparison of different measures. J. Phys. Chem. B **105**, 5515–5524 (2001)
23. Davidov, A.S.: Theory of Molecular Excitons. Springer, New York (1971)
24. D'Avino, G., Mothy, S., Muccioli, L., Zannoni, C., Wang, L., Cornil, J., Beljonne, D., Castet, F.: Energetics of electron–hole separation at p3ht/pcbm heterojunctions. J. Phys. Chem. C **117**(25), 12981–12990 (2013)
25. D'Avino, G., Muccioli, L., Castet, F., PoelkingC, Andrienko D., Soos, Z.G., CornilJ, Beljonne D.: Electrostatic phenomena in organic semiconductors: fundamentals and implications for photovoltaics. J. Phys.: Condens. Matter **28**(43), 433002 (2016)
26. D'Avino, G., Muccioli, L., Olivier, Y., Beljonne, D.: Charge separation and recombination at polymer-fullerene heterojunctions: Delocalization and hybridization effects. J. Phys. Chem. Lett. **7**(3), 536–540 (2016)
27. Dicke, R.H.: Coherence in spontaneous radiation processes. Phys. Rev. **93**, 99–110 (1954)
28. Dreuw, A., Head-Gordon, M.: Single-reference ab initio methods for the calculation of excited states of large molecules. Chem. Rev. **105**(11), 4009–4037 (2005)
29. Dreuw, A., Weisman, J.L., Head-Gordon, M.: Long-range charge-transfer excited states in time-dependent density functional theory require non-local exchange. J.Chem. Phys. **119**(6), 2943–2946 (2003)
30. Fedorov, D.G., Kitaura, K. (eds.): The Fragment Molecular Orbital Method: Practical Applications to Large Molecular Systems. CRC Press, Boca Raton, FL (2009)
31. Fedorov, D.G., Kitaura, K.: Many-body expansion of the fock matrix in the fragment molecular orbital method. J. Chem. Phys. **147**(10), 104106 (2017)
32. Few, S., Frost, J.M., Nelson, J.: Models of charge pair generation in organic solar cells. Phys. Chem. Chem. Phys. **17**, 2311–2325 (2015)
33. Fratini, S., Mayou, D., Ciuchi, S.: The transient localization scenario for charge transport in crystalline organic materials. Adv. Funct. Mater. **26**(14), 2292–2315 (2016)
34. Frisch, M.J., Trucks, G.W., Schlegel, H.B., Scuseria, G.E., Robb, M.A., Cheeseman, J.R., Scalmani, G., Barone, V., Petersson, G.A.,Nakatsuji, H., Li, X., Caricato, M., Marenich, A.V., Bloino, J., Janesko, B.G., Gomperts, R., Mennucci, B., Hratchian, H.P., Ortiz, J.V., Izmaylov, A.F., Sonnenberg, J.L., Williams-Young, D., Ding, F., Lipparini, F., Egidi, F., Goings, J., Peng, B., Petrone, A., Henderson, T., Ranasinghe, D., Zakrzewski, V.G., Gao, J., Rega, N., Zheng, G., Liang, W., Hada, M., Ehara, M., Toyota, K., Fukuda, R., Hasegawa, J., Ishida, M., Nakajima, T., Honda, Y., Kitao, O., Nakai, H., Vreven, T., Throssell, K., Montgomery, J.A. Jr., Peralta, J.E., Ogliaro, F., Bearpark, M.J., Heyd, J.J., Brothers, E.N., Kudin, K.N., Staroverov, V.N., Keith, T.A., Kobayashi, R., Normand, J., Raghavachari, K., Rendell, A.P., Burant, J.C., Iyengar, S.S., Tomasi, J., Cossi, M., Millam, J.M., Klene, M., Adamo, C., Cammi, R., Ochterski, J.W., Martin, R.L., Morokuma, K., Farkas, O., Foresman, J.B., Fox, D.J.: Gaussian 16 Revision B.01. Gaussian Inc. Wallingford CT (2016)

35. Fu, Y.T., Risko, C., Brédas, J.L.: Intermixing at the pentacene-fullerene bilayer interface: A molecular dynamics study. Adv. Mater. **25**(6), 878–882 (2013)
36. Fujimoto, K.J.: Transition-density-fragment interaction combined with transfer integral approach for excitation-energy transfer via charge-transfer states. J. Chem. Phys. **137**(3), 034101 (2012)
37. Fujita, T., Hoshi, T.: Fmo-based investigations of excited-state dynamics in molecular aggregates. In: Mochizuki, Y., Tanaka, S., Fukuzawa, K. (eds) Recent Advances of the Fragment Molecular Orbital Method - Enhanced Performance and Applicability, Springer (2020), in press
38. Fujita, T., Mochizuki, Y.: Development of the fragment molecular orbital method for calculating nonlocal excitations in large molecular systems. J. Phys. Chem. A **122**(15), 3886–3898 (2018)
39. Fujita, T., Noguchi, Y.: Development of the fragment-based cohsexmethod for large and complex molecular systems. Phys. Rev. B **98**, 205140 (2018). https://doi.org/10.1103/physrevb.98.205140
40. Fujita, T., Brookes, J.C., Saikin, S.K., Aspuru-Guzik, A.: Memory-assistedexciton diffusion in the chlorosome light-harvesting antenna of green sulfurbacteria. J. Phys. Chem. Lett. **3**, 2357–2361 (2012)
41. Fujita, T., Huh, J., Saikin, S.K., Brookes, J.C., Aspuru-Guzik, A.: Theoretical characterization of excitation energy transfer in chlorosome light-harvesting antennae from green sulfur bacteria. Photosynth. Res. **120**(3), 273–289 (2014)
42. Fujita, T., Atahan-Evrenk, S., Sawaya, N.P.D., Aspuru-Guzik, A.: Coherent dynamics of mixed frenkel and charge-transfer excitons in dinaphtho[2,3-*b*:2′ 3′ -*f*]thieno[3,2-*b*]-thiophene thin films: The importance of hole delocalization. J. Phys. Chem. Lett. **7**(7), 1374–1380 (2016)
43. Fujita, T., Haketa, Y., Maeda, H., Yamamoto, T.: Relating stacking structures and charge transport in crystal polymorphs of the pyrrole-based π-conjugated molecule. Org. Electron. **49**, 53–63 (2017)
44. Fujita, T., Alam, M.K., Hoshi, T.: Thousand-atom ab initio calculations of excited states at organic/organic interfaces: Toward first-principles investigations of charge photogeneration. Phys. Chem. Chem. Phys. **20**, 26443–26452 (2018). https://doi.org/10.1039/c8cp05574b
45. Fujita, T., Noguchi, Y., Hoshi, T.: Charge-transfer excited states in the donor/acceptor interface from large-scale gw calculations. J. Chem. Phys. **151**(11), 114109 (2019)
46. Gao, F., Inganas, O.: Charge generation in polymer-fullerene bulkheterojunctionsolar cells. Phys. Chem. Chem. Phys. **16**, 20291–20304 (2014)
47. Gao, W., Xia, W., Gao, X., Zhang, P.: Speeding up gw calculations to meet the challenge of large scale quasiparticle predictions. Sci. Rep. **6**, 36849 (2016)
48. Gélinas, S., Rao, A., Kumar, A., Smith, S.L., Chin, A.W., Clark, J., van der Poll, T.S., Bazan, G.C., Friend, R.H.: Ultrafast long-range charge separation in organic semiconductor photovoltaic diodes. Science **343**(6170), 512–516 (2014)
49. Giustino, F.: Electron-phonon interactions from first principles. Rev. Mod. Phys. **89**(1), 015003 (2017)
50. Gordon, M.S., Fedorov, D.G., Pruitt, S.R., Slipchenko, L.V.: Fragmentation methods: A route to accurate calculations on large systems. Chem. Rev. **112**(1), 632–672 (2012)
51. Govoni, M., Galli, G.: Large scalegw calculations. J. Chem. Theory Comput. **11**(6), 2680–2696 (2015)
52. Halasinski, T.M., Hudgins, D.M., Salama, F., Allamandola, L.J., Bally, T.: Electronic absorption spectra of neutral pentacene (c22h14) and its positive and negative ions in ne, ar, and kr matrices. J. Phys. Chem. A **104**(32), 7484–7491 (2000)
53. Hedin, L.: New method for calculating the one-particle green's function with application to the electron-gas problem. Phys. Rev. **139**(3A), A796 (1965)
54. Hestand, N.J., Spano, F.C.: Expanded theory of h- and j-molecular aggregates: The effects of vibronic coupling and intermolecular charge transfer. Chem. Rev. **118**(15), 7069–7163 (2018)

55. Hestand, N.J., Yamagata, H., Xu, B., Sun, D., Zhong, Y., Harutyunyan, A.R., ChenG, Dai H.L., Rao, Y., Spano, F.C.: Polarized absorption in crystallinepentacene: Theory versus experiment. J. Phys. Chem. C **119**(38), 22137–22147 (2015)
56. Hirata, S., Head-Gordon, M.: Time-dependent density functional theory within the tamm-dancoff approximation. Chem. Phys. Lett. **314**(3E), 291–299 (1999)
57. Hoshi, T., Imachi, H., Kuwata, A., Kakuda, K., Fujita, T., Matsui, H.: Numerical aspect of large-scale electronic state calculation for flexible device material. Japn. J. Indust. Appl. Math. **36**(2), 685–698 (2019)
58. Hsu, C.P.: The electronic couplings in electron transfer and excitation energy transfer. Acc. Chem. Res. **42**(4), 509–518 (2009)
59. Huh, J., Saikin, S.K., Brookes, J.C., Valleau, S., Fujita, T., Aspuru-Guzik, A.: Atomistic study of energy funneling in the light-harvesting complex of green sulfur bacteria. J. Am. Chem. Soc. **136**(5), 2048–2057 (2014)
60. Iikura, H., Tsuneda, T., Yanai, T., Hirao, K.: A long-range correction scheme for generalized-gradient-approximation exchange functionals. J. Chem. Phys. **115**(8), 3540–3544 (2001)
61. Kitaura, K., Ikeo, E., Asada, T., Nakano, T., Uebayasi, M.: Fragment molecular orbital method: an approximate computational method for large molecules. Chem. Phys. Lett. **313**(3–4), 701–706 (1999)
62. Kitoh-Nishioka, H., Ando, K.: Fmo3-lcmo study of electron transfer coupling matrix element and pathway: Application to hole transfer between two tryptophans through cis- and trans-polyproline-linker systems. J. Chem. Phys. **145**(11), 114103 (2016)
63. Koch, N., Vollmer, A., Salzmann, I., Nickel, B., Weiss, H., Rabe, J.P.: Evidence for temperature-dependent electron band dispersion in pentacene. Phys. Rev. Lett. **96**, 156803 (2006)
64. Körzdörfer, T., Brèdas, J.L.: Organic electronic materials: recent advances in the dft description of the ground and excited states using tuned rangeseparatedhybrid functionals. Acc. Chem. Res. **47**(11), 3284–3291 (2014)
65. Kresse, G., Furthmüller, J.: Efficient iterative schemes for ab initio totalenergycalculations using a plane-wave basis set. Phys. Rev. B **54**, 11169–11186 (1996)
66. Kurashige, Y., Yanai, T.: Theoretical study of the $\pi \rightarrow \pi*$ excited states of oligoacenes: A full π-valence dmrg-caspt2 study. Bull. Chem. Soc. Jpn. **87**(10), 1071–1073 (2014)
67. Lee, C., Yang, W., Parr, R.G.: Development of the colle-salvetticorrelationenergyformula into a functional of the electron density. Phys. Rev. B **37**(2), 785 (1988)
68. Li, J., D'Avino, G., Duchemin, I., Beljonne, D., Blase, X.: Accurate description of charged excitations in molecular solids from embedded many-body perturbation theory. Phys. Rev. B **97**(3), 035108 (2018)
69. Lin, Y.L., Fusella, M.A., Rand, B.P.: The impact of local morphology onorganic donor/acceptor charge transfer states. Adv. Energy Mater. (2018)
70. Lin, Y.L., Zhang, F., Kerner, R.A., Yang, T.C.J., Kahn, A., Rand, B.P.: Variable charge transfer state energies at nanostructured pentacene/c60 interfaces. Appl. Phys. Lett. **112**(21), 213302 (2018)
71. Liu, X., Rand, B.P., Forrest, S.R.: Engineering charge-transfer states for efficient, low-energy-loss organic photovoltaics. Trends Chem. (2019)
72. Ma, H., Troisi, A.: Direct optical generation of long-range charge-transfer states in organic photovoltaics. Adv. Mater. **26**(35), 6163–6167 (2014)
73. Martin, R.L.: Natural transition orbitals. J. Chem. Phys. **118**(11), 4775–4777 (2003)
74. Matsui, H., Mishchenko, A.S., Hasegawa, T.: Distribution of localized states from fine analysis of electron spin resonance spectra in organic transistors. Phys. Rev. Lett. **104**(5), 056602 (2010)
75. Minami, T., Nakano, M., Castet, F.: Nonempirically tuned long-range corrected density functional theory study on local and charge-transfer excitation energies in a pentacene/c60 model complex. J. Phys. Chem. Lett. **2**(14), 1725–1730 (2011)
76. Mochizuki, Y., Koikegami, S., Amari, S., Segawa, K., Kitaura, K., Nakano, T.: Configuration interaction singles method with multilayer fragment molecular orbital scheme. Chem. Phys. Lett. **406**, 283–288 (2005)

77. Nakano, T., Kaminuma, T., Sato, T., Akiyama, Y., Uebayasi, M., Kitaura, K.: Fragment molecular orbital method: use of approximate electrostatic potential. Chem. Phys. Lett. **351**, 475–480 (2002)
78. Nakayama, Y., Mizuno, Y., Hikasa, M., Yamamoto, M., Matsunami, M., Ideta, S., Tanaka, K., Ishii, H., Ueno, N.: Single-crystal pentacene valence-band dispersion and its temperature dependence. J. Phys. Chem. Lett. **8**(6), 1259–1264 (2017)
79. Neaton, J.B., Hybertsen, M.S., Louie, S.G.: Renormalization of molecular electronic levels at metal-molecule interfaces. Phys. Rev. Lett. **97**, 216405 (2006)
80. Nishi, T., Kanno, M., Kuribayashi, M., Nishida, Y., Hattori, S., Kobayashi, H., von Wrochem, F., Rodin, V., Nelles, G., Tomiya, S.: Impact of molecular orientation on energy level alignment at c60/pentacene interfaces. Appl. Phys. Lett. **113**(16), 163302 (2018)
81. Noguchi, Y., Sugino, O.: Molecular size insensitivity of optical gap of [n]cycloparaphenylenes (n = 3-16). J. Chem. Phys. **146**(14), 144304 (2017)
82. Noguchi, Y., Hirose, D., Sugino, O.: Optical properties of six isomers of three dimensionally delocalizedπ-conjugated carbon nanocage. Eur. Phys. J. B **91**(6), 125 (2018)
83. Ohkita, H., Ito, S. Exciton and charge dynamics in polymer solar cells studied by transient absorption spectroscopy. In: Organic Solar Cells, Springer (2013), pp 103–137
84. Onida, G., Reining, L., Rubio, A.: Electronic excitations: Densityfunctionalversus many-body green's-function approaches. Rev. Mod. Phys. **74**(2), 601 (2002)
85. Park, S.H., Roy, A., Beaupré, S., Cho, S., Coates, N., Moon, J.S., Moses, D., Leclerc, M., Lee, K., Heeger, A.J.: Bulk heterojunction solar cells with internal quantum efficiency approaching 100%. Nat. Photonics **3**(5), 297–302 (2009)
86. Perdew, J.P., Burke, K., Ernzerhof, M.: Generalized gradient approximation made simple. Phys. Rev. Lett. **77**(18), 3865 (1996)
87. Plasser, F.: Entanglement entropy of electronic excitations. J. Chem. Phys. **144**(19), 194107 (2016)
88. Poelking, C., Tietze, M., Elschner, C., Olthof, S., Hertel, D., Baumeier, B., WürthnerF, Meerholz K., Leo, K., Andrienko, D.: Impact of mesoscale order on open-circuit voltage in organic solar cells. Nat. Mater. **14**(4), 434 (2015)
89. Pope, M., Swenberg, C.E.: Electronic processes in organic crystals and polymers. Oxford University Press on Demand (1999)
90. Refaely-Abramson, S., Jain, M., Sharifzadeh, S., Neaton, J.B., Kronik, L.: Solid-state optical absorption from optimally tuned time-dependent rangeseparatedhybrid density functional theory. Phys. Rev. B **92**, 081204 (2015)
91. Reimers, J.R.: A practical method for the use of curvilinear coordinates in calculations of normal-mode-projected displacements and duschinsky rotation matrices for large molecules. J. Chem. Phys. **115**(20), 9103–9109 (2001)
92. Ren, X., Rinke, P., Blum, V., Wieferink, J., Tkatchenko, A., Sanfilippo, A., Reuter, K., Scheffler, M.: Resolution-of-identity approach to hartree–fock, hybrid density functionals, rpa, mp2 and gw with numeric atom-centered orbital basis functions. New J. Phys. **14**(5), 053020 (2012)
93. Rohlfing, M., Louie, S.G.: Electron-hole excitations and optical spectra from first principles. Phys. Rev. B **62**, 4927–4944 (2000)
94. RühleV, Lukyanov A., May, F., Schrader, M., Vehoff, T., Kirkpatrick, J., BaumeierB, Andrienko D.: Microscopic simulations of charge transport in disordered organic semiconductors. J. Chem. Theory Comput. **7**(10), 3335–3345 (2011)
95. Salzmann, I., Duhm, S., Opitz, R., Johnson, R.L., Rabe, J.P., Koch, N.: Structural and electronic properties of pentacene-fullerene heterojunctions. J. Appl. Phys. **104**(11), 114518 (2008)
96. Sawaya, N.P.D., Huh, J., Fujita, T., Saikin, S.K., Aspuru-Guzik, A.: Fast delocalization leads to robust long-range excitonic transfer in a large quantum chlorosome model. Nano Lett. **15**(3), 1722–1729 (2015)
97. Schiefer, S., Huth, M., Dobrinevski, A., Nickel, B.: Determination of the crystal structure of substrate-induced pentacene polymorphs in fiber structured thin films. J. Am. Chem. Soc. **129**(34), 10316–10317 (2007)

98. Schwarze, M., Schellhammer, K.S., Ortstein, K., Benduhn, J., Gaul, C., HinderhoferA, Toro L.P., Scholz, R., Kublitski, J., Roland, S., et al.: Impact of molecular quadrupole moments on the energy levels at organic heterojunctions. NatCommun **10**(1), 2466 (2019)
99. van Setten, M.J., Weigend, F., Evers, F.: The gw-method for quantum chemistry applications: Theory and implementation. J. Chem. Theory Comput. **9**(1), 232–246 (2013)
100. Sharifzadeh, S., Biller, A., Kronik, L., Neaton, J.B.: Quasiparticle and optical spectroscopy of the organic semiconductors pentacene and ptcda from first principles. Phys. Rev. B **85**(12), 125307 (2012)
101. Sharifzadeh, S., Darancet, P., Kronik, L., Neaton, J.B.: Low-energy chargetransferexcitons in organic solids from first-principles: The case of pentacene. J. Phys. Chem. Lett. **4**(13), 2197–2201 (2013)
102. Spano, F.C.: The spectral signatures of frenkelpolarons in H- And J-aggregates. Acc. Chem. Res. **43**(3), 429–439 (2010)
103. Szabo, A., Ostlund, N.: Modern Quantum Chemistry: Introduction to Advanced Electronic Structure Theory. Dover Publications, New York (1996)
104. Takeshi, I.: Ab initio quantum chemical calculation of electron density, electrostatic potential, and electric field of biomolecule based on fragment molecular orbital method. Int. J. Quantum Chem. **118**(8), e25535 (2018)
105. Takeya, J., Tsukagoshi, K., Aoyagi, Y., Takenobu, T., Iwasa, Y.: Hall effect of quasi-hole gas in organic single-crystal transistors. Jpn. J. Appl. Phys. **44**(46), L1393–L1396 (2005)
106. Tamura, H., Burghardt, I.: Ultrafast charge separation in organic photovoltaics enhanced by charge delocalization and vibronically hot exciton dissociation. J. Am. Chem. Soc. **135**(44), 16364–16367 (2013)
107. Tanaka, S., Mochizuki, Y., Komeiji, Y., Okiyama, Y., Fukuzawa, K.: Electron-correlated fragment-molecular-orbital calculations for biomolecular and nano systems. Phys. Chem. Chem. Phys. **16**, 10310–10344 (2014)
108. Tanaka, S., Miyata, K., Sugimoto, T., Watanabe, K., Uemura, T., Takeya, J., Matsumoto, Y.: Enhancement of the exciton coherence size in organic semiconductor by alkyl chain substitution. J. Phys. Chem. C **120**(15), 7941–7948 (2016)
109. Tiago, M.L., Northrup, J.E., Louie, S.G.: Ab initio calculation of the electronic and optical properties of solid pentacene. Phys. Rev. B **67**, 115212 (2003)
110. Troisi, A.: Charge transport in high mobility molecular semiconductors: classical models and new theories. Chem. Soc. Rev. **40**, 2347–2358 (2011)
111. Troisi, A., Orlandi, G.: Charge-transport regime of crystalline organic semiconductors: Diffusion limited by thermal off-diagonal electronic disorder. Phys. Rev. Lett. **96**, 086601 (2006)
112. Tsuneda, T., Song, J.W., Suzuki, S., Hirao, K.: On koopmans' theorem in density functional theory. J. Chem. Phys. **133**(17), 174101 (2010)
113. Tsuneyuki, S., Kobori, T., Akagi, K., Sodeyama, K., Terakura, K., Fukuyama, H.: Molecular orbital calculation of biomolecules with fragment molecular orbitals. Chem. Phys. Lett. **476**, 104–108 (2009)
114. Vandewal, K.: Interfacial charge transfer states in condensed phase systems. Annu. Rev. Phys. Chem. **67**, 113–133 (2016)
115. Verlaak, S., Beljonne, D., Cheyns, D., Rolin, C., Linares, M., Castet, F., CornilJ, Heremans P.: Electronic structure and geminate pair energetics at organic-organic interfaces: The case of pentacene/c60 heterojunctions. Adv. Funct. Mater. **19**(23), 3809–3814 (2009)
116. Vydrov, O.A., Scuseria, G.E.: Assessment of a long-range corrected hybrid functional. J. Chem. Phys. **125**(23), 234109 (2006)
117. Würthner, F., Kaiser, T.E., Saha-Möller, C.R.: J-aggregates: From serendipitous discovery to supramolecular engineering of functional dye materials. Angew. Chem. Int. Ed. **50**(15), 3376–3410 (2011)
118. Yamada, K., Yanagisawa, S., Koganezawa, T., Mase, K., Sato, N., Yoshida, H.: Impact of the molecular quadrupole moment on ionization energy and electron affinity of organic thin films: Experimental determination of electrostatic potential and electronic polarization energies. Phys. Rev. B **97**, 245206 (2018)

119. Yamamoto, M., Nakayama, Y., Uragami, Y., Kinjo, H., Mizuno, Y., Mase, K., Koswattage, K.R., Ishii, H.: Electronic structures of a well-defined organic hetero-interface: C60 on pentacene single crystal. E-J. Surf. Sci. Nanotechnol. **13**, 59–64 (2015)
120. Yanai, T., Tew, D.P., Handy, N.C.: A new hybrid exchange-correlation functional using the coulomb-attenuating method (cam-b3lyp). Chem. Phys. Lett. **393**(1–3), 51–57 (2004)
121. Yang, B., Yi, Y., Zhang, C.R., Aziz, S.G., Coropceanu, V., Brédas, J.L.: Impact of electron delocalization on the nature of the charge-transfer states in model pentacene/c60 interfaces: A density functional theory study. J. Phys. Chem. C **118**(48), 27648–27656 (2014)
122. Yao, J., Kirchartz, T., Vezie, M.S., Faist, M.A., Gong, W., He, Z., Wu, H., Troughton, J., Watson, T., Bryant, D., Nelson, J.: Quantifying losses in open-circuit voltage in solution-processable solar cells. Phys. Rev. Appl. **4**, 014020 (2015)
123. Yoshida, H., Yamada, K., Tsutsumi, J., Sato, N.: Complete description of ionization energy and electron affinity in organic solids: Determining contributions from electronic polarization, energy band dispersion, and molecular orientation. Phys. Rev. B **92**(7), 075145 (2015)
124. Zheng, Z., Brédas, J.L., Coropceanu, V.: Description of the charge transfer states at the pentacene/c60 interface: Combining range-separated hybrid functionals with the polarizable continuum model. J. Phys. Chem. Lett. **7**(13), 2616–2621 (2016)
125. Zheng, Z., Tummala, N.R., Fu, Y.T., Coropceanu, V., Brédas, J.L.: Chargetransferstates in organic solar cells: Understanding the impact of polarization, delocalization, and disorder. ACS Appl. Mater. Inter. **9**(21), 18095–18102 (2017)

Chapter 8
Open-Circuit Voltage in Organic Solar Cells

Seiichiro Izawa

8.1 Introduction

Organic solar cells (OSCs) have attracted considerable attention because of their potential advantages, which include low cost, lightweight, and environmental friendliness [1]. After more than 20 years of research, the power conversion efficiencies (PCEs) of single-junction OSCs based on blends of electron donor (D) and acceptor (A) semiconducting materials now exceed 17% [2]. This progress was mainly driven by the invention of bulk heterojunction device structure at an early stage [3, 4] and the subsequent development of low-bandgap polymer semiconductors [5]. Recently, the development of non-fullerene acceptors has brought about a rapid increase in the PCEs of OSCs [6].

The PCE of OSCs is lower compared to that of other SC devices: GaAs, single-crystal silicon, and perovskite SCs have PCE values of over 25% [7]. The next research target for OSCs is to achieve a PCE of a similar level in a highly efficient SC device. With regard to the current generation properties, the external quantum efficiencies (EQEs) of OSCs have already exceeded 85% [8], with internal quantum efficiencies from absorbed photons to electrons close to 100% [9]. Further increases in the PCE of OSCs is currently limited mainly by the open-circuit voltage (V_{OC}), and to achieve additional increases in the PCE, it is therefore very important to clarify the factors that determine V_{OC} in OSCs, and to find a strategy for improving it.

In this chapter, we first provide the theoretical background explaining how V_{OC} is determined in OSCs. We then describe recent findings on how the D/A interfacial properties largely affect V_{OC}, and how it can be improved by controlling the nanostructure at the D/A interface. Finally, we present our conclusions and describe future prospects for further improvement of V_{OC} in OSCs.

S. Izawa (✉)
National Institutes of Natural Science, Institute for Molecular Science, 5-1 Higashiyama Myodaiji, Aichi, Okazaki 444-8787, Japan
e-mail: izawa@ims.ac.jp

M. Hiramoto and S. Izawa (eds.), *Organic Solar Cells*,
https://doi.org/10.1007/978-981-15-9113-6_8

8.2 Theoretical Background of Open-Circuit Voltage in Organic Solar Cells

This section summarizes the theoretical background of V_{OC} in OSCs, beginning with an explanation of the charge generation and recombination mechanism. Subsequently, the empirical understanding of V_{OC}, and the models explaining how V_{OC} in OSCs is determined are described.

8.2.1 Photoconversion Mechanisms in Organic Solar Cells

The photoconversion process in OSCs is described below, and a schematic image of the working principle is depicted in Fig. 8.1. Under solar irradiation, organic semiconductors absorb light, and excitons that are stabilized by strong Coulomb attraction are formed. The exciton then migrates to the D/A interface, where it dissociates into a free electron and a hole. These electrons and holes are transported in the acceptor and the donor materials, respectively, and collected at the electrodes.

8.2.1.1 Exciton Formation and Diffusion

In the formation of an exciton, light irradiation excites an electron from the highest occupied molecular orbital (HOMO) to the lowest unoccupied molecular orbital (LUMO) in organic semiconductors. In organic semiconductors, the exciton is stabilized by coulombic attraction forces owing to the low dielectric constants of these materials ($\varepsilon_r \approx 2$–4) [10]. This is referred to as a Frenkel-type exciton. The typical exciton binding energy in organic semiconductors is 0.1–1 eV [11]. The exciton is

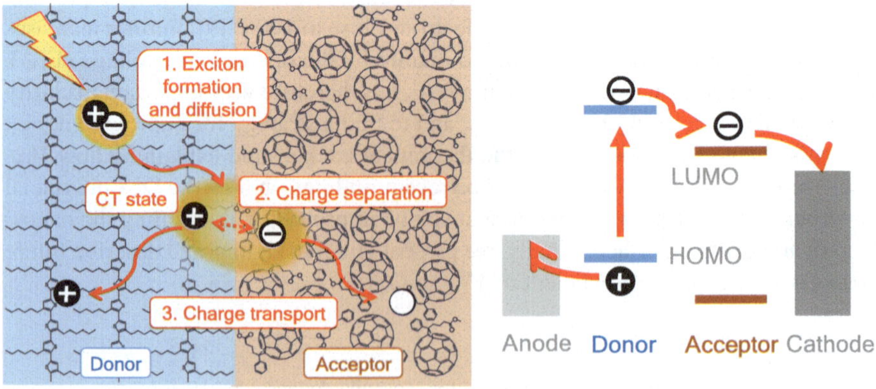

Fig. 8.1 Schematic image of the working principle in OSCs

required to be dissociated into free charges within its lifetime, and in organic photovoltaics, the heterojunction of the D/A is used to promote this dissociation. In organic semiconductors, the exciton moves around by the Förster energy transfer process, with a typical exciton diffusion length of approximately 5–10 nm [12]. Thus, the size of the donor and acceptor domains should be controlled at the ~10 nm scale, and otherwise, the exciton will relax to the ground state without being separated from the free charges before it reaches the D/A interface.

8.2.1.2 Charge Separation

The exciton may be formed in either the donor or the acceptor material. In the donor, the exciton is then quenched by electron transfer from donor to acceptor, and in the acceptor, it is quenched by hole transfer from acceptor to donor. The charge transfer does not necessarily generate free charges, because the coulombic attraction force between the charge pairs is large and is effective over a long range, owing to the low dielectric constants of organic semiconductors. It has been reported that the estimated Coulomb capture length is approximately 4 nm [10]. Thus, the charge transfer (CT) state, which is defined as an electron-hole pair stabilized by the Coulomb attraction force at the D/A interface, is potentially formed before complete charge separation [13]. This CT state could decay to the ground state (geminate recombination), resulting in the loss of the photocurrent. The charge separation process requires dissociation of the CT state into free charges.

8.2.1.3 Charge Recombination

After charge dissociation, free electrons and holes are transported in the donor and acceptor domains, respectively. Finally, free charges are collected at the electrodes.

When the free charge pairs meet at the D/A interface, a CT state is formed and can decay to the ground state (non-geminate, or bimolecular, recombination: Fig. 8.2). Charge recombination is further classified according to the mechanism of the process. Bimolecular recombination with light emission is called radiative recombination, whereas if it occurs with thermal relaxation, it is called non-radiative recombination [14]. The recombination processes are sometimes explained using the theory of the *pn* junction in inorganic SCs. Recombination between free electrons in the conduction band and holes in the valence band is called band-to-band recombination. On the other hand, recombination with a trapped charge in the trap state inside the bandgap is called trap-assisted recombination [15]. The kinetics of these processes are strongly related to the V_{OC} in OSCs.

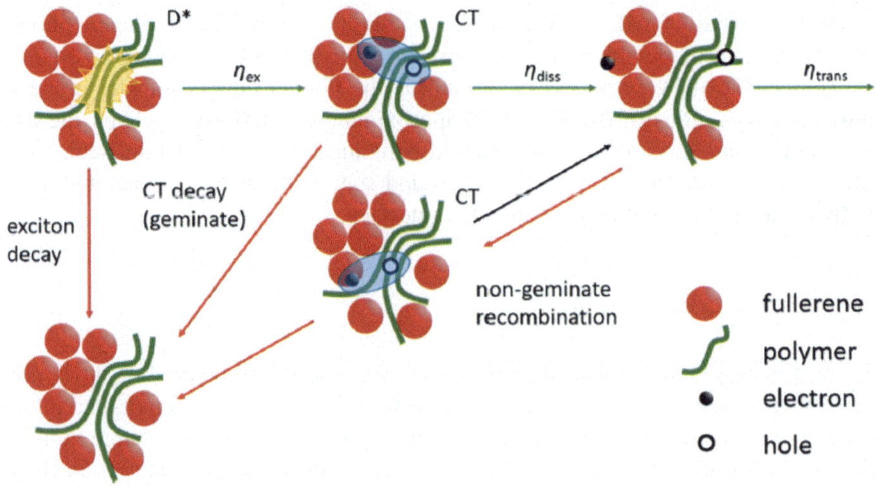

Fig. 8.2 Schematic representation of charge separation and recombination processes. Reprinted with permission from [13]. Copyright 2013, American Chemical Society

8.2.2 *Empirical Understanding of Open-Circuit Voltage in Organic Solar Cells*

In OSCs, V_{OC} is found empirically to have a linear dependence on the energy difference between the HOMO of the donor and the LUMO of the acceptor materials. Scharber et al. reported a linear correlation between V_{OC} and the onset of the oxidation potential (HOMO level) of various donor polymers combined with the same PCBM acceptor [16]. Using –4.3 eV for the LUMO of PCBM, they reported that V_{OC} followed the empirical equation

$$eV_{OC} = E_{Donor}^{HOMO} - E_{Acceptor}^{LUMO} - 0.3\,\text{eV} \tag{8.1}$$

where e is the elementary charge. However, the values of the HOMO and LUMO energy levels of the donor and acceptor materials in this report were not very accurate because they were determined by cyclic voltammetry. Measurement by ultraviolet photoemission spectroscopy and inverse photoemission spectroscopy is desirable [17]. Further, devices with the same D/A combination sometimes show different V_{OC} values owing to structural or electrical interaction at the D/A interface after junction formation [18]. Thus, V_{OC} should be compared with the state energy of the devices.

In a recent report, it was found that V_{OC} was determined by the energy of the CT state, which consists of coulombically bound charge pairs at the D/A interface [19]. The energy of the CT state (E_{CT}) is expressed by the following equation, assuming point charges [20]:

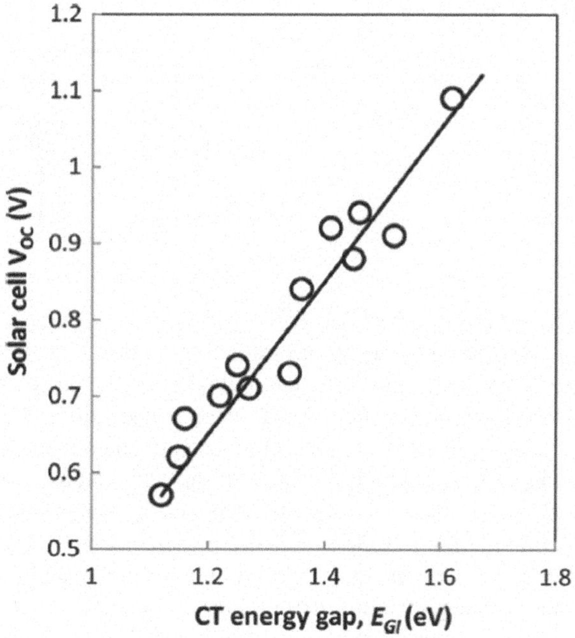

Fig. 8.3 Plot of measured V_{OC} *vs.* energy of the CT state. Reprinted with permission from [19]. Copyright 2014, American Chemical Society

$$E_{CT} = E_{DA} - \frac{e^2}{4\pi\varepsilon_0\varepsilon_r d} \tag{8.2}$$

where E_{DA} is the energy-level difference between the HOMO of the donor and the LUMO of the acceptor, ε_0 is the vacuum dielectric constant, ε_r is the relative dielectric constant, and d is the distance between electron and hole at the D/A interface. Street et al. reported a linear correlation between the energy of the CT state and the measured V_{OC} of the organic photovoltaic devices (Fig. 8.3) [19]. The empirical energy difference between the CT state and eV_{OC} was 0.55 eV. A strategy to minimize this energy loss should be devised in order to enhance the PCE of OSCs.

8.2.3 Equivalent Circuit Model

An equivalent circuit model of SCs has been developed in inorganic semiconductor theory. The model in Fig. 8.4 gives the relations between the current density and voltage in SCs as follows [21]:

$$J = \frac{1}{1 + R_S/R_P}\left[J_0\left\{\exp\left(\frac{V - JR_SA}{nkT/e}\right) - 1\right\} + \frac{V}{R_PA}\right] - J_{ph} \tag{8.3}$$

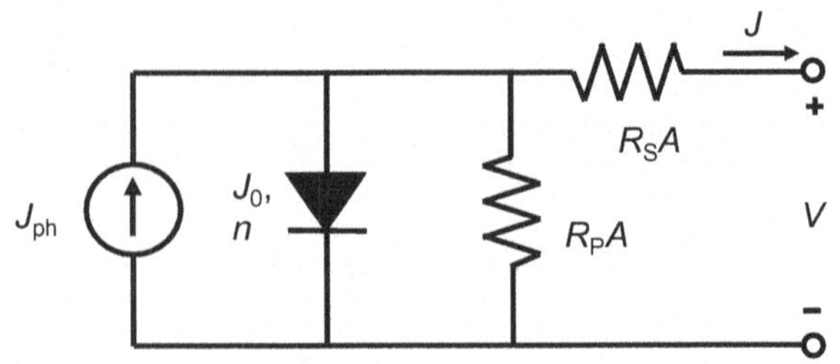

Fig. 8.4 Equivalent circuit model for SCs

where J is the current density, V is the applied voltage, J_{ph} is the photocurrent density, J_0 is the dark saturation current density, n is the ideality factor of the diode, R_S is the series resistance, R_P is the parallel (shunt) resistance, A is the area of the device, k is the Boltzmann constant, and T is the temperature.

This equivalent circuit is composed of four parts. First is the photocurrent generation part, J_{ph}, which depends on the light intensity and becomes zero under dark conditions. J_{ph} sometimes also depends on the applied voltage in OSCs, because the recombination processes are voltage dependent. The second part is represented by a diode, in which the important parameters are J_0 and n. J_0 is the dark current density under the negative infinity condition, i.e., the current density generated by thermal activation in the dark. Under dark conditions with $V = 0$ and no current flowing, the thermal-activation current and recombination current are balanced. Therefore, another meaning of J_0 is the recombination current density under zero-bias conditions. J_0 is a pre-exponential factor; thus, the onset of the J–V curve shifts to the lower voltage side if J_0 becomes larger. The parameter n is equal to unity in ideal *pn* junctions, where recombination is dominated by band-to-band recombination. However, in classical *pn* junctions, n has been reported to increase owing to Shockley–Read–Hall (SRH) recombination. An n value greater than unity also indicates the presence of trap-assisted recombination in OSCs [15]. R_S is the resistance at the electrodes or active-layer/electrode interface. The short-circuit current density (J_{SC}) and fill factor (FF) become small if R_S is large; thus, a small R_S is desirable. R_P is a resistance related to the leak current. V_{OC} and FF become small if R_P is small; thus, a large R_P is desirable.

In an ideal situation, where R_S is small enough, R_P is large enough, and J_{ph} is sufficiently larger than J_0, Eq. 8.3 can be simplified and solved under open-circuit conditions ($J = 0$, $V = V_{OC}$) as follows:

$$eV_{OC} = nkT\ln\left(\frac{J_{ph}}{J_0}\right) \tag{8.4}$$

Equation 8.4 gives the important insight that V_{OC} is mainly determined by the type of recombination (n), generated current density (J_{ph}), and recombination current density (J_0). The origin of J_0 in OSCs is the thermal-activation current at the D/A interface; thus, J_0 is expressed by the Arrhenius equation [22],

$$J_0 = J_{00}\exp\left(-\frac{\varphi}{nkT}\right) \tag{8.5}$$

where the pre-exponential term J_{00} is an electronic interaction term in Marcus theory related to orbital overlap or area at the D/A interface and φ is the activation energy. The ground state and CT state are in equilibrium under open-circuit conditions; thus, φ can be identified with E_{CT}. Substitution of J_0 from Eqs. 8.5 into 8.4 gives Eq. 8.6.

$$eV_{OC} = E_{CT} - nkT\ln\left(\frac{J_{00}}{J_{ph}}\right) \tag{8.6}$$

The reason for the linear dependence of V_{OC} on E_{CT} in Fig. 8.3 can be understood from Eq. 8.6. The most effective ways to enhance V_{OC} are to increase E_{CT}, realize ideal band-to-band recombination ($n = 1$), and reduce electronic interactions leading to charge recombination (hence, a smaller J_{00}).

8.2.4 Detailed Balance Theory

The detailed balance between absorption and emission in SCs gives the following relationship [23]:

$$J_0 = \frac{e}{EQE_{EL}}\int EQE_{PV}(E)\varphi_{BB}(E)dE \tag{8.7}$$

where EQE_{EL} is the external quantum efficiency of electroluminescence (EL) in an SC device, EQE_{PV} is the external quantum efficiency for photocurrent generation, and φ_{BB} is the black body radiation spectrum at the temperature of the SC device. The essence of this relation is that recombination emission under light irradiation is black body radiation from the SC device, and it balances the light absorption. The well-known Shockley and Queisser limit is calculated from Eqs. 8.4 and 8.7 by assuming that $EQE_{EL} = 1$ and EQE_{PV} is a step function [24]. This is an ideal case, meaning that all the recombination is radiative and all the photons with energies above the bandgap are collected as current. However, real SC devices cannot satisfy this assumption. φ_{BB} increases significantly at longer wavelengths, as explained by Planck's law; thus, J_0 in Eq. 8.7 is dominated by the absorption in the state that has the smallest bandgap in an SC device. In the case of OSCs, it is the CT state. Thus, EQE_{PV} is assumed to be the efficiency of CT state absorption, given by the following equation [25]:

$$EQE_{PV}(E) = \frac{f}{E\sqrt{4\pi\lambda kT}}\exp\left(\frac{-(E_{CT} + \lambda - E)^2}{4\lambda kT}\right) \tag{8.8}$$

where λ is the reorganization energy and f is an electronic coupling matrix represented in the Marcus theory. Replacing J_0 in Eq. 8.4 by the expression in Eq. 8.7, EQE_{PV} in Eqs. 8.7 by 8.8, and φ_{BB} by the Planck's law relation gives the following relationship:

$$V_{OC} = \frac{E_{CT}}{e} + \frac{kT}{e}\ln\left(\frac{J_{SC}h^3c^2}{fe2\pi(E_{CT} - \lambda)}\right) + \frac{kT}{e}\ln(EQE_{EL}) \quad (8.9)$$

where h is Planck's constant and c is the speed of light. Equation 8.9 gives another explanation for how V_{OC} in OSC is determined. Equation 8.9 is composed of three terms: first the energetic term, second the radiative recombination term, and third the non-radiative recombination term. Radiative recombination is reduced if the CT state absorption is small (small f) and the absorption edge is sharp. Non-radiative recombination is suppressed if the EQE_{EL} of the SC device is large. This relation leads to a useful maxim: "An efficient SC device is also an efficient LED" (light-emitting diode). However, EQE_{EL} of OSCs from the CT state is generally very small (10^{-5} – 10^{-7}) [14] because the oscillator strength between the CT state and ground state is very small. Thus, suppression of non-radiative recombination is very important for increasing the V_{OC} and PCE of OSCs.

8.3 Increasing Open-Circuit Voltage in Organic Solar Cells by Modifying Donor/Acceptor Interface

The Sect. 8.2 led to the important result that charges separation and recombination processes from the CT state occurring at the D/A interface determines V_{OC} in OSCs. These processes are closely related to the properties of the D/A interface. This section summarizes our recent results on how V_{OC} in OSCs can be increased by modifying the D/A interface. The section starts with an explanation of the way that energy-level alignment at the D/A interface affects V_{OC} in OSCs and can be controlled by doping. The way in which the CT state energy is affected by the energetic structure at the D/A interface is subsequently explained. Finally, a strategy to suppress charge recombination and thereby increase V_{OC} in OSCs is presented.

8.3.1 Effect of Energy-Level Alignment at Donor/Acceptor Interface on Open-Circuit Voltage in Organic Solar Cells

When different layers are contacted, charge transfer occurs between them until thermodynamic equilibrium is attained. The generated charges form an electrostatic field near the interface and induce a vacuum-level shift to align the Fermi level (E_F). This

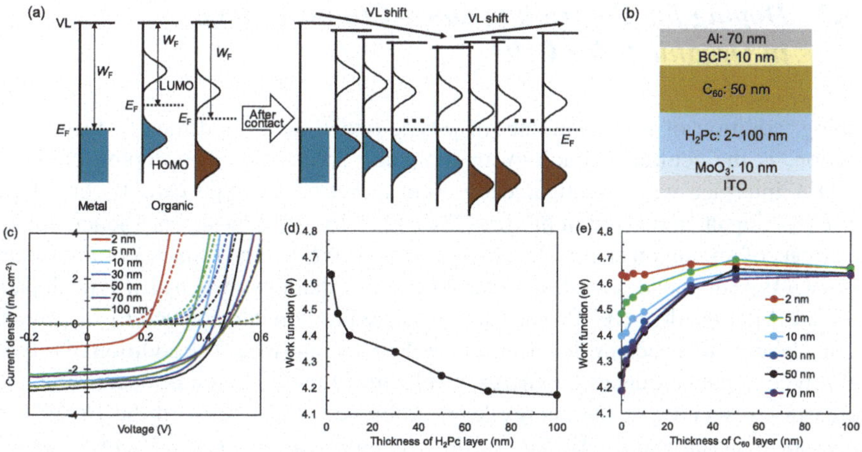

Fig. 8.5 **a** Schematic of band bending near interfaces. **b** Schematic diagram of device structure. **c** *J-V* curves of OSCs with different H_2Pc layer thicknesses. **d** Work function of the H_2Pc film on the MoO_3/ITO substrate, as measured by the Kelvin probe method and plotted as a function of layer thickness. **e** Work function of C_{60} films on H_2Pc (2–70 nm)/MoO_3/ITO substrate as a function of C_{60} layer thickness. Reprinted with permission from [18]. Copyright 2018, American Chemical Society

is called energy-level alignment at the interface (Fig. 8.5a) [26]. This effect was revealed by the thickness dependence of the donor layer in bilayer OSCs (Fig. 8.5b) with phthalocyanine (H_2Pc) as the donor and fullerene (C_{60}) as the acceptor [18]. The V_{OC} of the device was significantly affected by the H_2Pc layer thickness, starting from 0.20 V at a thickness of 2 nm and gradually increasing to 0.49 V at 100 nm (Fig. 8.5c). The reason for the large difference in V_{OC} was explained by the band-bending phenomenon. The work function measured by a Kelvin probe of H_2Pc films changed considerably with changes in the layer thickness, leading to band bending because of the large E_F difference between the MoO_3 hole-transporting layer and H_2Pc (Fig. 8.5d). Thus, different thicknesses of the H_2Pc layer lead to variations in the degree of band bending in the C_{60} layer due to the E_F alignment at the D/A interface (Fig. 8.5e). This difference in the vacuum-level shift near the D/A interface induces a large V_{OC} change with the donor layer thickness. This finding gives the important insight that V_{OC} could change even though the same D/A materials are used. If E_F can be controlled by doping, V_{OC} will be increased by the energy-level alignment. E_F control by doping is discussed in the Sect.8.3.2.

8.3.2 Doping for Controlling Open-Circuit Voltage in Organic Solar Cells

Doping, addition of trace amounts of *p*-type and *n*-type impurities to form a *pn* junction, is the central technology in inorganic solar cells. The doping effect at the D/A interface was investigated by co-evaporating a *p*-type (MoO_3) or *n*-type (Cs_2CO_3) dopant and H_2Pc in bilayer OSCs [27]. Figure 8.6a shows a schematic of the effects of doping on energy-level alignment. The E_F of the films becomes deeper when a *p*-type dopant is added, whereas it becomes shallower when an *n*-type dopant is added. After the doped H_2Pc and C_{60} layers are brought into contact, E_F alignment occurs, leading to a vacuum-level shift near the D/A interface. The addition of *p*-type and *n*-type dopants induces an energy-level shift in the direction of decreasing and increasing, respectively, the energy-level difference (E_{DA}) between the HOMO of the donor material and the LUMO of the acceptor material. The *J-V* curves of the devices are shown in Fig. 8.6b. In the devices with MoO_3-doped H_2Pc, V_{OC} gradually decreased from 0.43 to 0.40 and 0.36 V for devices with dopant concentrations of 200, 2,000, and 20,000 ppm, respectively. In the devices with Cs_2CO_3-doped H_2Pc, V_{OC} increased from 0.47 V in the undoped film to 0.51 and 0.52 V in films with dopant concentrations of 200 and 2,000 ppm, respectively.

To estimate the energy-level shift near the D/A interface, energy-level mapping using the Kelvin probe method was performed. The vacuum level shifted steeply within 10 nm near the D/A interface as a result of the charge transfer, with the direction of the shift depending on the dopant. MoO_3 doping induced an energy-level shift that decreased the E_{DA}, whereas Cs_2CO_3 doping increased the E_{DA}, as expected and as shown in Fig. 8.6a. The E_{DA} change explains the large V_{OC} difference produced by doping.

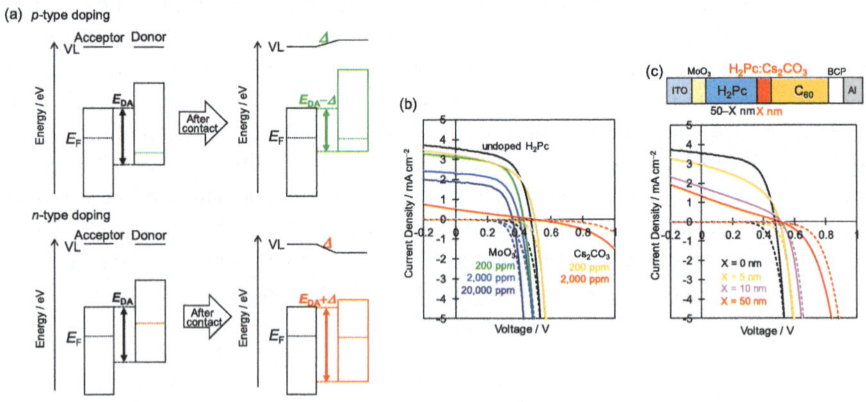

Fig. 8.6 **a** Schematic of E_F alignment at the D/A interface when the H_2Pc layer is doped with *p*-type and *n*-type dopants. **b** *J-V* curves of undoped and doped H_2Pc/C_{60} devices. **c** *J-V* curves of the devices with 0, 5, 10, and 50 nm of the 2,000 ppm Cs_2CO_3-doped H_2Pc layer. Reprinted with permission from [27]. Copyright 2018, American Chemical Society

To investigate the effect of doping near the D/A interface, trilayer devices with an active layer consisting of H_2Pc/doped-H_2Pc/C_{60} were fabricated. The device with a 5 nm Cs_2CO_3-doped layer showed a V_{OC} of 0.50 V, and the device with a 10 nm Cs_2CO_3-doped layer showed an increased V_{OC} of 0.52 V; these values are approximately the same as those of the bilayer device with a 50 nm Cs_2CO_3-doped layer (0.52 V), as shown in Fig. 8.6c. These results indicate that 10-nm-thick doped layers determined the magnitude of the increase in V_{OC}, which is consistent with the band-mapping results. Notably, the device with a thinner Cs_2CO_3-doped layer showed larger J_{SC} and FF values. These results indicate that the addition of *n*-type dopants only near the D/A interface could result in an increase in V_{OC} while maintaining the J_{SC} and FF values.

8.3.3 Organic pn Homojunction Solar Cells

D/A heterojunctions are generally used in OSCs to split coulombically bound excitons formed in the semiconductor. As discussed in the Sect. 8.3.2, energy-level alignment creates an electric field at the interface, and this electric field can separate excitons if they are formed at a single organic semiconductor interface. Doping was applied to create a *pn* homojunction at an organic semiconductor interface, similar to the *pn* junction in an inorganic SC [28]. A bilayer *pn* homojunction device was created using diindenoperylene (DIP) as the host material, and MoO_3 and Cs_2CO_3 as *p-type* and *n-type* dopants, respectively (Fig. 8.7a, b). DIP is well known to have ambipolar characteristics [29], which are necessary for the host material in *pn* homojunction SCs, where electrons and holes are transported in a single material.

The *J-V* curves of the *pn* homojunction device with different dopant concentrations are displayed in Fig. 8.7c. The undoped device showed a very low J_{SC}, as the device could not split the coulombically bound excitons to generate a photocurrent. In contrast, the *pn* homojunction device with higher dopant concentrations showed a larger J_{SC}. The J_{SC} value became more than eight times larger with an increase in dopant concentration from 0 to 5%. This is due to the large electric field (built-in-potential) near the *pn* homojunction interface formed by doping, as shown in Fig. 8.7d. The internal quantum efficiency of the 5% doped device was found to be 30%, indicating that the *pn* homojunction created by doping can efficiently separate excitons to form free charges.

However, V_{OC} decreased with increasing dopant concentration, as shown in the *J-V* curves. Analysis of the V_{OC} variation helps to expose the critical difference between organic and inorganic semiconductors with respect to how V_{OC} is determined [30]. Plots of V_{OC} of the *pn* homojunction devices as a function of temperature are shown in Fig. 8.7e. The temperature dependence of V_{OC} follows Eq. 8.6 in the Sect. 8.2.3, and the magnitude of V_{OC} is determined mainly by two parts: the energetic term (E_g^{eff} is used here instead of E_{CT} because the devices do not have a D/A interface), and the recombination loss term, which is dependent on the temperature. The E_g^{eff} values of the devices with 5% and 1% doped DIP, calculated from the intercepts of the plots,

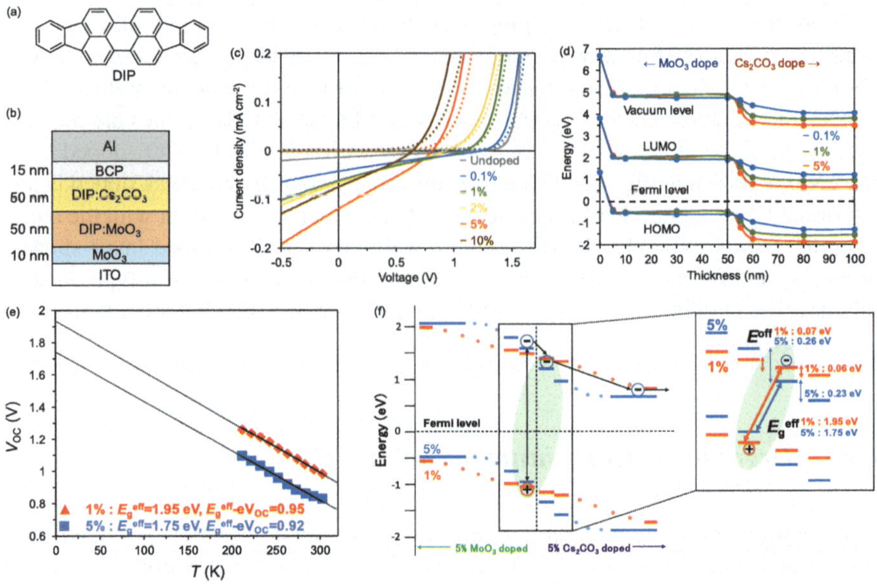

Fig. 8.7 **a** Chemical structure of DIP. **b** Schematic of the *pn* homojunction device with the MoO_3 and Cs_2CO_3 doped layers. **c** *J-V* curves of homojunction devices. **d** Vacuum level, HOMO, and LUMO energy levels relative to E_F in DIP films with different dopant concentrations. **e** Temperature-dependent V_{OC} for the 1% and 5% doped *pn* homojunction devices. **f** Schematic energy diagram and charge recombination mechanisms in organic *pn* homojunction SCs. Reprinted with permission from [28]. Copyright 2019, Elsevier

are 1.75 and 1.95 eV, respectively. On the other hand, the V_{OC} loss induced by the recombination ($E_{CT}/e-V_{OC}$) of these two devices showed values of 0.95 and 0.92 V, respectively. These results indicate that the decrease in V_{OC} due to an increase in dopant concentration is due to a reduction in the effective energy gap at the interface, rather than to the difference in the recombination processes. In the case of inorganic SCs, E_g^{eff} corresponds to the energy of the bandgap of the semiconductor material, because charge recombination occurs from delocalized charges in the conduction and valence bands [31]. In contrast, E_g^{eff} in conventional D/A type OSCs corresponds to the CT state energy, because the localized electrons in the LUMO of the acceptor and holes in the HOMO of the donor recombine [32]. In the case of the organic *pn* homojunction SCs, E_g^{eff} decreased with an increase in built-in-potential. The E_g^{eff} difference between the devices with 1% and 5% doping is 0.20 eV, which is almost equal to the difference between the 1% and 5% cases in energy offset for adjacent molecules at the *pn* interface (Fig. 8.7f). This result indicates that charge recombination in organic *pn* homojunction SCs occurs from localized holes and electrons in the two adjacent molecules. The mechanism is similar to that of the CT state recombination in conventional D/A type OSCs, even though the devices were fabricated using a single host material and a doping technique resembling that for the inorganic *pn* homojunction SCs. These results reveal the important insight that the localized nature of charges in organic semiconductors has a dominant role in the recombination processes and in how V_{OC} is determined.

8.3.4 *Enhancement of Open-Circuit Voltage in Organic Solar Cells by Monolayer Cascade Energy Structure at Donor/Acceptor Interface*

The effect of energy-level alignment at the D/A interface and the strategy to control it by doping were presented in the Sect. 8.3.3. The dominant effect of the energy structure at the D/A interface on E_{CT} and V_{OC} in OSCs is discussed in this section [20].

To tune the energetics at the D/A interface, a new surface modifier based on fullerene bisadducts with a fluoroalkyl chain (F-bisPCBM; Fig. 8.8a) has been synthesized. Fullerenes with fluoroalkyl chains automatically segregate at the surface of the film after spin coating because of its low surface energy [33]; thus, the nanostructure at the D/A interface can be controlled at a monolayer scale if the donor layer is transferred onto it. Further, fullerene bisadducts have a higher-lying LUMO than monoadducts because the π-conjugation is broken [34]. By combining two surface modifiers with different LUMO energy levels (F-PCBM and F-bisPCBM) with two

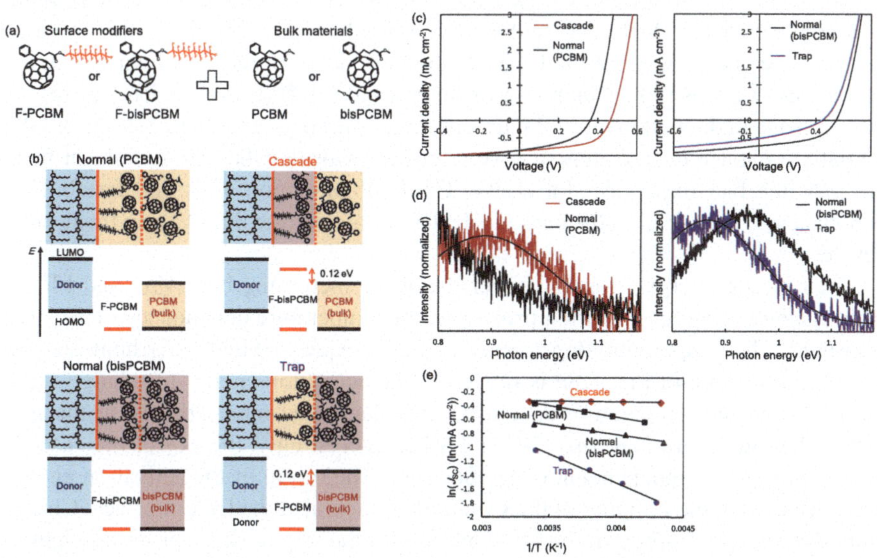

Fig. 8.8 **a** Chemical structures of the surface modifiers and bulk materials. **b** Schematic images of the four different energy landscapes at the D/A interface in bilayer OSCs. **c** *J-V* curves, **d** EL spectra, and **e** temperature dependence of J_{SC} of the four bilayer OSCs. Reprinted with permission from [20]. Copyright 2015, Wiley-VCH

fullerene acceptors (PCBM and bisPCBM), four energy landscapes (cascade, normal with PCBM, normal with bis PCBM, and trap) were constructed at the D/A interface in the bilayer OSCs (Fig. 8.8b). These four devices differ only in the energy level of the monolayer at the D/A interface.

The *J*-*V* characteristics of the four devices with Poly(3-hexylthiophene-2,5-diyl) (P3HT) as the donor layer are depicted in Fig. 8.8c. The cascade device exhibited V_{OC} and FF values that were larger than those of the normal (PCBM) device. The changes in V_{OC} and FF were +0.09 V and +0.13, respectively. In contrast, the trap device exhibited much lower V_{OC}, FF, and J_{SC} values than the normal (bisPCBM) device. The changes in V_{OC} and FF were −0.12 V and −0.18, respectively.

To explain the large V_{OC} changes caused by different energy landscapes at the D/A interface, E_{CT} of the four types of devices was measured by electroluminescence (EL). The EL spectra from the CT state emission of the bilayer devices are shown in Fig. 8.8d. A shift of approximately +0.1 eV in the EL spectra was observed for the cascade device compared to the case of the normal (PCBM) device. In contrast, the peak top for the trap device shifted by approximately −0.1 eV compared to the case of the normal (bisPCBM) device. Shifts in E_{CT} were also observed in the temperature dependence of V_{OC}. The energy changes in E_{CT}, extrapolated from the low-temperature limits for the cascade (+0.09 eV) and trap (−0.12 eV) compared to the corresponding normal devices, generally agreed with those from EL measurements.

The differences in E_{CT} were shown to be the main reason for the change in V_{OC} in the cascade and trap devices. E_{CT} is expressed as the energy-level difference between the HOMO of the donor and the LUMO of the acceptor, and the Coulomb binding is dominated by the electron-hole distance, as explained by Eq. 8.2 in the previous section. The energy level of the LUMO at the first layer of the D/A interface in the cascade device was higher than that of the normal (PCBM) device. The energy of the electrons in the first layer of the acceptor was raised by a higher LUMO, and the spatially separated charge pairs that were further away than the first monolayer from the D/A interface experienced a weaker Coulomb attraction. E_{CT} was destabilized in both cases; thus, V_{OC} in the cascade device became larger than V_{OC} in the normal device.

The cascade energy structure is beneficial for increasing both V_{OC} and FF. To investigate the charge separation processes, the temperature dependence of J_{SC} was measured. The temperature dependence of J_{SC} is expressed by the Arrhenius equation, and the slope in Fig. 8.8e is related to the activation energy for charge separation. When the temperature decreased, J_{SC} decreased more sharply in the trap device than in the normal (bisPCBM) device. Interestingly, the J_{SC} of the cascade device was almost independent of temperature. The origin of the activation energy was attributed to the trapping of the Coulomb attraction at the D/A interface. Charge pairs were deeply trapped at the D/A interface in the trap devices, promoting geminate recombination. In contrast, the temperature independence of J_{SC} in the cascade devices indicated free charge generation at the D/A interface without thermal activation. These mechanisms of promotion and suppression of geminate recombination produced changes in FF. This result suggests that the Coulomb binding in the CT state was weak enough to form free charges in the cascade device.

In conclusion, the monolayer cascade energy landscape weakens the Coulomb binding energy at the interface; thus, destabilization of E_{CT} and charge generation without thermal activation are realized. Therefore, the monolayer cascade energy landscape at the D/A interface is an ideal structure for achieving higher V_{OC} and FF.

8.3.5 Importance of Donor/Acceptor Interfacial Crystallinity to Reduce Open-Circuit Voltage Loss in Organic Solar Cells

The relationship between the energetic structure at the D/A interface and V_{OC} has been discussed in the Sect. 8.3.4. In the following Sects. 8.3.5 and 8.3.6 , strategies to increase V_{OC} by suppressing charge recombination are presented.

Bilayer OSCs have been fabricated using highly crystalline 2,7-dioctyl[1]benzothieno[3,2-*b*][1]benzothiophene (BTBT) and *N*,*N*′-dialkyl-3,4,9,10-perylenedicarboximides (C_n-PTCDI) as the donor and acceptor materials, respectively (Fig. 8.9a) [35]. BTBT and C_8-PTCDI are common materials for organic field-effect transistors, giving very high hole and electron mobility values, 43 cm^2 $V\ s^{-1}$ [36] and 1.3 cm^2 $V\ s^{-1}$ [37], respectively. The influence of crystallinity at the D/A interface on the OSC performance was investigated using three C_n-PTCDI analogs with different alkyl side-chain lengths (C_3, C_6, and C_8). The PTCDI derivatives with longer side-chain lengths have higher crystallinity.

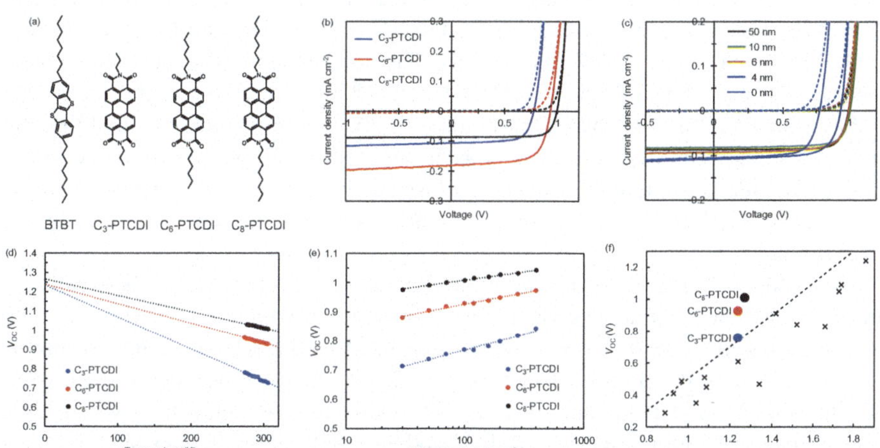

Fig. 8.9 **a** Chemical structures of the donor and acceptor materials. *J*-*V* curves of **b** three bilayer OSCs with C_3-PTCDI, C_6-PTCDI, or C_8-PTCDI as the acceptor, and **c** the trilayer (BTBT/C_8-PTCDI/C_3-PTCDI) devices. **d** Temperature dependence and **e** light-intensity dependence of V_{OC} for OSCs. **f** V_{OC} under 1 sun illumination of OSCs as a function of E_{CT}. Reprinted with permission from [35]. Copyright 2019, AIP Publishing

The *J*-*V* characteristics of the devices under dark conditions and on irradiation with simulated solar light are presented in Fig. 8.9b. The V_{OC} was found to increase as the side-chain length of C_n-PTCDI increased, with values of 0.76 V, 0.93 V, and 1.01 V for C_3-PTCDI, C_6-PTCDI, and C_8-PTCDI, respectively. The FF of the BTBT/C_8-PTCDI device was 0.79, which is higher than those of the C_3-PTCDI and C_6-PTCDI devices, and is the highest FF value reported so far for OSCs fabricated by thermal evaporation [38].

To elucidate the effect on device performance of crystallinity near the D/A interface and in the bulk, trilayer devices were fabricated with the active layer composed of BTBT/C_8-PTCDI/C_3-PTCDI. The D/A interface of the trilayer devices consisted of BTBT and thin C_8-PTCDI layers, with the majority of the layer, apart from the interface, made up of the less crystalline C_3-PTCDI. The trilayer devices based on 10 and 6 nm C_8-PTCDI layers exhibited almost the same V_{OC} values, 1.01 and 0.99 V, as the BTBT/C_8-PTCDI bilayer device, whereas the device with the 4 nm C_8-PTCDI layer exhibited a decreased V_{OC} of 0.94 V, as shown in Fig. 8.9c. The *J*-*V* characteristics of the devices with the C_8-PTCDI layer thicker than 6 nm were almost the same, indicating that the device parameters were not affected by the less crystalline C_3-PTCDI layer except at the D/A interface. This result suggests that the performance of the BTBT/C_8-PTCDI bilayer device was determined by the region within 6 nm (i.e., three molecular layers of C_8-PTCDI) of the D/A interface, and that highly ordered donor and acceptor materials in this region are very important for obtaining high V_{OC} and FF.

To quantitatively evaluate the energy loss in the output voltage, E_{CT} was determined from measurements of the temperature dependence of V_{OC} (Fig. 8.9d). E_{CT}/e can be identified with the intercept of the linear plot of V_{OC} versus temperature at $T = 0$ (Eq. 8.6). The E_{CT} values of C_3-PTCDI, C_6-PTCDI, and C_8-PTCDI were 1.24 eV, 1.24 eV, and 1.27 eV, respectively. The difference in eV_{OC} at room temperature between C_3-PTCDI and C_8-PTCDI was 0.25 eV; in contrast, the difference in E_{CT} (i.e., eV_{OC} at 0 K) was very small, 0.03 eV. These results indicated that the energy loss between E_{CT} and eV_{OC} was strongly affected by the difference in the C_n-PTCDI side-chain length. The energy loss exhibited the smallest value, 0.26 eV, in the BTBT/C_8-PTCDI device, which contained the D/A interface with the highest crystallinity.

According to Eq. 8.6, the origin of the large energy-level difference between E_{CT} and eV_{OC} for the three devices is a change in either *n*, which is related to the type of charge recombination, or J_{00}, which is related to the rate of charge recombination. To clarify the dominant factors, the ideality factor *n* was determined from the light-intensity dependence of V_{OC}. The value of *n* is equal to 1 when ideal band-to-band recombination occurs [15]. When additional pathways such as trap-assisted recombination are also involved, V_{OC} displays a stronger dependence on light intensity and *n* is greater than 1. Figure 8.9e shows the light-intensity dependence of V_{OC} for the BTBT/C_n-PTCDI devices. The values of *n* for C_3-PTCDI, C_6-PTCDI, and C_8-PTCDI were 1.90, 1.37, and 1.00, respectively. As the side-chain length of C_n-PTCDI increased, *n* decreased, and the BTBT/C_8-PTCDI device was an ideal diode. Only band-to-band recombination occurred in the BTBT/C_8-PTCDI device; in contrast, trap-assisted recombination took place in the BTBT/C_3-PTCDI and BTBT/C_6-PTCDI devices. The values of J_{00} can be calculated from the measured n and the slope of the temperature dependence of V_{OC}. The calculated J_{00} values were not very different for the three devices, indicating that the origin of the decreased V_{OC} loss in the devices using C_n-PTCDI with longer side chains was the smaller n.

Figure 8.9f shows a plot of the V_{OC} values under 1 sun illumination versus the E_{CT} values for various kinds of OSCs reported in the literature, as well as the values for the BTBT/C_n-PTCDI devices. The reported V_{OC} values for the OSCs were smaller than $E_{CT}/e - 0.5$ V. The V_{OC} loss for the BTBT/C_8-PTCDI device was very small, less than 0.3 V, because trap-assisted recombination was greatly suppressed and ideal band-to-band recombination was realized, as mentioned above in the quantitative analysis of V_{OC} loss. The origin of the trap states in organic semiconductor films is under debate, but a possible candidate is a tailing in the density of states induced by the thermal motion of the molecules [26]. The D/A materials used in this work have high charge mobility (BTBT: 43 $cm^2\ V\ s^{-1}$, C_8-PTCDI: 1.3 $cm^2\ V\ s^{-1}$). Recently, the octyl side chains of BTBT have been reported to assist in molecular packing and to suppress molecular vibration of the π-conjugated cores, which are the origin of the high hole mobility of BTBT [39]. The origin of the extraordinarily small V_{OC} loss in the BTBT/C_8-PTCDI device could be explained by analogy with the origin of the high hole mobility of BTBT: the highly ordered nature of the molecules at the D/A interface suppressed the thermal motion of the π-conjugated cores and reduced the energetic disorder near the D/A interface.

To summarize this section, the model OSC device with high mobility and highly crystalline donor and acceptor materials was able to reduce the V_{OC} loss to less than 0.3 V and to attain an FF of approximately 0.8. The high crystallinity of the several molecular layers near the D/A interface is important for reducing the energy loss in the output voltage and realizing ideal band-to-band recombination.

8.3.6 Controlling Energy Level of Acceptor Dye Molecule to Reduce Open-Circuit Voltage Loss in Organic Solar Cells

Tetra-substituted perylenediimide (4SubPDI) with four electron-donating groups as substituents was synthesized with the aim of developing a novel acceptor molecule (Fig. 8.10a) [40]. The molecule has a higher-lying LUMO level (~ 0.1 eV) than the non-substituted reference PDI molecule (RefPDI). The effect of the LUMO difference on the V_{OC} loss was investigated.

Figure 8.10b depicts the J -V curves of mixed bulk heterojunction OSC devices with PTB7-Th as the donor and 4SubPDI or RefPDI as the acceptor. The devices with 4SubPDI showed a very high V_{OC}, reaching 1.0 V, comparable to the highest value ever found for PTB7-Th based devices [41]. The optical HOMO-LUMO gap (E_g^{opt}) of the device was estimated to be 1.57 eV from the absorption spectral edges

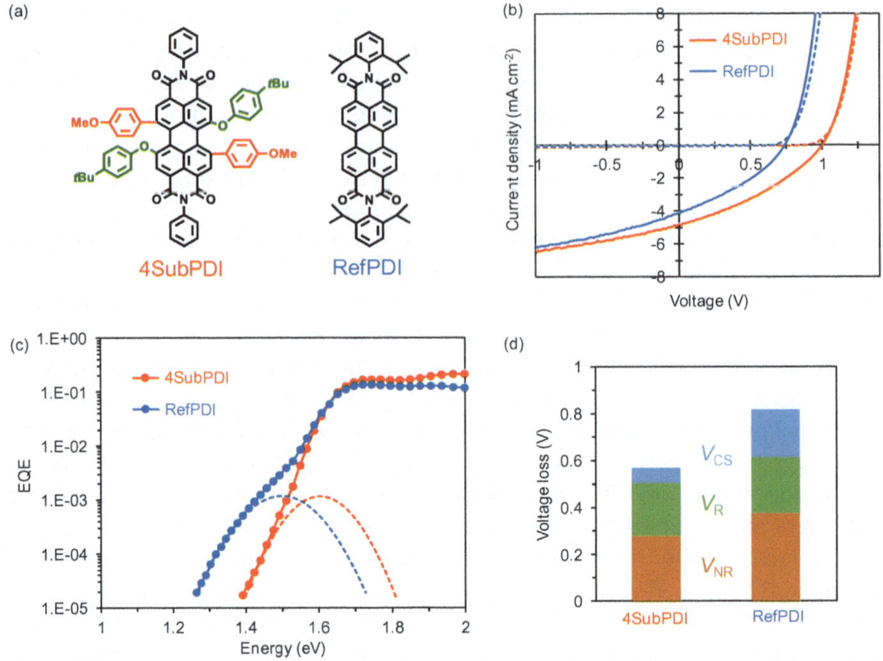

Fig. 8.10 **a** Chemical structures of 4SubPDI and RefPDI. **b** *J-V* curves. **c** High resolution EQE spectra. **d** Breakdown of open-circuit voltage loss. Reprinted with permission from [40]. Copyright 2020, Wiley-VCH

of PTB7-Th. The voltage loss (V_{loss}), calculated from the equation $V_{loss} = E_g^{opt}/e$-V_{OC}, was found to have the remarkably small value of 0.57 V. On the other hand, the devices with RefPDI gave a much smaller V_{OC} of 0.75 V, which produced a large V_{loss} of 0.82 V.

V_{loss} is divided into contributions from charge separation (V_{CS}), radiative recombination (V_r), and non-radiative recombination (V_{nr}), represented in the equation $V_{loss} = V_{CS} + V_r + V_{nr}$. Generally, V_{CS} is closely related to the LUMO-LUMO offset between the donor and the acceptor. V_r is the inevitable voltage loss in all types of photovoltaic devices. Importantly, OSCs usually have a large V_{nr} compared to inorganic SCs, and the large V_{nr} is the primary reason for the small PCE [14]. To further investigate the small V_{loss} in the device based on 4SubPDI, high resolution EQE spectra were measured to observe the CT state absorption, shown in Fig. 8.10c. The device based on RefPDI showed a clear shoulder in the CT state absorption at an energy of ca. 1.5 eV, whereas the absorption edge of the CT state and S_1 state of PTB7-Th could not be distinguished in the device based on4SubPDI. E_{CT} was determined by Gaussian fitting of the spectra. From the equation $V_{CS} = (E_g^{opt} - E_{CT})/e$, the V_{CS} in the PTB7-Th/4SubPDI device was determined to be only 0.06 V and was substantially smaller than that of RefPDI (0.20 V) owing to the elevated LUMO level. Efficient charge separation could be achieved even with a negligibly small LUMO-LUMO offset between the donor and the acceptor. V_r and V_{nr} were also quantified using Eq. 8.9 and are shown in the bar graph (Fig. 8.10d). The device based on 4SubPDI showed a V_{nr} value of 0.28 V, significantly smaller than that of RefPDI (0.38 V), while both devices exhibited similar V_r values. The reason for the suppression of V_{nr} was the enhancement of radiative efficiency from the CT state. This was realized by an intensity borrowing mechanism between CT state emission and S_1 emission by hybridizing these states [42] by getting E_{CT} close to E_g^{opt} as a result of the bay-functionalization, which increases the LUMO of PDI.

8.4 Conclusion and Future Prospects

In this chapter, we began by describing the theoretical background of V_{OC} in OSCs, including the equivalent circuit model and a detailed balance of thermodynamics in the devices. The important outcome of these theories is that they reveal the factors determining V_{OC} in OSCs, namely the energetics, types, and rates of charge recombination, both radiative and non-radiative. These processes occur at the D/A interface and are thus closely related to each other. Our recent study on the relationship between the D/A interfacial structure and V_{OC}, and how V_{OC} in OSCs can be increased by modifying the D/A interface, was then presented. A brief summary of our results is as follows. First, energy-level alignment is very important to determine the energetics at the D/A interface and the value of V_{OC}. V_{OC} could change even though the same D/A materials are used, because of the vacuum-level shift near the D/A interface induced by E_F alignment. E_F can be controlled by doping, resulting in the control of V_{OC} and successful realization of *pn* homojunction OSCs. In particular, doping near the D/A interface (within less than 10 nm) has a dominant effect on the device performance. In the Sect. 8.3.4, the energy structure difference of the monolayer at the D/A interface was investigated. The research revealed that a monolayer cascade can weaken the Coulomb binding energy at the interface; in this way, both reduction in E_{CT} and thermal-activation-free charge generation are realized. Thus, a monolayer cascade is an ideal structure for achieving high V_{OC} and FF. The suppression of charge recombination was presented in the Sects. 8.3.5 and 8.3.6. A model OSC device with high mobility and highly crystalline donor and acceptor materials was demonstrated and was able to reduce the V_{OC} loss to less than 0.3 V and attain an FF of almost 0.8. The results revealed that high crystallinity in several molecular layers near the D/A interface is important for reducing the energy loss in the output voltage and realizing ideal band-to-band recombination. In the Sect. 8.3.6, the use of bay-functionalization of PDI for lifting the LUMO of the acceptor molecule was described. The energy-level control available leads to an E_{CT} close to E_g^{opt}, leading to a reduction in energy loss for charge separation, and to the suppression of non-radiative recombination as a result of hybridization of the CT and S_1 states.

These results reveal that how we can maximize V_{OC} in OSCs. First, the CT state should be destabilized by weakening the Coulomb binding energy at the D/A interface. Second, E_{CT} should be close to the E_g^{opt} of the smaller D/A bandgap material, thereby reducing the LUMO-LUMO or HOMO-HOMO offset between the donor and the acceptor. Third, crystallinity should be increased to reduce the trap state near the D/A interface. In Sect. 8.3.6, it was shown that the energy difference between E_g^{opt} and E_{CT} could be reduced to 0.06 eV, and in Sect. 8.3.5 that the energy difference between E_{CT} and eV_{OC} could be reduced to 0.26 eV. By combining these two effects, a very small V_{OC} loss of less than 0.3 eV, similar to that in inorganic SCs, can be achieved. Our research has revealed that controlling the D/A interfacial structure can reduce V_{OC} loss in OSCs, and we expect that this research will provide a route to achieving high power conversion efficiencies in OSCs, close to those in the efficient inorganic SCs.

Acknowledgments The research works presented in this chapter were conducted with by Prof. Hiramoto, Dr. Shintaku, Dr. Kikuchi, Mr. Lee, and Mr. Perrot in Institute for Molecular Science, and Dr. Tajima and Dr. Nakano in Riken, and Prof. Takahashi and Prof. Fujimoto in Shizuoka University. They were partly supported by the JSPS KAKENHI (Grant-in-Aid for Research Activity Start-up, No. 16H07421 and for Young Scientist, 18K14115), the foundation of Chubu Science and Technology Center and the Mazda foundation.

References

1. Dou, L., You, J., Hong, Z., Xu, Z., Li, G., Street, R.A., Yang, Y.: 25th Anniversary article—a decade of organic/polymeric photovoltaic research. Adv. Mater. **25**, 6642–6671 (2013)
2. Liu, L., Kan, Y., Gao, K., Wang, J., Zhao, M., Chen, H., Zhao, C., Jiu, T., Jen, A.K.Y., Li, Y.: Graphdiyne derivative as multifunctional solid additive in binary organic solar cells with 17.3% efficiency and high reproductivity. Adv. Mater. **32**, 1907604 (2020)
3. Hiramoto, M., Fujiwara, H., Yokoyama, M.: Three-layered organic solar cell with a photoactive interlayer of codeposited pigment. Appl. Phys. Lett. **58**, 1062–1064 (1991)
4. Yu, G., Gao, J., Hummelen, J.C., Wudl, F., Heeger, A.J.: Polymer photovoltaic cells: enhanced efficiencies via a network of internal donor-acceptor heterojunctions. Science **270**, 1789–1791 (1995)
5. Lu, L., Zheng, T., Wu, Q., Schneider, A. M., Zhao, D., Yu, L.: Recent advances in bulk heterojunction polymer solar cells. Chem. Rev. **115**, 12666–12731 (2015)
6. Hou, J., Inganäs, O., Friend, R., Gao, F.: Organic solar cells based on non-fullerene acceptors. Nat. Mater. **17**, 119–128 (2018)
7. National Renewable Energy Laboratory, Best Research-Cell Efficiency Chart (2019)
8. Baran, D., Gasparini, N., Wadsworth, A., Tan, C.H., Wehbe, N., Song, X., Hamid, Z., Zhang, W., Neophytou, M., Kirchartz, T., Brabec, C.J., Durrant, J.R., McCulloch, I.: Robust nonfullerene solar cells approaching unity external quantum efficiency enabled by suppression of geminate recombination. Nat. Commun. **9**, 2059 (2018)
9. Park, S.H., Roy, A., Beaupré, S., Cho, S., Coates, N., Moon, J.S., Moses, D., Leclerc, M., Lee, K., Heeger, A.J.: Bulk heterojunction solar cells with internal quantum efficiency approaching 100%. Nat. Photon. **3**, 297–295 (2009)
10. Clarke, T.M., Durrant, J.R.: Charge photogeneration in organic solar cells. Chem. Rev. **110**, 6736–6767 (2010)
11. Knupfer, M.: Exciton binding energies in organic semiconductors. Appl. Phys. A **77**, 623–626 (2003)
12. Kroeze, J.E., Savenije, T.J., Vermeulen, M.J.W., Warman, J.M.: Contactless determination of the photoconductivity action spectrum, exciton diffusion length, and charge separation efficiency in polythiophene-sensitized TiO_2 bilayers. J. Phys. Chem. B **107**, 7696–7705 (2003)

13. Vandewal, K., Himmelberger, S., Salleo, A.: Structural factors that affect the performance of organic bulk heterojunction solar cells. Macromolecules **46**, 6379–6387 (2013)
14. Benduhn, J., Tvingstedt, K., Piersimoni, F., Ullbrich, S., Fan, Y., Tropiano, M., McGarry, K.A., Zeika, O., Riede, M.K., Douglas, C.J., Barlow, S., Marder, S.R., Neher, D., Spoltore, D., Vandewal, K.: Intrinsic non-radiative voltage losses in fullerene-based organic solar cells. Nat. Energy **2**, 17053 (2017)
15. Shintaku, N., Hiramoto, M., Izawa, S.: Effect of trap-assisted recombination on open-circuit voltage loss in phthalocyanine/fullerene solar cells. Org. Electron. **55**, 69–74 (2018)
16. Scharber, M.C., Wuhlbacher, D., Koppe, M., Denk, P., Waldauf, C., Heeger, A.J., Brabec, C.L.: Design rules for donors in bulk-heterojunction solar cells—towards 10% energy-conversion efficiency. Adv. Mater. **18**, 789–794 (2006)
17. Yoshida, H.: Near-ultraviolet inverse photoemission spectroscopy using ultra-low energy electrons. Chem. Phys. Lett. **539–540**, 180–185 (2012)
18. Izawa, S., Shintaku, N., Hiramoto, M.: Effect of band bending and energy level alignment at the donor/acceptor interface on open-circuit voltage in organic solar cells. J. Phys. Chem. Lett. **9**, 2914–2918 (2018)
19. Street, R.A., Hawks, S.A., Khlyabich, P.P., Li, G., Schwartz, B.J., Thompson, B.C., Yang, Y.: Electronic structure and transition energies in polymer-fullerene bulk heterojunctions. J. Phys. Chem. C **118**, 21873–21883 (2014)
20. Izawa, S., Nakano, K., Suzuki, K., Hashimoto, K., Tajima, K.: Dominant effects of first monolayer energetics at donor/acceptor interfaces on organic photovoltaics. Adv. Mater. **27**, 3025–3031 (2015)
21. Sze, S.M.: Physics of semiconductor devices (2005)
22. Zhong, Y., Tada, A., Izawa, S., Hashimoto, K., Tajima, K.: Enhancement of V_{OC} without loss of J_{SC} in organic solar cells by modification of donor/acceptor interfaces. Adv. Energy Mater. **4**, 1301332 (2014)
23. Vandewal, K., Tvingstedt, K., Gadisa, A., Inganas, O., Manca, J.V.: On the origin of the open-circuit voltage of polymer–fullerene solar cells. Nature Mater. **8**, 904–909 (2009)
24. Vandewal, K., Tvingstedt, K., Gadisa, A., Inganas, O., Manca, J.V.: Relating the open-circuit voltage to interface molecular properties of donor:acceptor bulk heterojunction solar cells. Phys. Rev. B **81**, 125204 (2010)
25. Shockley, W., Queisser, H.J.: Detailed balance limit of efficiency of pn junction solar cells. J. Appl. Phys. **32**, 510–519 (1961)
26. Yang, J.-P., Bussolotti, F., Kera, S., Ueno, N.: Origin and role of gap states in organic semiconductor studied by UPS: as the nature of organic molecular crystals. J. Phys. D Appl. Phys. **50**, 423002 (2017)
27. Shintaku, N., Hiramoto, M., Izawa, S.: Doping for controlling open-circuit voltage in organic solar cells. J. Phys. Chem. C **122**, 5248–5253 (2018)
28. Izawa, S., Perrot, A., Lee, J., Hiramoto, M.: Organic pnhomojunction solar cell. Org. Electron. **71**, 45–49 (2019)
29. Horlet, M., Kraus, M., Brütting, W., Opitz, A.: Diindenoperylene as ambipolar semiconductor:influence of electrode materials and mobility asymmetry in organicfield-effect transistors. Appl. Phys. Lett. **98**, 233304 (2011)
30. Lee, J., Perrot, A., Hiramoto, M., Izawa, S.: Photoconversion mechanism at the pn-HomojunctionInterface in single organic semiconductor. Materials **13**, 1727 (2020)
31. Hegedus, S.S., Shafarman, W.N.: Thin-film solar cells: device measurementsand analysis. Prog. Photovolt. Res. Appl. **12**, 155–176 (2004)
32. Burke, T.M., Sweetnam, S., Vandewal, K., McGehee, M.D.: Beyond Langevin recombination: How equilibrium between free carriers and charge transfer states determines the open-circuit voltage of organic solar Cells. Adv. Energy Mater. **5**, 1500123 (2015)
33. Wei, Q., Tajima, K., Tong, Y., Ye, S., Hashimoto, K.: Surface-segregated monolayers: a new type of ordered monolayer for surface modification of organic semiconductors. J. Am. Chem. Soc. **131**, 17597 (2009)

34. Lenes, M., Wetzelaer, G.-J.A.H., Kooistra, F.B., Veenstra, S.C., Hummelen, J.C., Blom, P.W.M.: Fullerene bisadducts for enhanced open-circuit voltages and efficiencies in polymer solar cells. Adv. Mater. **20**, 2116 (2008)
35. Izawa, S., Shintaku, N., Kikuchi, M., Hiramoto, M.: Importance of interfacial crystallinity to reduce open-circuit voltage loss in organic solar cells. Appl. Phys. Lett. **115**, 153301 (2019)
36. Yuan, Y., Giri, G., Ayzner, A.L., Zoombelt, A.P., Mannsfeld, S.C., Chen, J., Nordlund, D., Toney, M.F., Huang, J., Bao, Z.: Ultra-high mobility transparent organic thin film transistors grown by an off-centre spin-coating method. Nat. Commun. **5**, 3005 (2014)
37. Chesterfield, R.J., McKeen, J.C., Newman, C.R., Ewbank, P.C., da Silva, D.A., Bredas, J.L., Miller, L.L., Mann, K.R., Frisbie, C.D.: Organic thin film transistors based on N-alkyl perylenediimides: Charge transport kinetics as a function of gate voltage and temperature. J. Phys. Chem. B **108**, 19281–19292 (2004)
38. Wagner, J., Gruber, M., Hinderhofer, A., Wilke, A., Bröker, B., Frisch, J., Amsalem, P., Vollmer, A., Opitz, A., Koch, N., Schreiber, F., Brütting, W.: High fill factor and open circuit voltage in organic photovoltaic cells with Diindenoperylene as donor material. Adv. Funct. Mater. **20**, 4295–4303 (2010)
39. Illig, S., Eggeman, A., Troisi, A., Jiang, L., Warwick, C., Nikolka, M., Schweicher, G., Yeates, S.G., Geerts, Y.H., Anthony, J.E., Sirringhaus, H.: Reducing dynamic disorder in small-molecule organic semiconductors by suppressing large-amplitude thermal motions. Nat. Commun. **7**, 10736 (2016)
40. Fujimoto, K., Izawa, S., Arikai, Y., Sugimoto, S., Oue, H., Inuzuka, T., Uemura, N., Sakamoto, M., Hiramoto, M., Takahashi, M.: Regioselective bay-functionalization of perylenes toward tailor-made synthesis of acceptor materials for organic photovoltaics. ChemPlusChem **85**, 285–293 (2020)
41. Qian, D., Zheng, Z., Yao, H., Tress, W., Hopper, T.R., Chen, S., Li, S., Liu, J., Chen, S., Zhang, J., Liu, X.-K., Gao, B., Ouyang, L., Jin, Y., Pozina, G., Buyanova, I.A., Chen, W.M., Inganäs, O., Coropceanu, V., Bredas, J.-L., Yan, H., Hou, J., Zhang, F., Bakulin, A.A., Gao, F.: Design rules for minimizing voltage losses in high-efficiency organic solar cells. Nature Mater. **17**, 703–709 (2018)
42. Eisner, F.D., Azzouzi, M., Fei, Z., Hou, X., Anthopoulos, T.D., John, T., Dennis, S., Heeney, M., Nelson, J.: Hybridization of local exciton and charge-transfer states reduces nonradiative voltage losses in organic solar cells. J. Am. Chem. Soc. **141**, 6362–6374 (2019)

Chapter 9
Parts-Per-Million-Level Doping Effects and Organic Solar Cells Having Doping-Based Junctions

Masahiro Hiramoto

9.1 Introduction

9.1.1 Motivation

Controlling the *p*- and *n*-type behavior, i.e., whether holes or electrons are responsible for electric conduction, respectively, by adding an extremely small quantity of impurity (doping) has been a central part of inorganic semiconductor electronics since the 20th century, such as silicon integrated circuits, solar cells, and light-emitting diodes. However, the genuine potential of organic semiconductors (OSCs) has been obscured for a long time by external contaminants such as oxygen. We strongly believe that the *p*- and *n*-type behavior of OSCs should be controlled by parts-per-million (ppm)-level doping, similar to inorganic semiconductors such as silicon. We expect that veiled physical properties unique to OSCs would be found during such studies to achieve conduction-type control of OSCs by ppm doping.

We propose that the points described below are essential to developing the principles underlying the doping of OSCs.

(i) A ppm-doping strategy should be performed on sub-ppm purified OSCs together with the total removal of oxygen from the air, which acts as an external dopant.
(ii) Perfect *pn*-control, namely, any single or blended OSC should exhibit either *n* or *p*-type behavior only by impurity doping.
(iii) To precisely clarify the nature of the doping effects, ppm doping in the bulk of OSC single crystals with few grain boundaries should be performed.

M. Hiramoto (✉)
National Institutes of Natural Sciences, Institute for Molecular Science, 5-1 Higashiyama, Myodaiji, Okazaki 444-8787, Aichi, Japan
e-mail: hiramoto@ims.ac.jp

M. Hiramoto and S. Izawa (eds.), *Organic Solar Cells*,
https://doi.org/10.1007/978-981-15-9113-6_9

9.1.2 History

Organic semiconductors were previously thought to be insensitive to doping. Now, researchers recognize that external impurities due to imperfect refinement and contaminants from the ambient atmosphere have hidden the true properties of OSCs for a long time. A representative contaminant is oxygen. Because oxygen acts as an acceptor and is doped in various kinds of OSCs, such as phthalocyanines, these OSCs always behave as *p*-type semiconductors. Completely eliminating oxygen under ultra-high vacuum throughout film fabrication and measurements has shown that phthalocyanines are essentially *n*-type [1, 2]. A few kinds of OSCs, such as perylene pigments, are not influenced by oxygen and behave as *n*-type semiconductors [3].

Though impurity doping in OSCs has a long history, very few dopants were previously available. Halogen vapors such as I_2 and Br_2 have been typical acceptor dopants [4, 5]. A perylene pigment was converted from *n*- to *p*-type by Br_2 doping, and *pn*-homojunction behaviors were observed by forming a junction between *n*- and *p*-type regions [6, 7]. Moreover, available donor dopants were limited to alkaline metals such as Na and Ca, which are easily deteriorated by oxidation in the ambient air.

In the last ten years, however, because of the research outcomes on organic electroluminescence, many novel dopants have been developed. Organic acceptor dopants such as F_4-TCNQ [8, 9], F6-TCNNQ [10, 11], HAT-CN [12], and $F_{36}C_{60}$ [13] and inorganic acceptor dopants such as MoO_3, V_2O_5, and Fe_2Cl_6 [14, 15] have been found. Ru- [16–18], Co- [19], and W-complexes [10, 20] and Cs_2CO_3 [21, 22] have been discovered to be relatively stable donors in air.

Simultaneously, important physics were reported, for example, the realization of a *pin*-homojunction fabricated by connecting *p*- and *n*-doped zinc phthalocyanines (ZnPc) [16] and *p*- and *n*-type pentacenes [18] and the formation of an ohmic junction for carrier injection [15].

Recently, even an extremely small amount of dopant reaching the ppm level was revealed to affect the electrical characteristics of OSC films [10, 23, 24] and organic single crystals [25]. Some papers reported significantly higher doping efficiency [10, 25, 26]. Such recent progress on doping strongly suggests that the field of organic electronics is now shifting from conventional intrinsic organic semiconductor electronics to doped organic semiconductor electronics [27–30]. Today, organic semiconductors are very sensitive to doping.

9.2 Principles

In this section, the processes behind doping, including charge transfer and ionization to create carriers, are described.

9.2.1 Charge Transfer

In the case of inorganic semiconductors such as silicon, the behavior of dopants is determined by their valence. For *p*-type doping, a boron atom, which has three valence electrons, acts as an acceptor in a crystal of Si atoms, which have four valence electrons. A hole near the negatively ionized boron atom (B^-) can be generated with a significantly small activation energy of 45 meV [31]. The holes are released thermally, and the ionization rate at room temperature (RT) can reach 100%.

In the case of OSCs, the behavior of dopants is not determined by their valence, as solid-state molecules are only bound by van der Waals forces, and no chemical bonds exist. Doping is caused by the charge transfer between the dopant molecule and the OSC molecule. As shown in Fig. 9.1a (left), the energy of the highest occupied

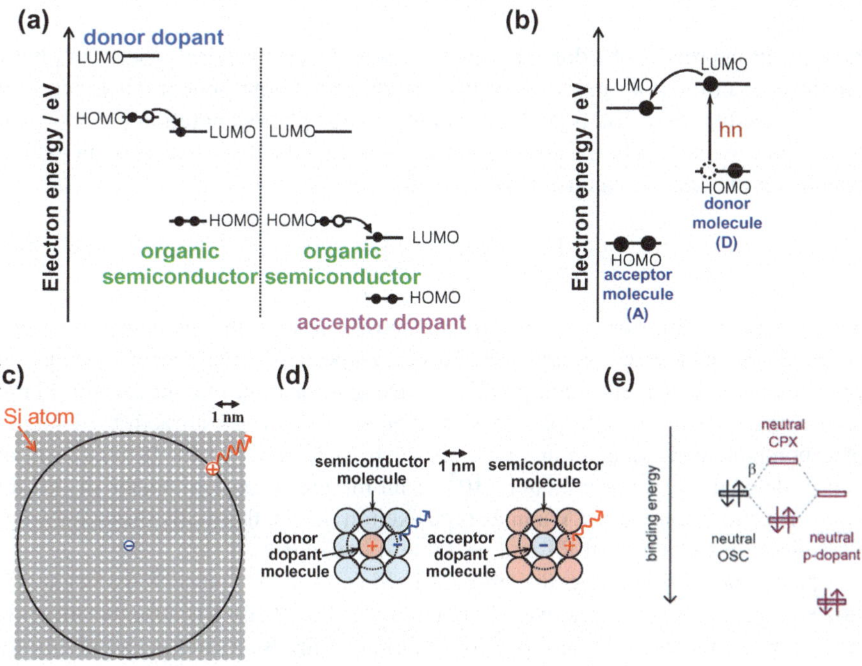

Fig. 9.1 **a** Energetic relationships between an organic semiconductor molecule and donor dopant molecule (left) or acceptor dopant molecule (right). **b** Energetic relationship between donor and acceptor molecules for D/A sensitization to generate photocarriers. Photoexcitation of the donor molecules is shown. **c** A positive charge bound loosely around the negative charge on Si. Such an e^--h^+ pair resembles a Wannier exciton and a loosely bound hole around a negatively ionized acceptor dopant atom (B^-). **d** CT state between donor molecules (pink circles) and an acceptor molecule (gray circle). This CT state can be considered a CT exciton and a weakly bound electron around a positively ionized donor dopant molecule (left) and vice versa (right). The solid circle represents the size of the C_{60} molecule. Parts (c) and (d) have the same scale. **e** Formation of ground-state charge transfer complex. Reproduced with permission from M. Hiramoto et al., *Adv. Mater.*, Copyright 2018 John Wiley and Sons

molecular orbital (HOMO) of a donor dopant should be less than that of the lowest unoccupied molecular orbital (LUMO) of the OSC molecule. Upon electron transfer (ET) from the donor molecule to the OSC molecule, a charge transfer (CT) state is formed (Fig. 9.1d, left). The negative electric charge on the OSC molecule is thermally released at RT. Consequently, the OSC shows *n*-type behavior. In contrast, as shown in Fig. 9.1a (right), the energy of the LUMO of an acceptor dopant molecule should be higher than the energy of the HOMO of the host OSC molecule. Upon ET from the OSC molecule to the acceptor molecule, a CT state is produced (Fig. 9.1d, right). The positive electric charge on the OSC molecule is released thermally at RT. Consequently, the OSC shows *p*-type behavior.

9.2.2 Ionization

The ionization process of a dopant is analogous to the dissociation process of photogenerated electron–hole pairs (excitons). Here, the charge generation process by doping is analogously discussed based on the charge photogeneration process in an organic photovoltaic cell. Coulomb's law dictates that the dissociation of an exciton depends on the dielectric constant of the solid (ε):

$$F = (1/4p\varepsilon\varepsilon_0)\left(q_1 q_2/r^2\right) \tag{9.1}$$

where ε_0, q_1, q_2, and r are the absolute dielectric constant, the elementary charges, and the distance between elementary charges, respectively. For a small ε value, the opposite charge pair feels a comparatively strong attraction. In contrast, for a large ε value, the opposite charge pair feels a comparatively weak attraction. A large ε corresponds to inorganic semiconductors such as Si (ε = 11.9). In this case, an exciton delocalizing approximately 10^4 Si atoms has a large diameter of 9.0 nm (Fig. 9.1c) and is a so-called Wannier-type exciton, which thermally dissociates into a free electron and a hole at RT and generates photocarriers [31].

A pair of positive and negative charges in Si can also be considered a negatively ionized dopant atom (B^-) and weakly bound hole (Fig. 9.1c). Because the attractive force between the negative and positive charges in the Si crystal is essentially the same, a hole weakly bound to the negatively ionized dopant atom can thermally dissociate into a free hole at RT, thereby creating the majority carrier. Unlike an exciton, the negative charge of the ionized acceptor atom (B^-) cannot move in the crystal lattice of Si.

In contrast, a small ε corresponds to OSCs such as fullerenes (C_{60}) (ε = 4.4) [32]. In this case, an exciton localizing on one molecule has a small diameter of 0.50 nm (not shown) and is a so-called Frenkel-type exciton, which hardly dissociates thermally into a free electron and hole at RT and generates few photocarriers.

To dissociate Frenkel excitons, donor/acceptor (D/A) sensitization is utilized. Electron-donating (D) and electron-accepting (A) molecules are blended (Fig. 9.3, lower right). The energy positions of their HOMO and LUMO are shifted in parallel with each other. The photoexcitation of the donor molecule (Fig. 9.1b) and that of the acceptor molecule (not shown) separate the positive and negative charges on the adjacent donor and acceptor molecules and create a CT exciton (Fig. 9.1d) via photoinduced ET. Thus, the CT exciton can dissociate thermally into a free electron and a hole at RT. By utilizing D/A sensitization, OSCs can dissociate excitons and generate photocurrent.

A pair of positive and negative charges separated to adjacent molecules (Fig. 9.1d) can be considered a positively ionized donor dopant molecule and a negatively charged neighboring semiconductor molecule (Fig. 9.1d, left) or as a negatively ionized acceptor dopant molecule and a positively charged neighboring semiconductor molecule (Fig. 9.1d, right). Because the attractive force between negative and positive charges in an organic solid is similar, a hole comparatively weakly bound to a negatively ionized dopant molecule and vice versa can thermally dissociate into a free hole at RT and become the majority carrier. In contrast with a CT exciton, the charge of an ionized dopant molecule is spatially fixed in the molecular solid and cannot move.

Since the distance between the orbital of the bound charge and the dopant ion (Fig. 9.1d) is shorter than that in Si (Fig. 9.1c), the interaction between the dopant molecule and semiconductor molecule may significantly influence the ionization process. By forming the ground-state charge transfer complex by hybridization (Fig. 9.1e), the ionization rate is expected to be lower than that of the simple ion pair formation (Fig. 9.1a) [33, 34].

9.3 ppm-Level Doping Method

The keys to ppm-level doping are rigorous purification and an extremely low deposition speed of 10^{-9} nm s^{-1}.

9.3.1 Purification

To clarify the effects of an extremely small doping amount reaching 1 ppm, the organic semiconductor samples must be thoroughly purified to be free from residual impurities, and external impurities from air must be avoided. Therefore, the air exposure of organic films has been completely avoided by using an evaporation chamber built in a glove-box filled with circulating nitrogen gas ($O_2 < 0.2$ ppm, $H_2O < 0.5$ ppm) throughout depositions and measurements. The observed values of the Fermi level (E_F) of organic films were changed irreversibly upon air exposure, and reproducible and accurate values could not be obtained.

The organic semiconductors were purified through the formation of single crystals by using train sublimation under nitrogen convection conditions at a pressure between

1 and 0.1 atm (Fig. 9.2a) [35]. C_{60} single crystals were confirmed to have at least seven-nines (7 N) purity (99.99999%) by secondary ion mass spectrometry [36]; 7 N purity is equivalent to 0.1 ppm by volume.

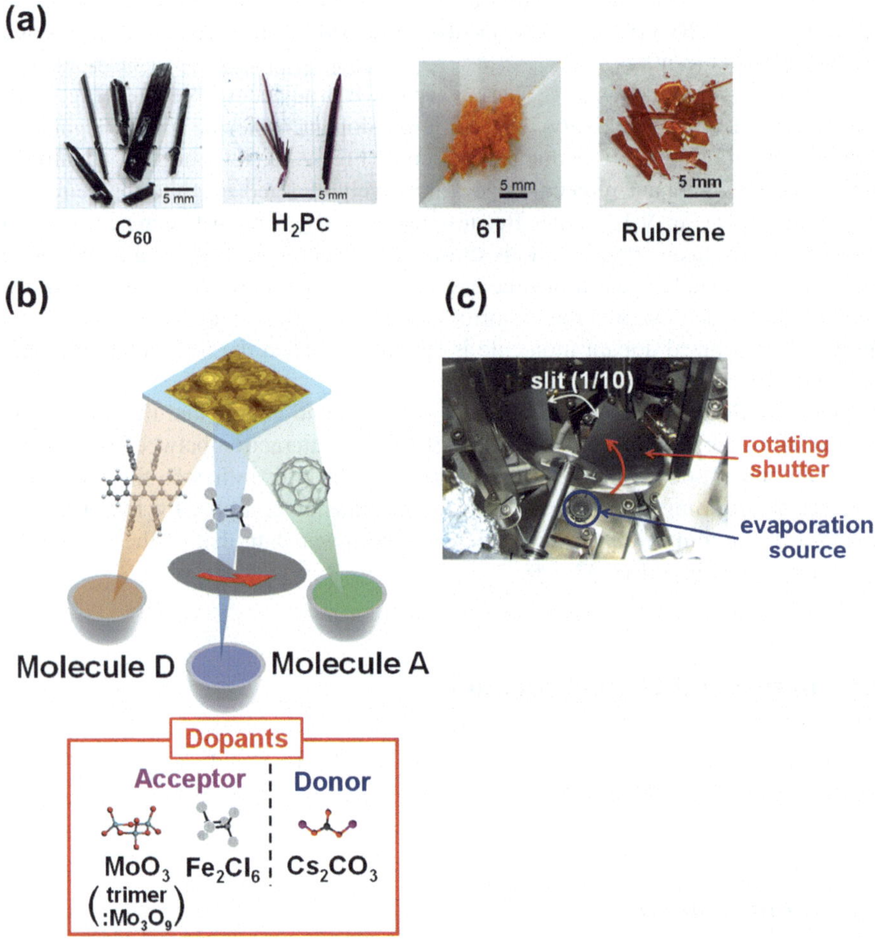

Fig. 9.2 **a** Photographs of single crystals of C_{60}, H_2Pc (metal-free phthalocyanine), 6 T (sexithiophene), and rubrene. **b** Three-component evaporation system with a rotating shutter for dopants. The chemical structures of rubrene (molecule D), C_{60} (molecule A), and dopant molecules (MoO_3, Fe_2Cl_3, and Cs_2CO_3) are also shown. AFM image (5 μm × 5 μm) of a 100 ppm Fe_2Cl_3-doped rubrene single crystal is shown at the substrate position. **c** A photograph of a rotating shutter having a slit with an aperture ratio of 1/10. Reproduced with permission from M. Hiramoto et al., *Adv. Mater.*, Copyright 2018 John Wiley and Sons

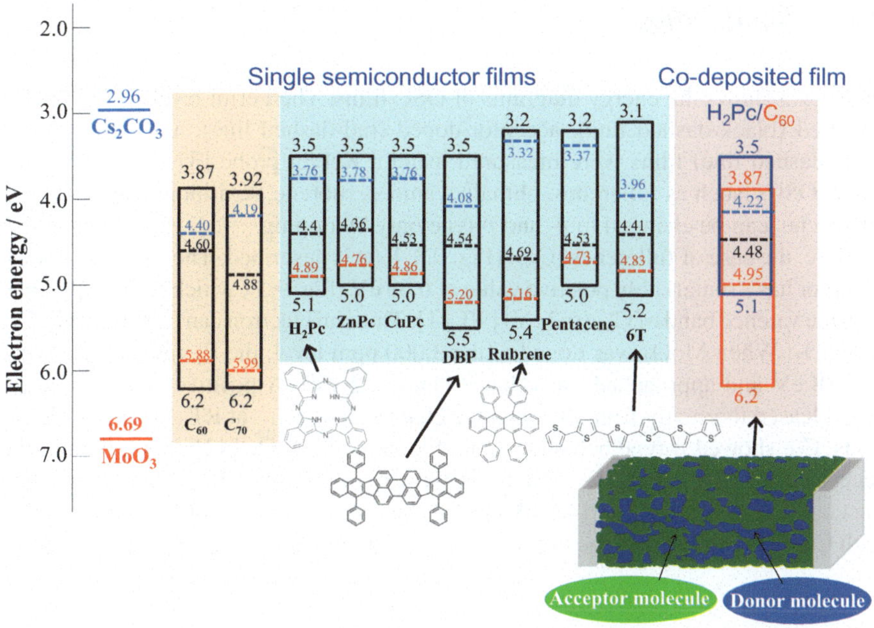

Fig. 9.3 Fermi level (E_F) of the undoped (black-dashed line), MoO_3-doped (3,000 ppm) (red-dashed line), and Cs_2CO_3-doped (3,000 ppm) (blue-dashed line) OSC films. In the case of the H_2Pc:C_{60} blended film, the overlapped diagram of H_2Pc (blue) and C_{60} (red) is shown. Solid red and blue lines on the left side show the work functions of films of MoO_3 and Cs_2CO_3, respectively. Reproduced with permission from M. Hiramoto et al., *Adv. Mater.*, Copyright 2018 John Wiley and Sons

9.3.2 *Ultra-Slow Deposition at 10^{-9} nm s^{-1}*

Single OSC films were doped by the co-evaporation of an OSC and a dopant (Fig. 9.2b). The blended OSC films were doped by the co-evaporation from three evaporation sources for two kinds of OSCs and a dopant. The acceptor dopants molybdenum oxide (MoO_3) and iron chloride (Fe_2Cl_6) and donor dopant cesium carbonate (Cs_2CO_3) were used. The slow deposition rate (1.8×10^{-6} nm s^{-1}) enabled doping of the materials at an extremely low concentration of 9 ppm by volume [36]. To further reduce the evaporation rate to 10^{-9} nm s^{-1}, rotary shutters with apertures having ratios ranging from 1/10 to 1/1,000 were used (Fig. 9.2b, c).

9.4 *pn*-Control by Doping

Complete *pn*-control was demonstrated for the single and blended OSC films. A series of homojunctions was designed and fabricated in the bulk of the OSC films.

9.4.1 Single Films

Figure 9.3 shows the energy diagrams of OSC films. The Fermi levels (E_Fs) for the undoped (black-dashed line), acceptor-doped (red-dashed line), and donor-doped (blue-dashed line) films were measured using a Kelvin probe [37–39]. Generally, single OSCs, such as fullerenes, phthalocyanines, rubrene, sexithiophene (6 T), and pentacene, can be changed to *n*- and *p*-type only by doping.

Here, the case of fullerenes (C_{60}) (Fig. 9.3, left) is described. The E_F of the MoO_3 acceptor has a remarkably positive value of 6.69 eV (Fig. 9.3), which is more positive than the valence band of C_{60} (6.2 eV) [40, 41]. Thus, an electron can transfer from C_{60} to MoO_3. When MoO_3 was doped at the 3,000 ppm level, the E_F moved positively to 5.88 eV and approached the valence band (6.2 eV). A positive charge on C_{60}^+ was released thermally from the negative charge on MoO_3^- at RT (Fig. 9.1d, right). Thus, C_{60} showed *p*-type behavior upon doping with MoO_3 [42]. Upon doping the acceptor (3,000 ppm Cs_2CO_3), the E_F negatively moved to 4.40 eV and approached the conduction band of C_{60} (4.0 eV) [39]. The *n*-type behavior of C_{60} upon doping with Cs_2CO_3 results from the opposite mechanism of MoO_3 doping (Fig. 9.1d, left).

9.4.2 Blended Films

pn-control by doping can be extended to blended films fabricated by co-deposition, which consists of D:A (donor:acceptor) organic semiconductors, such as $H_2Pc:C_{60}$ and 6 T:C_{60}, having the energy relationship shown in Fig. 9.1b. Blended films are indispensable in device applications, especially for organic photovoltaic cells. The organic blended junction (Fig. 9.3, lower right) was proposed by Hiramoto in 1991 [43, 44]. By doping a $H_2Pc:C_{60}$ co-deposited film with a donor (Cs_2CO_3), the E_F moved negatively to 4.22 eV (blue-dashed line) and approached the conduction band of C_{60} (Fig. 9.3, right). In contrast, by acceptor (V_2O_5) doping, the E_F moved to 4.95 eV (red-dashed line) and approached the valence band of H_2Pc. For *pn*-control of the blended films [45, 46], the E_F shift occurs in "the bandgap of the blended film," i.e., between the C_{60} conduction band and H_2Pc valence band.

9.5 Doping Efficiency

The doping efficiency of single organic semiconductor films is less than 10%. With D/A sensitization, the doping efficiency of blended films can be increased to 97%.

9.5.1 Band-Mapping

Band-mapping using a Kelvin probe can estimate the carrier concentrations created by doping [48, 49]. For example, a contact between *p*-type OSC and ITO electrode is shown (Fig. 9.4a). Due to the E_F alignment, the vacuum level (E_{VAC}) is bent upward. Since the work function is the energetic difference between E_{VAC} and E_F (red double arrows), it depends on the thickness of films. Thus, we can map the band-bending (Fig. 9.4a, lower) from the dependence of the work function on the film thickness (Fig. 9.4a, middle). The depletion layer width (W_{dep}) is related to the built-in potential (V_{bi}) and the carrier concentration (N) by Eq. (9.2).

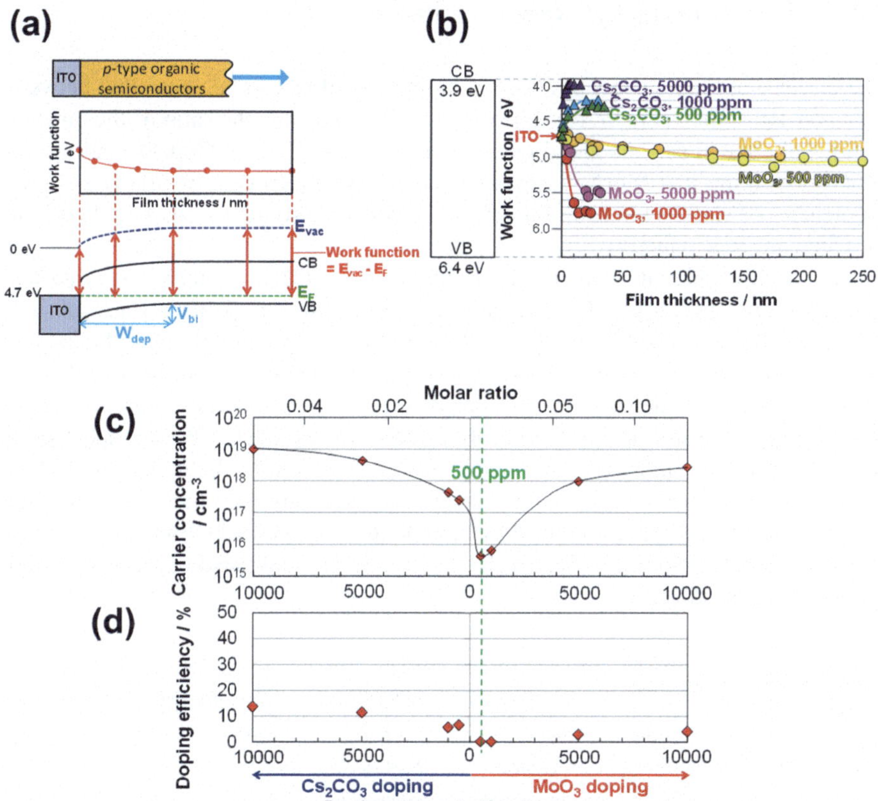

Fig. 9.4 **a** Mapping of band-bending using a Kelvin probe. The ITO electrode and a *p*-type semiconductor film in contact. Double red arrows (lower figure) indicate the work function values (middle figure), which depend on the OSC film thickness. **b** Dependence of the work function on C_{60} film thickness. Triangular and circular markers correspond to Cs_2CO_3 and MoO_3 doping. Solid curves are fit to Poisson's equation. **c** The carrier concentration versus the doping concentration. **d** The doping efficiency vs. the doping concentration. Reproduced with permission from M. Hiramoto et al., *Adv. Mater.*, Copyright 2018 John Wiley and Sons

$$W_{dep} = (2\varepsilon\varepsilon_0 V_{bi}/eN)^{1/2} \tag{9.2}$$

Band-mapping of the doped C_{60} films is shown in Fig. 9.4b. Upon doping with the donor (Cs_2CO_3) and acceptor (MoO_3), the work function moved negatively (triangular dots) and positively (circular dots), respectively. For both dopants, increasing the doping concentration decreased the depletion layer width (W_{dep}) decreased and increased the built-in potential (V_{bi}). The band-bending profile obeyed the quadratic curve of Poisson's equation (Fig. 9.4b, solid curves), thus providing precise values of W_{dep} and V_{bi}.

9.5.2 Low Doping Efficiency in Single Films

The electron and hole concentrations in single C_{60} films can be controlled between 10^{15} and 10^{19} cm^{-3} (Fig. 9.4c). The doping efficiency, i.e., the ratio of the concentrations of created carriers and dopant molecules, was 10% for Cs_2CO_3 and 3% for MoO_3, assuming that MoO_3 forms a trimer (Mo_3O_9) [50] (Fig. 9.4d). The doping efficiencies of various kinds of dopants in C_{60} films are summarized in Table 9.1, all of which are less than 10%.

The doping efficiency equals the product of the CT state formation rate and ionization rate (Fig. 9.1d). Because all the dopants listed in Table 9.1 are substantial molecules, all dopant molecules can be assumed to evaporate individually, and the CT state formation rate with C_{60} is nearly unity. Therefore, we believe that the obtained doping efficiencies can be considered rough estimations of the ionization rates. Because all the observed ionization rates are less than 10%, irrespective of the kind of dopant (inorganic or organic, acceptor or donor), the nature of the host organic semiconductor (C_{60}) mainly determines the ionization rate. Note that the doping efficiency is also influenced by other factors such as trap filling at extremely low doping concentrations (Sect. 8.1.1.) [10, 23, 51] and carrier reserves at high doping concentrations [10, 51].

Table 9.1 Doping efficiency for various dopants in vacuum-deposited C_{60} films. Reproduced with permission from M. Hiramoto et al., *Adv. Mater.*, Copyright 2018 John Wiley and Sons

	Dopants	Doping efficiency (%)
Donor	Cs_2CO_3	10
Acceptors	MoO_3	3
	V_2O_5	1
	Fe_2Cl_6	1
	F_4-TCNQ	7

9.6 Junction Formation

9.6.1 Ohmic Junctions

9.6.1.1 Organic/Metal Ohmic Junction

It is very important to make the two organic/metal contacts in a photovoltaic cell ohmic. When the region in the vicinity of a metal electrode is heavily doped, even if there is a Schottky barrier, its width becomes extremely thin allowing charge-carrier tunneling, and as a result, an ohmic contact is expected to be formed similar to that on heavily p^+- or n^+-doped inorganic semiconductors [52–54]. Here, + means heavily doped. Moreover, an ohmic contact can be formed irrespective of which electrode material is used, since tunneling is less dependent on the metal work function, enabling the cell structure to be inverted. This technique would allow flexibility in the design of the cell structure.

As a test case, two-layer cells consisting of C_{60} and H_2Pc (Fig. 9.5a, b) were examined [55]. The electrode materials were ITO and Ag. Heavy doping of the order of 10,000 ppm (1%) and 50,000 ppm (5%) was applied to thin 10 nm regions close to the C_{60} and H_2Pc/metal interfaces. Figure 9.5c, d shows the current–voltage (J-V) characteristics for the cells with heavily doped regions in Fig. 9.5a, b (red curves). For the cell in Fig. 9.5a, the fill factor (FF) reaches a value of 0.59, and clear rectification characteristics can be seen with heavily doped regions (red broken curve). Without the heavily doped regions (blue curves), however, FF is only 0.29, and the forward current is significantly suppressed. For the inverted cell (Fig. 9.5b), without heavily doped regions (Fig. 9.5d, blue curves), photovoltaic and rectification behavior are scarcely perceptible. However, with the heavily doped regions (red curves), the FF recovers, reaching a value of 0.49, and rectification is clearly observed. Clearly, the photovoltaic properties of the cells with thin heavily doped regions at the interfaces are independent of the type of electrode material used. Thus, H_2Pc/C_{60} cells are invertible using this interfacial heavy-doping technique.

Since electron extraction from C_{60} to the ITO electrode is crucial for the operation of the inverted cell (Fig. 9.5d (red curves)), we estimated the interfacial energy band structure of ITO/10,000 ppm Cs_2CO_3-doped C_{60} by Kelvin band-mapping (Fig. 9.6) (Sect. 5.1.). There is a distinct barrier to electrons with a height of 0.34 eV from the conduction band of C_{60} to the ITO. However, since the band bends down steeply within 5 nm of the interface, photogenerated electrons can tunnel through this barrier. Heavily doped C_{60} acts as an n^+-type semiconductor and makes the n^+-C_{60}/ITO junction ohmic. Organic/metal ohmic junctions can be fabricated by making tunneling contacts with heavy interfacial doping.

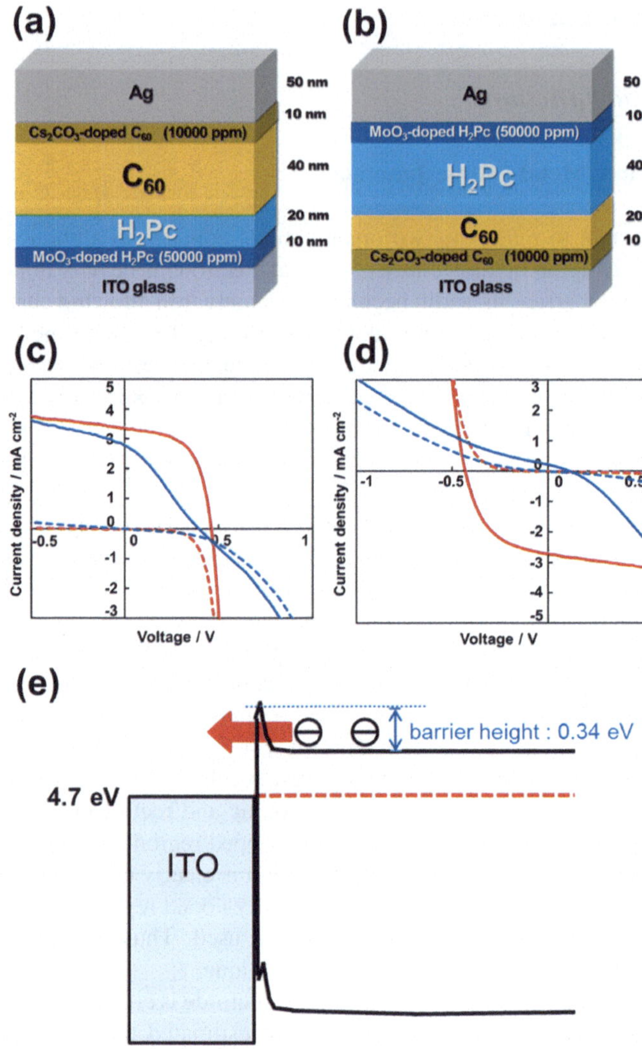

Fig. 9.5 (**a**) (**b**) Structures of invertible two-layered H_2Pc/C_{60} cells with heavily doped organic/metal interfaces. For the cells in (**a**) and (**b**), photogenerated holes and electrons are extracted to ITO and Ag, and to Ag and ITO, respectively. The total thicknesses of the H_2Pc and C_{60} films are kept the same. (**c**) (**d**) Current–voltage (J-V) characteristics for cells (**a**) and (**b**) (red curves). The blue curves are for the cells without heavily doped interfaces. The photocurrent and the dark current are shown by solid and broken curves, respectively. The ITO electrode was irradiated with simulated solar light (AM1.5, 100 mWcm^{-2}). Cell performances: (**a**) J_{sc}: 3.38 mAcm^{-2}, V_{oc}: 0.46 V, FF: 0.59, Efficiency: 0.91%. (**b**) J_{sc}: 2.70 mAcm^{-2}, V_{oc}: 0.43 V, FF: 0.49, Efficiency: 0.57%. (**e**) Energy structure for an ITO/n^+-C_{60} contact measured by Kelvin band-mapping. A tunneling junction for photogenerated electrons is formed. Reproduced with permission from [55]. Copyright 2012 The Japan Society of Applied Physics

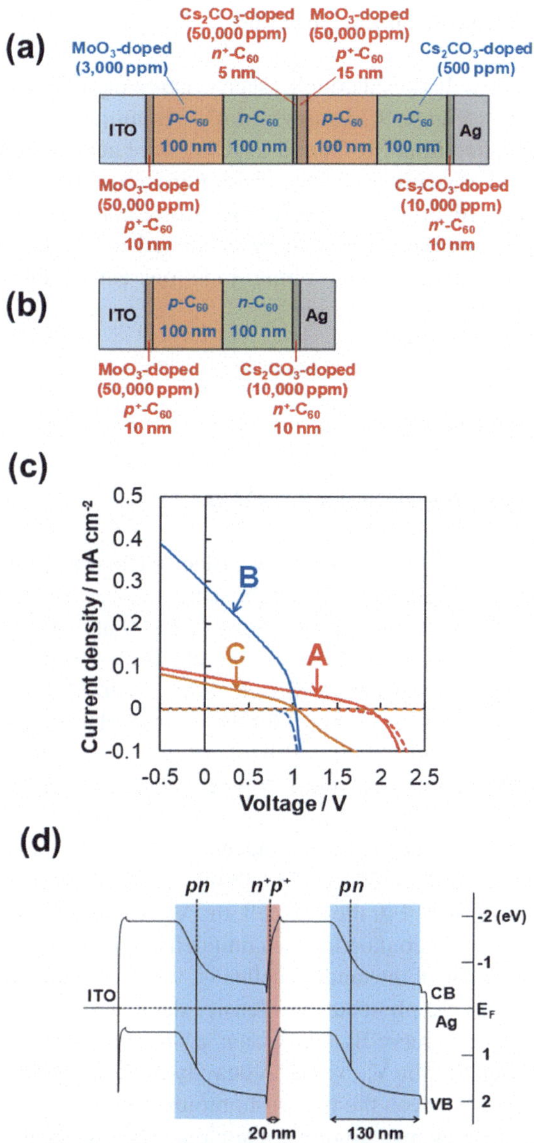

Fig. 9.6 **a** Structure of a tandem cell incorporating two *pn*-homojunction C_{60} cells connected by an n^+p^+-homojunction. **b** Structure of a C_{60} unit cell incorporating a *pn*-homojunction. **c** Current–voltage (*J*-*V*) characteristics for the tandem cell (curves A)(Fig. 9.6a) and the unit cell (curves B) (Fig. 9.6b). Curves C are for a tandem cell without an n^+p^+-interconnecting homojunction. The photo and dark currents are shown by solid and broken curves, respectively. The ITO electrode was irradiated with simulated solar light (AM1.5, 100 mWcm^{-2}). The short-circuit photocurrent, open-circuit photovoltage (V_{oc}), fill factor, and efficiency of the cells: (A) 0.09 mAcm^{-2}, 1.90 V, 0.29, 0.05%. (B) 0.29 mAcm^{-2}, 1.03 V, 0.34, 0.10%. (C) 0.06 mAcm^{-2}, 1.03 V, 0.32, 0.02%. **d** Energy band diagram of a tandem cell based on Kelvin probe measurements. The blue and red-shaded regions correspond to the depletion layers for the *pn*- and n^+p^+-homojunctions, respectively. Reproduced with permission from [39]. Copyright 2012 AIP Publishing

9.6.1.2 Organic/Organic Ohmic Junction

By making an n^+p^+ heavily doped double layer, organic/organic ohmic junctions can be fabricated [47, 56]. Figure 9.7d shows an n^+p^+-homojunction device fabricated in a C_{60}:6 T co-deposited film. MoO_3 and Cs_2CO_3 were heavily doped (50,000 ppm (5%)) for the p^+ and n^+-regions, respectively. Obviously, the n^+p^+-homojunction showed good ohmic properties (Fig. 9.7e). We think that the origin of the ohmic behavior of the n^+p^+-homojunction can be attributed to carrier recombination or tunneling. Similarly, organic/organic ohmic junctions can be fabricated also for a C_{60} single film (Fig. 9.6).

9.6.2 *Homojunctions*

9.6.2.1 *pn-Homojunction in Single C_{60}Films*

Since both *p*- and *n*-type C_{60} were formed, we tried to fabricate *pn*-homojunctions in single C_{60} films [56, 57]. Figure 9.6b shows the structure of a *pn*-homojunction. MoO_3 and Cs_2CO_3 dopants in volume concentrations of 3,000 and 500 ppm, respectively, were used. Ohmic contacts to the ITO and Ag electrodes were formed by heavily doping 10 nm thick layers adjacent to the electrodes with MoO_3 (50,000 ppm) for the p^+ contact and Cs_2CO_3 (10,000 ppm) for the n^+ contact, respectively. Figure 9.6a shows the structure of a tandem cell consisting of two *pn*-C_{60} unit cells (Fig. 9.6b). To make the n^+p^+-homojunction, both doping concentrations were increased to 50,000 ppm.

Curves A and B in Fig. 9.6c show the current–voltage (*J*-*V*) characteristics for a tandem cell (Fig. 9.6a) and a unit cell (Fig. 9.6b), respectively. For the tandem cell, the V_{oc} value reaches 1.90 V. For the unit cell, the V_{oc} value is 1.03 V. V_{oc} has almost doubled (84% increase) by connecting two unit cells. When the n^+p^+-homojunction between the two unit cells of the tandem cell (Fig. 9.6a) was removed, V_{oc} remained the same (1.03 V), but the photocurrent density decreased significantly (curve C) compared to the unit cell (curve B). Thus, an n^+p^+-interconnecting homojunction is indispensable for doubling the V_{oc} value. A heavily doped n^+p^+-homojunction acted as an ohmic interlayer between the two *pn*-homojunction cells.

Figure 9.6d shows the energy structure of tandem cell in which n^+p^+-homojunction (Fig. 9.6d, red-shaded region) connecting two *pn*-homojunctions (Fig. 9.6d, blue-shaded regions). The width of the depletion layers of the *pn*-homojunctions (blue-shaded) and the n^+p^+-homojunction (red-shaded) is 130 and 20 nm, respectively. The former and the latter act as the active regions for photocurrent generation and the ohmic interlayer, respectively. Thus, the depletion layer width, which is closely related to the behavior of the homojunction, was intentionally controlled by the doping concentration.

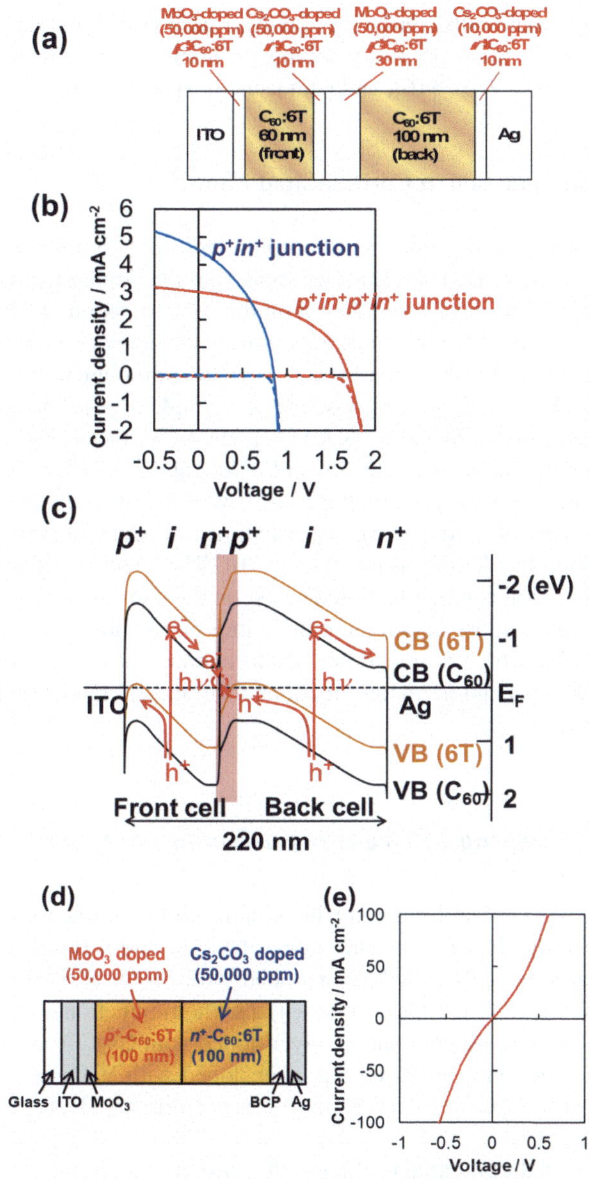

Fig. 9.7 **a** A C_{60}:6 T blended film with a $p^+in^+p^+in^+$-homojunction formed by doping alone. The p^+-, i-, and n^+-junctions are the MoO_3 (50,000 ppm)-doped, intrinsic (undoped), and MoO_3 (50,000 ppm)-doped C_{60}:6 T layers. **b** Current–voltage characteristics of the p^+in^+- (blue curves) and $p^+in^+p^+in^+$-homojunctions (red curves). The solid and dashed curves correspond to photo and dark currents, respectively. An ITO electrode was irradiated with simulated solar light (AM 1.5, 100 mW cm^{-2}). **c** Kelvin-probe-mapped energetic structure for a $p^+in^+p^+in^+$-homojunction. **d** Organic/organic ohmic n^+p^+-homojunction in C_{60}:6 T layer, and **e** current–voltage characteristics. (**a**) (**b**) (**c**) Reproduced with permission from [47]. Copyright 2013 Elsevier. Reproduced with permission from M. Hiramoto et al., *Adv. Mater.*, Copyright 2018 John Wiley and Sons. **e** Reproduced with permission from [47]. Copyright 2013 Elsevier

Tandem photovoltaic cells consisting of two *pn*-homojunctions were fabricated in single fullerene films only by impurity doping. *pn*-homojunctions were also reported in C_{60} [57], H_2Pc [62], ZnPc [16], and pentacene [18] single films.

9.6.2.2 *pn*-Homojunction in Co-Deposited Films

Blended film consisting of donor and acceptor molecules is indispensable in the modern organic solar cells. Therefore, we think that the doping technology for the blended films should be developed. Based on this consideration, we have designed and fabricated a variety of organic homojunctions in the bulk of co-deposited organic films by doping only. As an example, a $p^+in^+p^+in^+$-homojunction was fabricated only by doping. The homojunction consisted of a couple of p^+in^+-homojunctions in a uniform 6 T:C_{60} blended film (Fig. 9.7a) [47]. Here, "+" means heavily doped. By comparing the photovoltaic characteristics of p^+in^+- and $p^+in^+p^+in^+$-homojunctions (Fig. 9.7a), the open-circuit photovoltage (V_{oc}) was found to double from 0.85 V (blue curve) to 1.69 V (red curve), representing a tandem organic photovoltaic cell [58, 59] made by doping alone (Fig. 9.7b). The precise potential profile of the n^+p^+-homojunction can be illustrated based on the conventional junction theory (Fig. 9.6c) [60, 61]. An organic/organic tunneling ohmic junction is formed at the n^+p^+-homojunction, which is doped heavily to connect two p^+in^+-homojunctions. Heavily doped p^+/ITO and n^+/metal junctions form metal/organic ohmic junctions via tunneling [55].

9.6.3 *Band-Mapping of Organic pn-Homojunctions*

In this section, precise band-mapping in organic *pn*-homojunctions with various balances between the doping concentrations of the *p*- and *n*-type layers, i.e., (i) a p^+n^+-homojunction in 6 T:C_{60} (p^+ layer doped with MoO_3 at 50,000 ppm and n^+ layer doped with Cs_2CO_3 at 50,000 ppm), (ii) a *pn*-homojunction in 6 T:C_{60} (p-layer doped with MoO_3 at 3,000 ppm and n^+ layer doped with Cs_2CO_3 at 500 ppm), and (iii) a pn^+-homojunction in H_2Pc:C_{60} (p-layer doped with V_2O_5 at 1,000 ppm and n^+ layer doped with Cs_2CO_3 at 10,000 ppm) was performed [61]. Here, " + " means heavily doped.

In order to obtain the potential profiles of the *pn*-homojunctions, the film thickness dependence of the work function both in *p*-type layers on *n*-type layers and vice versa were measured using a Kelvin probe. The work function shift for *p*-side can be observed by repeat of deposition and work function measurement of *p*-type layer on 100 nm-thick *n*-type layer. Similarly, the work function shift for *n*-side can be observed by repeat of deposition and work function measurement of *n*-type layer on 100 nm-thick *p*-type layer.

For the heavily doped p^+n^+-homojunction (i), the bulk values of E_F in 100 nm-thick p^+ and n^+ layers were determined to be 5.58 eV and 4.38 eV, respectively. The variations of the work function (ϕ) with film thickness in the n^+- and p^+-type regions of the junction are shown in Fig. 9.8a, b, respectively. On the n^+-type side, ϕ becomes rapidly less positive moving away from the interface, converging towards a value of 4.44 eV at a depth of about 3 nm, which is close to the value of E_F for 100 nm-thick n^+-6 T:C_{60} (4.38 eV). This indicates that the space charge layer (SCL) is formed up to 3 nm into the n^+-type region. On the other hand, on the p^+-type side, ϕ becomes more positive moving away from the interface, converging toward a value of 5.63 eV at a depth of about 15 nm, which is close to the value of E_F for 100 nm-thick p^+-6 T:C_{60}

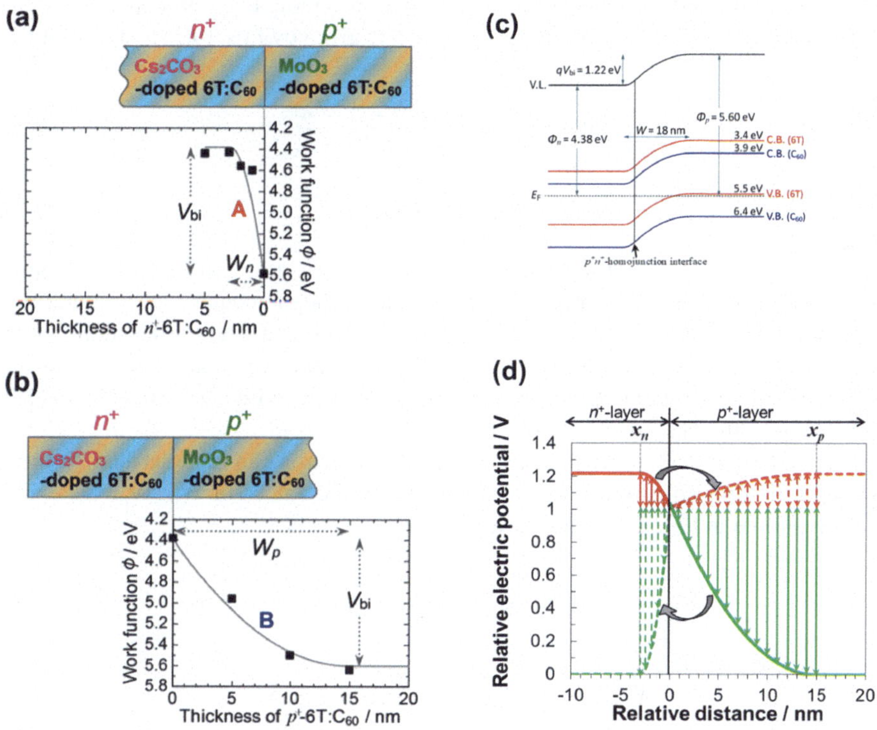

Fig. 9.8 Work functions in the p^+n^+-homojunctions. **a** n^+ on p^+. **b** p^+ on n^+. The black squares and solid curves are the observed points and the simulated curves, respectively. W_n, W_p, and V_{bi} denote the depletion layer widths in the n^+- and p^+-regions, and the built-in potential, respectively. **c** Energy band diagram of the p^+n^+-homojunction illustrated. V.L., E_F, V.B., and C.B. denote the vacuum level, Fermi level, the valence band, and the conduction band, respectively. Φ_n and Φ_p are the bulk work functions of n^+- and p^+-6 T:C_{60}. The bands for 6 T and C_{60} are shown by the red and blue lines, respectively. **d** Electric potential profile (solid curve) in the p^+n^+-homojunction in Figs. 9.7c, 9.8a–c illustrated based on Poisson's equation. The point $x = 0$ corresponds to the p^+n^+ interface. The work function change in Fig. 9.8a(solid curve) is calculated by adding the potential change that occurred in n^+-layer (red solid double-headed arrow) and that which occurred in p^+-layer (green broken double-headed arrow). The work function change in Fig. 9.8b (solid curve) is calculated by adding the potential change that occurred in p^+-layer (green solid double-headed arrow) and that which occurred in p^+-layer (red broken double-headed arrow). Reproduced with permission from [61]. Copyright 2015 AIP Publishing

(5.58 eV). Thus, the SCL is formed up to 15 nm into the p^+-type region. Since the values toward which ϕ converge and the values of E_F in the 100 nm films agree well, we conclude that, in thermal equilibrium, full alignment of the Fermi level between the p^+- and n^+-type films is realized. Thus, we can say that the depletion layer widths (W_n) for the n^+-type region and (W_p) for the p^+-type region are 3 and 15 nm, respectively, i.e., the total depletion layer width (W) is 18 nm, and the built-in potential (V_{bi}), which is the difference between the bulk E_F of n^+-6 T:C_{60} and that of p^+-6 T:C_{60}, is 1.22 eV. Using these parameters, the ionized dopant concentrations for the n^+- and p^+-type regions, $N_D{}^+N_D^+$ and $N_A{}^-N_A^-$, respectively, can be estimated. According to the standard theory for uniformly doped inorganic semiconductors, the total depletion layer width (W) of a *pn*-junction is given by Eqs. (9.3) and (9.4) [60].

$$W = \sqrt{\frac{2\varepsilon_r\varepsilon_0 V_{bi}(N_A^- + N_D^+}{qN_A^- N_D^+}} \tag{9.3}$$

$$qN_A^- x_p = qN_D^+ x_n \tag{9.4}$$

Here, ε_r, ε_0, q, x_p, and x_n denote the relative dielectric constant of the semiconductor, the dielectric permittivity in a vacuum, the elementary charge, and the depletion layer width in the *p*- and *n*-type regions, respectively. Furthermore, from Eq. (9.4), since $x_n = W_n = 3$ nm and $x_p = W_p = 15$ nm, we obtain the relationship $N_D{}^+ = 5\ N_A{}^-$. Together with $V_{bi} = 1.22$ eV, we find from Eq. (9.3) that $N_D{}^+ = 1.1 \times 10^{19}$ cm^{-3} and $N_A{}^- = 2.2 \times 10^{18}$ cm^{-3}.

The electric potential distribution in the p^+n^+-junction can be described by integrating Poisson's equation twice (5) and (6) with the values of W_n, W_p, $N_D{}^+$, and $N_A{}^-$, as shown in Fig. 9.9 (solid curve).

$$V(x) = -\frac{qN_D^+}{2\varepsilon_r\varepsilon_0}(x + x_n)^2 + V_{bi}(-x_n \le x \le 0) \tag{9.5}$$

$$V(x) = \frac{qN_A^-}{2\varepsilon_r\varepsilon_0}(x - x_p)^2(0 \le x \le x_p) \tag{9.6}$$

Here, the red and green parts of the curve are the electric potentials in the n^+- and p^+-type regions, respectively.

Based on this distribution, we attempted to reproduce the observed work function change (Fig. 9.8, black squares). When an n^+-type layer is deposited step-wise on the p^+-type layer, an energy band is bent not only in the n^+-type layer but also simultaneously in the underlying p^+-type layer. Taking account of the fact that the total amounts of charge due to ionized dopant in the n^+-region ($N_D{}^+x_n$) and p^+-region ($N_A{}^-x_p$) are identical, by depositing x_n-thick n^+-layer on the p^+-layer, band-bending up to a depth of x_n in the n^+-layer and that up to a depth of x_p in the underlying p^+-layer develop simultaneously. So, in order to calculate the work function change, potential change occurred in n^+-layer (red solid double-headed arrow) and that occurred in

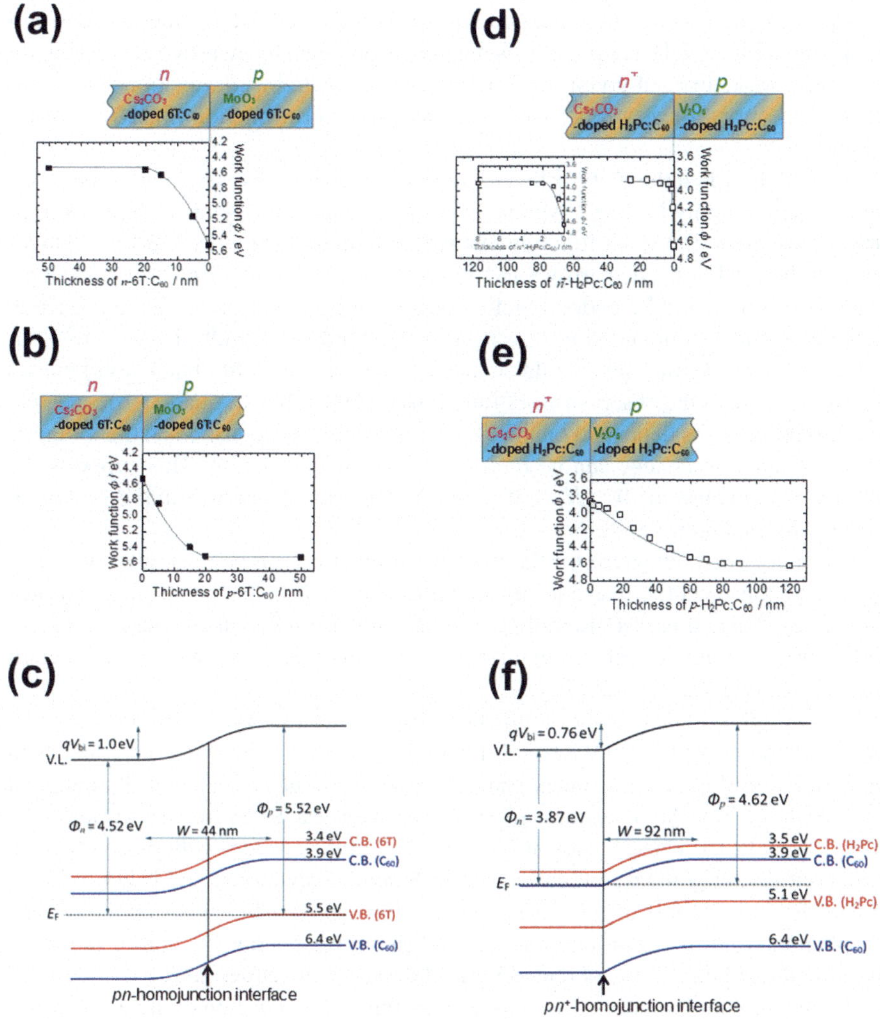

Fig. 9.9 (**a**) (**b**) Work functions in the *pn*-homojunction. (**a**) *n* on *p*. (**b**) *p* on *n*. The black squares and solid curves are the observed points and simulated curves, respectively. (**c**) Energy band diagram of the *pn*-homojunction for Fig. 9.9a, b. (**d**) (**e**) Work functions in the pn^+-homojunctions. (**d**) n^+ on *p*. (**e**) *p* on n^+. Inset: magnified view of the work function shifts within the first 8 nm for n^+ on *p*. The open squares and solid curves are the observed points and simulated curves, respectively. (**f**) Energy band diagram of the pn^+-homojunction for Fig. 9.9d, e. Reproduced with permission from [61]. Copyright 2015 AIP Publishing

p^+-layer (green broken double-headed arrow) should be added together as shown in n^+-region in Fig. 9.9. Here, green broken curve is obtained by inverting and shrinking in width from x_p to x_n of green solid curve. The calculated work function (Fig. 9.8a, black solid curve) clearly reproduces the observed work function (Fig. 9.8a, black squares).

Oppositely, when the p^+-type layer was deposited step-wise on the n^+-type layer, band is bent not only in the p^+-type layer but also in the underlying n^+-type layer. In order to calculate the work function change, potential change occurred in p^+-layer (green solid double-headed arrow) and that occurred in n^+-layer (red broken double-headed arrow) should be added together as shown in p^+-region in Fig. 9.8d. Here, red broken curve is obtained by inverting and extending in width from x_n to x_p of red solid curve. Again, the calculated work function (Fig. 9.8b, black solid curve) clearly reproduces the observed work function (Fig. 9.8b, black squares). Irrespective of whether the n^+- is deposited on the p^+-layer or the p^+- is deposited on the n^+-layer, the work function change can be reproduced by the calculation. This supports the validity of the values of W_p, N_A^-, W_n, and N_D^+ obtained and potential distribution in Fig. 9.8b (solid curve).

The energy band diagram for the p^+n^+-homojunction can be illustrated precisely by turning the potential distribution shown in Fig. 9.8d (solid curve) upside down (Fig. 9.8c). Since this p^+n^+-homojunction has an extremely narrow depletion layer width (W) of 18 nm, it confirms that it behaves as an organic/organic ohmic junction due to tunneling (Sect. 6.1.2., Fig. 9.7).

For the lightly doped *pn*-homojunction (ii), as was the case for the first sample, the simulated curve (Fig. 9.9a, b, solid curves) reproduces the observed changes in work function (Fig. 9.9a, b, black squares), and the values of W_n and W_p obtained were each 22 nm. As a result, N_D^+ and N_A^- were each determined to be 5.0×10^{17} (Fig. 9.9a, b). These values of N_D^+ and N_A^- almost accord with those obtained from band-bending measurements made in Schottky junctions, i.e., ITO/*n*-6 T:C_{60} (6.0×10^{17} cm^{-3})[63] or *p*-6 T:C_{60} (3.0×10^{17} cm^{-3}). The consistency between the parameters in the *pn*-homojunction and the corresponding Schottky junctions supports the validity of the obtained ionized dopant concentrations.

For the pn^+-homojunction (iii), since the doping concentration of the n^+-region is considerably larger than that of the *p*-region, it is assumed that in the n^+p-junction, which forms a one-sided abrupt junction, the SCL spreads predominantly into the *p*-region. As shown in Fig. 9.9d, e, as expected, the band-bending mainly occurred in the *p*-region (Fig. 9.9e). On the other hand, a sharp decrease in work function was observed to a depth of 2 nm in the n^+-region. This means that positive charge within 2 nm of the interface is sufficient to compensate for the negative charge in the depletion layer of the *p*-region. Conversely, deposition of an n^+-layer of only 2 nm thick on the *p*-layer can induce band-bending of 92 nm width in the underlying *p*-layer. The values of W_p and W_n obtained were 92 nm and 2 nm (see inset in Fig. 9.9d), respectively. As a result, N_D^+ and N_A^- were determined to be 1.9×10^{18} and 4.2×10^{16} cm^{-3}, respectively (Fig. 9.9d, e). The values of N_D^+ and N_A^- accord with those obtained from band-bending measurements made on a Schottky junction, i.e.,

ITO/p-$H_2Pc:C_{60}$ (4.3×10^{16} cm^{-3}), which has a similar energy structure to an n^+p-junction (Fig. 9.9f). Again, the simulated curves (Fig. 9.9e, solid curves) precisely reproduce the observed changes in work function (Fig. 9.9e, open squares).

The energy band diagram in the present n^+p-homojunction based on the simulated curve is shown in Fig. 9.9f. Since the ionized donor concentration is significantly larger than the ionized acceptor concentration, a one-sided abrupt junction, in which the SCL spreads predominantly into the *p*-type region, is formed.

We have demonstrated precise mapping of the band-bending in three cases, a p^+n^+-homojunction (i), a *pn*-homojunction (ii), and a pn^+-homojunction (iii), irrespective of the doping concentrations of the *n*- and *p*-layers and the balance between the doping concentrations of these layers. It was revealed that consideration of the change in band-bending in the underlying *n*-(or *p*-) type layer when a *p*-(or *n*-) type layer is gradually accumulated on it is necessary. The validity of the conventional theory for the space charge layer suggests that the dopants are spatially fixed even in the organic semiconductor films. The present results clearly show that the precise configuration of the built-in electric field in organic photovoltaic cells is fundamentally identical to that in inorganic ones.

9.7 Doping Sensitization in Blended Films

In this section, the doping sensitization observed for blended films ($H_2Pc:C_{60}$) is described [26]. The donor dopant (Cs_2CO_3) donates an electron to the conduction bands (LUMOs) of both H_2Pc of 3.2 eV and C_{60} of 3.9 eV [64, 65] (Fig. 9.10c), giving them *n*-type character. The carrier concentration evaluated by band-mapping (Fig. 9.4a) is about five times higher for the blended films (Fig. 9.10a, red dots) than for the corresponding single component films (Fig. 9.10a, yellow diamonds, blue triangles). Thus, a doping efficiency of approximately 50% was attained for the blended films (Fig. 9.10b, red dots), while that of the corresponding single films was approximately 10% (Fig. 9.10b, yellow diamonds, blue triangles). For $H_2Pc:C_{60}$ blended films (Fig. 9.10a), we confirmed that the specific conductivity increased 2,000 times from 7.5×10^{-6} (undoped) to 1.5×10^{-2} Scm^{-1} (MR = 0.05).

Doping sensitization can be explained by the charge separation superlattice model. Before contact (Fig. 9.10c), the corresponding single films show ionization rates of 10%, which correspond to ΔE_D (activation energy) = 0.12 eV. After contact (blending) (Fig. 9.10d), ET from H_2Pc to C_{60} (Fig. 9.10c, red arrow) accelerates electron release from the donor levels. As a result, the donor ionization rate increases for only H_2Pc. The ionization rates in the H_2Pc region of 100% and in the C_{60} region of 10% reproduce the observed total rate of 55%, i.e., $100\% \times 0.5 + 10\% \times 0.5 = 55\%$ (Table 9.2). The ionization rate increases from 10 to 100% indicates a sensitization factor of 10.

Identical ET from donor (H_2Pc) to acceptor (C_{60}) (Fig. 9.10c, red arrow) can be utilized to cause exciton dissociation, i.e., D/A sensitization (Fig. 9.1b). Therefore,

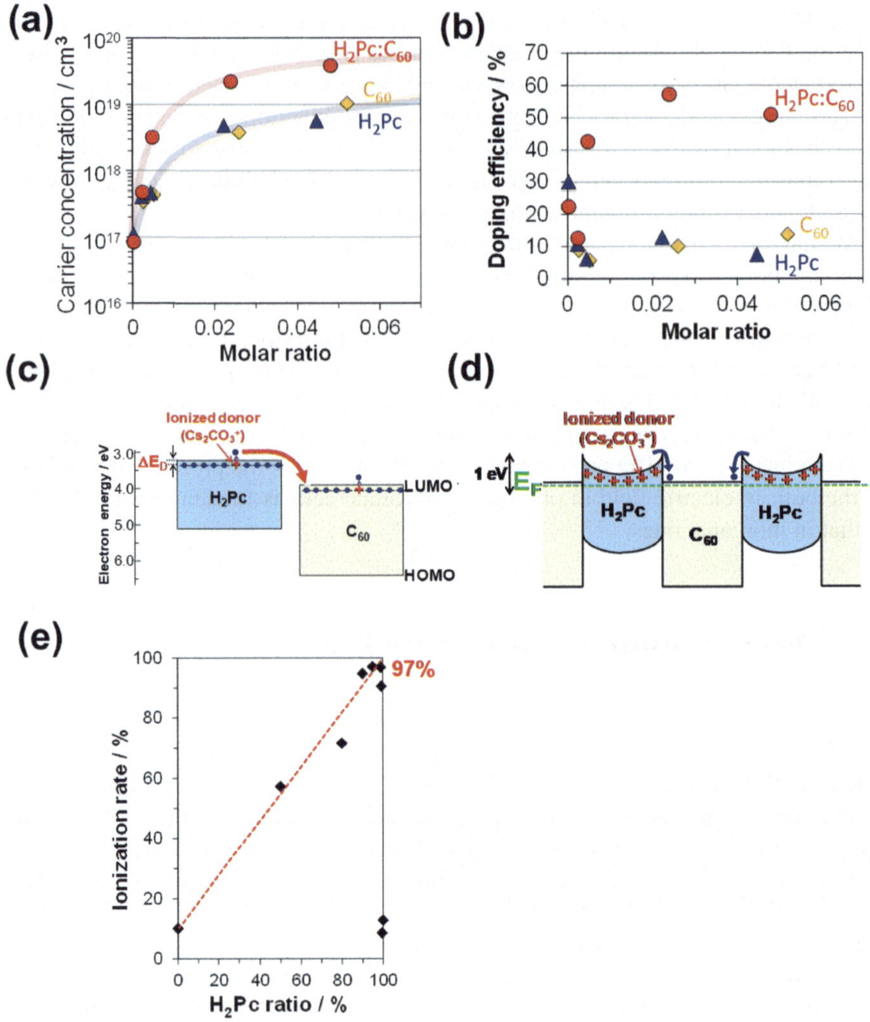

Fig. 9.10 Dependences of the carrier concentration (**a**) and the doping efficiency (**b**) on the molar doping ratio (Cs_2CO_3). (**c**) Energetic structures of H_2Pc and C_{60} films doped with Cs_2CO_3 before contact. (**d**) Energetic structure of a superlattice composed of H_2Pc and C_{60} doped with Cs_2CO_3 after contact. (**e**) Ionization rate versus H_2Pc ratio for Cs_2CO_3-doped H_2Pc:C_{60} films. Reproduced with permission from [26]. Copyright 2014 AIP Publishing. Reproduced with permission from M. Hiramoto et al., *Adv. Mater.*, Copyright 2018 John Wiley and Sons

the doping sensitization can be considered to be D/A sensitization, which causes dopant ionization.

This model predicts that the ionization rate of blended films can be further increased by increasing the H_2Pc ratio. In fact, the ionization rate (Fig. 9.10e) increases proportionally to the H_2Pc ratio and approached 97% at 99:1 (H_2Pc:C_{60}).

Table 9.2 Ionization rates of blended and composed single films and the sensitization factor for H_2Pc:C_{60} doped with a donor and H_2Pc:Me-PTC doped with an acceptor. Reproduced with permission from [26] Copyright 2014 AIP Publishing. Reproduced with permission from M. Hiramoto et al., *Adv. Mater.*, Copyright 2018 John Wiley and Sons

	Ionization rate (Doping efficiency) (%)				
		Single films			
	Co-deposited films	H_2Pc	C_{60}	Me-PTC	Sensitization factor
Donor doping (H_2Pc:C_{60} system)	55	10	10	–	10
Acceptor doping (H_2Pc:Me-PTC system)	30	2	–	0.2	300

To increase the specific conductivity by utilizing this high doping efficiency, the 1% C_{60} region should transport the created electrons. The same doping sensitization was observed for the acceptor (Fe_2Cl_6) doped in the H_2Pc:Me-PTC (perylene derivative) blended films (Table 9.2). The sensitization factor was increased 300-fold. Thus, doping sensitization is a universal phenomenon.

9.8 ppm-Level Doping Effects

9.8.1 Organic Semiconductor Films

A series of doping effects occurs depending on the doping level: trap filling (0–1 ppm), majority carrier appearance (1–10 ppm), homojunction formation (10–100 ppm), and mobility decrease (100–1,000 ppm).

9.8.1.1 Trap Filling

Many trap levels exist within the bandgap of OSC films due to grain boundaries, molecular defects, etc. Tieze et al. reported that hole traps can be passivated by doping [51]. Later, Olthof et al. [23] reported that electron trap levels can be passivated by doping at an ultra-low molar doping ratio (MR) of 10^{-4}. Because the carriers created by doping can fill and passivate such traps, the conductivity of organic semiconductor films increases due to the mobility increase.

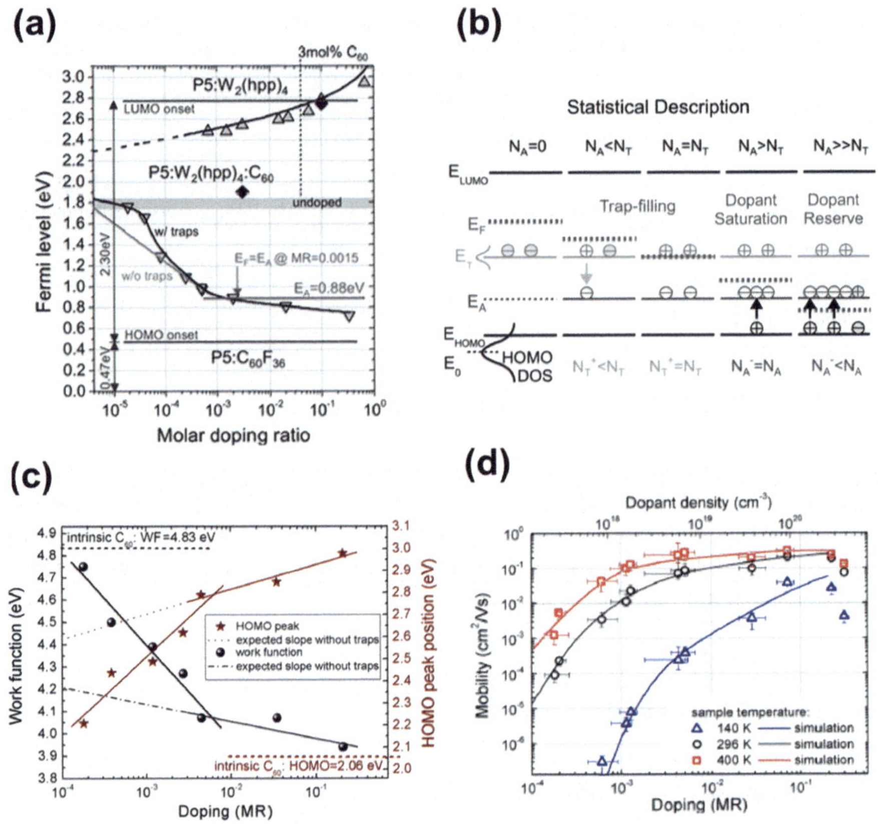

Fig. 9.11 **a** Fermi level (E_F) vs. molar doping ratio (MR). Pentacene (P5) films are doped with an acceptor ($C_{60}F_{36}$) (downward triangles) or donor ($W_2(hpp)_4$) (upright triangles). **b** Statistical description of acceptor doping. Activation of the dopant occurs at an effective acceptor level E_A. Trap levels E_T prevent doping at low concentrations. Free holes appear under the doping condition, $N_A > N_T$. **c** Dependences of the work function (black dots) and HOMO peak position (stars) on E_F by doping. The top- and bottom-dashed horizontal lines represent the work function and HOMO of the intrinsic C_{60} film, respectively. **d** Dependences of the mobility of a C_{60} layer on the doping concentration. (**a**) Reproduced with permission from [10]. Copyright 2015 John Wiley and Sons. (**b**) Reproduced with permission from [10]. Copyright 2015 John Wiley and Sons. Reproduced with permission from [51]. Copyright 2012 American Physical Society. (**c**) (**d**) Reproduced with permission from [23]. Copyright 2012 American Physical Society

Trap Filling by Acceptor Doping

Trap filling was discovered by Tietze et al. through acceptor doping for hole transporting organic films [51] and for pentacene films [10]. Pentacene films were doped with $C_{60}F_{36}$ acting as an acceptor. At doping concentrations below 4×10^{-5}, the E_F measured by ultraviolet photoelectron spectroscopy (UPS) was pinned near its intrinsic condition ($E_F \approx 1.8$ eV) (Fig. 9.11a). A slight increase in the doping ratio to 8 × 10^{-5} caused the rapid shift in the E_F to 1.29 eV, which corresponds to the trap filling region. When the concentration of holes (N_A) created by acceptor doping is smaller than that of traps (N_T), all created holes are captured by traps ($N_A < N_T$, Fig. 9.11b). This situation is maintained until all traps are filled by the created holes ($N_A = N_T$, Fig. 9.11b). After trap filling (MR > 10^{-4}), i.e., when the concentration of holes (N_A) exceeds that of traps (N_T), free holes appear ($N_A > N_T$, Fig. 9.11b). This situation corresponds to the dopant saturation region, i.e., the appearance of majority carriers, and the doping efficiency at MR = 2 × 10^{-4} reaches 0.98 [10, 66]. Recently, Tieze et al. reported that an amorphous hole transporting film (MeO-TPD) showed systematically lower activation energies (9.1 meV) than crystalline ZnPc (21 meV) or pentacene (19 meV) for F_6-TCNNQ doping [67]. The energetic disorder is essential for the dissociation of integer charge transfer complexes with low activation energy.

Trap Filling by Donor Doping

Trap filling was further confirmed by donor doping [23]. C_{60} films can be doped with a Ru-complex [16] acting as a donor dopant. By increasing the doping concentration, the work function measured by UPS decreases (black dots) (Fig. 9.11c), i.e., the Fermi level (E_F) negatively shifts and approaches the LUMO. At a high doping ratio from 6 × 10^{-3} to 10^{-1}, a slow negative shift is observed. In contrast, at a low doping ratio from 10^{-4} to 6 × 10^{-3}, a very rapid negative shift is observed, which is attributed to the filling of deep traps by electrons created by *n*-doping. Simultaneously, a very rapid increase in both conductivity and mobility (Fig. 9.11d) is observed at the same doping ratio from 10^{-4} to 6 × 10^{-3}. The electron mobility clearly increases after filling the traps by electrons created by donor dopants.

9.8.1.2 Majority Carrier, Homojunction, Mobility Decrease

In this section, the effects of 1 to 1,000 ppm doping on practical organic photovoltaic devices are described.

The formation of a typical junction, an abrupt and one-sided n^+p-homojunction formed in a photovoltaic composed of 6 T:C_{60} co-deposited films was investigated (Fig. 9.12a, b) [23]. The acceptor dopant (Fe_2Cl_6) concentration in the *p*-layer, which was in contact with the heavily doped n^+-layer (10,000 ppm Cs_2CO_3), was varied from 0, 1, 10, 100 to 1,000 ppm using a rotary shutter (Fig. 9.2b, c); 1 ppm is equivalent to a MR of 3.7 × 10^{-6}.

Both the short-circuit photocurrent (J_{sc}) and fill factor (FF) significantly increased upon direct ppm-level doping in the bulk of the photocarrier-generating co-deposited layer (Fig. 9.12a). Even at 1 ppm doping, slight effects can be seen (green curves). Clear effects appeared at 10 ppm doping (orange curves). The photocurrent reached its maximum at 100 ppm. However, a further increase to 1,000 ppm reduced both the photocurrent (blue solid curve) and forward current (blue-dashed curve).

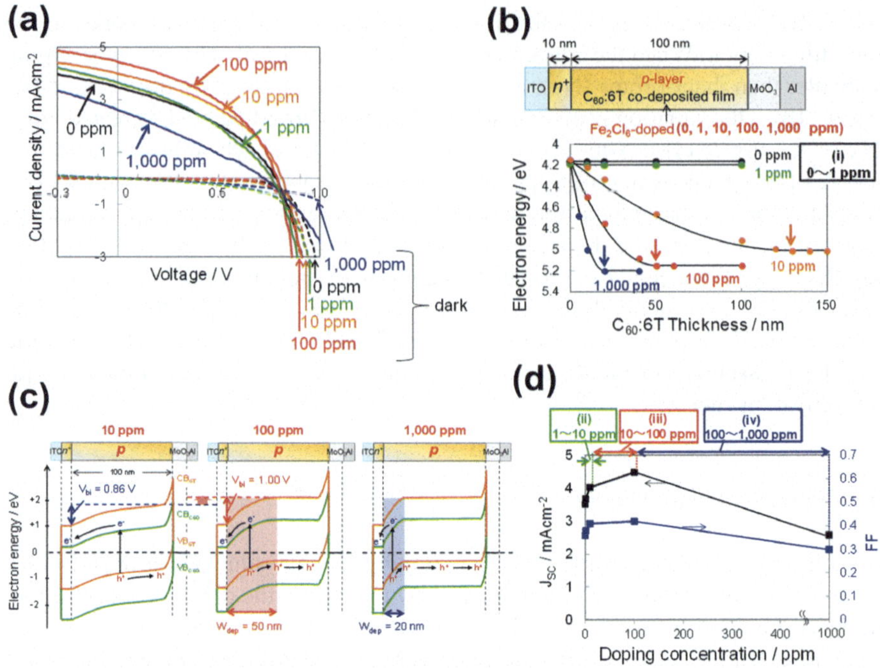

Fig. 9.12 **a** Current–voltage curves of $n^{+}p$-homojunctions for acceptor doping concentrations from 0 to 1,000 ppm. **b** $n^{+}p$-homojunction cell (upper). The work function vs. p-layer thickness (lower). **c** Energetic structures for 10, 100, and 1,000 ppm doping concentrations, which were fully mapped using a Kelvin probe. **d** J_{sc} (black) and FF (blue) versus doping concentration. Reproduced with permission from [24]. Copyright 2015 Elsevier

Band-mapping using a Kelvin probe (Sect. 5.1.) can reveal what occurred. The lack of change in the work function at 1 ppm doping (Fig. 9.12b, green dots) suggests trap filling, as the created carriers seemed to disappear (Sect. 8.1.1.) [10, 23]. Because band-bending occurs above 10 ppm (Fig. 9.12b, orange, red, and blue dots), the energetic structures of the $n^{+}p$-homojunctions (Fig. 9.12c) are illustrated by flipping Fig. 9.12b curves.

Doping effects on J_{sc} and FF were clearly observed and can be divided into three regions (Fig. 9.12d). Together with the trap filling region (0–1 ppm), a series of four regions can be distinguished from 0 to 1,000 ppm.

(i) **Trap filling (0–1 ppm)**: Slight increase in J_{sc}. The absence of band-bending (Fig. 9.12b) suggests that very few holes were created by 1 ppm doping (MR $= 3.7 \times 10^{-6}$), which were all captured by the traps and compensated by the negatively ionized acceptor molecules. The slight increase in photocurrent may be attributed to the conductivity increase due to trap filling [23].

(ii) **Majority carrier appearance (1–10 ppm)**: Steep increases in FF and J_{sc}. The FF increase is clearly synchronized with the cell resistance decrease. Thus, the

FF increase at 10 ppm can be attributed to the emergence of majority carriers after all traps in the *p*-region are filled (0–1 ppm).

(iii) **Built-in potential formation (10–100 ppm)**: J_{sc} increase continues, while FF becomes constant. The J_{sc} increase is clearly synchronized with the built-in potential (V_{bi}) increase (Fig. 9.12c, left and center). Thus, the J_{sc} increase at 100 ppm can be attributed to the creation of a built-in field, i.e., n^+p-homojunction formation.

(iv) **Mobility decrease (100–1,000 ppm)**: Decreases in FF and J_{sc}. Both decreases are clearly synchronized with the hole mobility decrease and the drastic increase in cell resistance from 2 to 50 Ωcm^2. Thus, the FF and J_{sc} decreases can be attributed to the hindered carrier transport due to the negatively ionized and neutral accepter dopant molecules. In addition, the decrease in the depletion layer thickness (W_{dep}) causes J_{sc} to decrease (Fig. 9.12c, center and right).

High doping concentrations reaching 1,000 ppm have harmful side effects, such as a mobility decrease caused by adding dopant molecules. Harada et al. [68, 69] reported decreased mobility in C_{60} films at high concentrations of doping. These doping-induced side effects must be clarified to develop effective doping technology for organic semiconductors. From this point of view, doping has been applied to organic single crystals with strictly defined spatial and energetic structures.

9.9 Bulk-Doped Organic Single Crystals

In the research on organic electronics, doping bulk organic single crystals and measuring their Hall effect have not been attempted. In addition, for vacuum-deposited films, it is difficult to exclude the possibility that the dopant molecules exist at the grain boundaries and that the essential nature of doping is hidden by the defects. In this section, ppm-level doping effects on bulk-doped rubrene single crystals measured by the Hall effect are described. Rubrene single crystals, which have inherent band conduction, showed a very high doping efficiency of 24%.

9.9.1 Doped Homoepitaxy

The Hall voltage of undoped rubrene single crystals was measured with the help of trap healing [70] and charge accumulation by a field-effect transistor (FET) [71, 72]. The bulk-doped rubrene single crystal was obtained by homoepitaxy, i.e., a homoepitaxial layer of rubrene was grown on a single-crystal rubrene substrate (Fig. 9.13a) [73]. When evaporated at a conventional speed of 0.2 nm s^{-1}, rubrene formed an amorphous film even on the single-crystal rubrene substrate, but homoepitaxial growth could be achieved at a low speed of 3.3×10^{-3} nm s^{-1} (Fig. 9.13b). An extremely low speed of 10^{-9} nm s^{-1} is needed for 1 ppm Fe_2Cl_6 doping due to the low

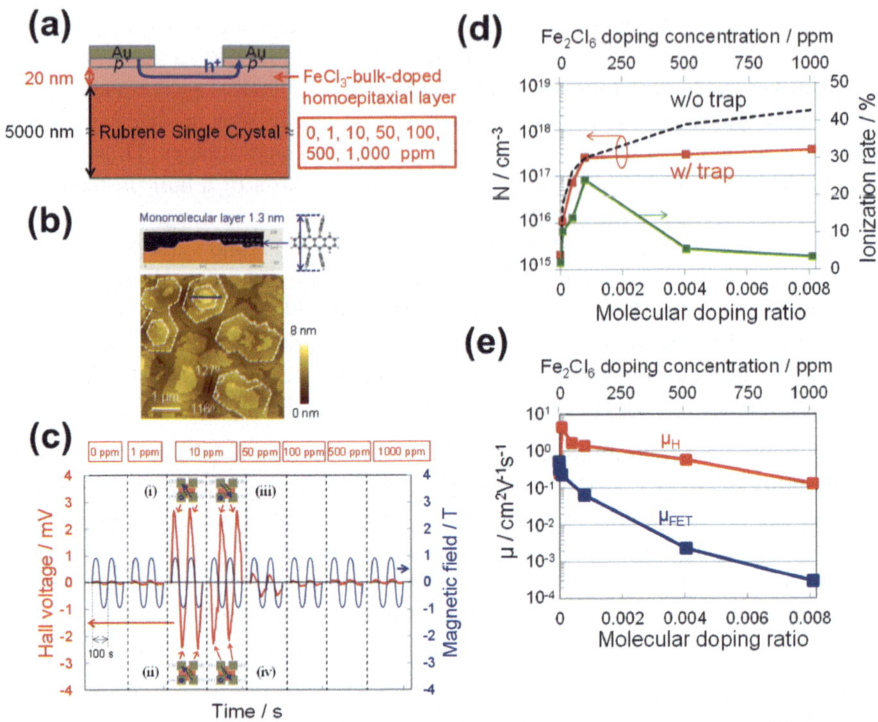

Fig. 9.13 **a** Side view of the bulk-doped homoepitaxial single-crystal layer used for Hall measurements. **b** AFM image (5 μm × 5 μm) of a homoepitaxial rubrene single crystal (20 nm) doped with Fe_2Cl_6 (10 ppm). The cross section of the AFM image is also shown. **c** Hall voltage (V_H) signals (red curves) synchronized with the magnetic field (B) (blue curves). **d** Hole concentration (N) (red curve) and ionization rate (green curve) versus doping concentration. The black-dashed curve was calculated assuming no traps were formed by doping. **e** Hall mobility (red curve) and FET mobility (blue curve) versus doping concentration. Reproduced with permission from [25]. Copyright 2017 John Wiley and Sons. Reproduced with permission from M. Hiramoto et al., *Adv. Mater.*, Copyright 2018 John Wiley and Sons

deposition rate of 10^{-3} nm s^{-1} for rubrene. Because 1,000 ppm Fe_2Cl_6 corresponds to a rate of 3.3×10^{-6} nm s^{-1}, 500, 100, 50, 10 and 1 ppm doping can be realized by rotary shutters having apertures with 1:5, 1:10, 1:50, 1:100 and 1:1,000 ratios (Sect. 3.2., Fig. 9.2b, c) [74, 75], and 1 ppm is equivalent to an MR of 3.7×10^{-6}.

AFM images of the surfaces of the Fe_2Cl_6-doped homoepitaxial rubrene films show many hexagonal layers aligned in the same direction with shapes identical to those of rubrene single crystals (001) having 116° and 127° angles (Fig. 9.13b). The observed step height of 1.3 nm (Fig. 9.13b, upper) corresponds to a monomolecular step [73, 76]. Under these hexagonal structures, uniform homoepitaxial layers are grown. The dopant molecules obviously exist in the rubrene single-crystal lattice.

9.9.2 Hall Effects

Fe_2Cl_6-doped rubrene single crystals show Hall voltage signals (Fig. 9.13c). Although the voltage signals (V_H) were low for undoped crystals (0.059 mV) and those doped at 1 ppm (0.065 mV), after a sharp increase to 2.5 mV at 10 ppm, the voltage gradually decreased from 50 ppm (0.37 mV), 100 ppm (0.11 mV) to 1,000 ppm (0.073 mV). Holes were the dominant carrier because Fe_2Cl_6 acts an acceptor dopant.

9.9.3 High Ionization Rate

The hole concentration (N) rapidly increased from 1×10^{15} to 3×10^{17} cm^{-3} upon 100 ppm doping (Fig. 9.13d, red curve). A surprising feature is the high ionization rate, which reached 24% at 100 ppm doping (Fig. 9.13d, green curve) and corresponds to the activation energy of the acceptor dopant (ΔE_A) of 72 meV, which is close to the ΔE_A of 45 mV for the acceptor dopant (B) in Si [31].

Recently, a very low activation energy (9.1 meV) was reported for an amorphous film [67]. Energetic disorder is essential for the dissociation of integer charge transfer complexes with low activation energy. Actually, an amorphous hole transporting film (MeO-TPD) shows systematically lower activation energies than crystalline ZnPc or pentacene. A systematic trend can be observed wherein smaller energetic disorder causes a higher activation energy. Though the activation energy for a band-conductive organic single crystal with low disorder should be large, a significantly low activation energy of 72 meV was observed. Therefore, we supposed that a significantly small activation energy obtained from the hole concentration, which was directly observed by the Hall effect for a band-conductive single crystals, has a different origin.

9.9.4 Scattering

The Hall mobility (μ_H) decreased from 4.6 to 0.7 cm^2V^{-1} s^{-1} upon increasing the dopant concentration from 10 to 500 ppm (Fig. 9.13e, red curve). Because holes in the traps cannot move due to the Lorentz force, holes captured in the traps did not contribute to the Hall mobility (μ_H). Therefore, the observed μ_H decrease is assigned to hole scattering by the negatively ionized neutral dopant Fe_2Cl_6 and the doping-induced lattice disturbances. Research on the scattering mechanism in bulk-doped organic single crystals is an unexplored field and should be clarified, especially the temperature dependence of μ_H [76–80].

9.9.5 Doping-Induced Trap Formation

The FET mobility (μ_{FET}) (Fig. 9.13e, blue curve), which represents the drift mobility, including the capture and release processes of traps, is one hundred times lower than the Hall mobility (μ_H), which represents the microscopic mobility free from trapping processes. Thus, the larger decrease in μ_{FET} than that in μ_H upon increasing the doping concentration is caused by hole traps. The activation energy of traps (ΔE_{trap}) determined from the temperature dependence of μ_{FET} was 0.18 eV for a 100 ppm bulk-doped single crystal, which is significantly larger than the activation energy of the acceptor (Fe_2Cl_6) (ΔE_A) of 0.072 eV (see Fig. 9.14a).

The trap concentration increased with the doping concentration. Without traps, the hole concentration increased without saturation above 100 ppm (Fig. 9.13d, black-dashed curve) [25]. Thus, the observed decrease in the ionization rate (Fig. 9.13d, green curve) above 100 ppm can be attributed to doping-induced trap formation.

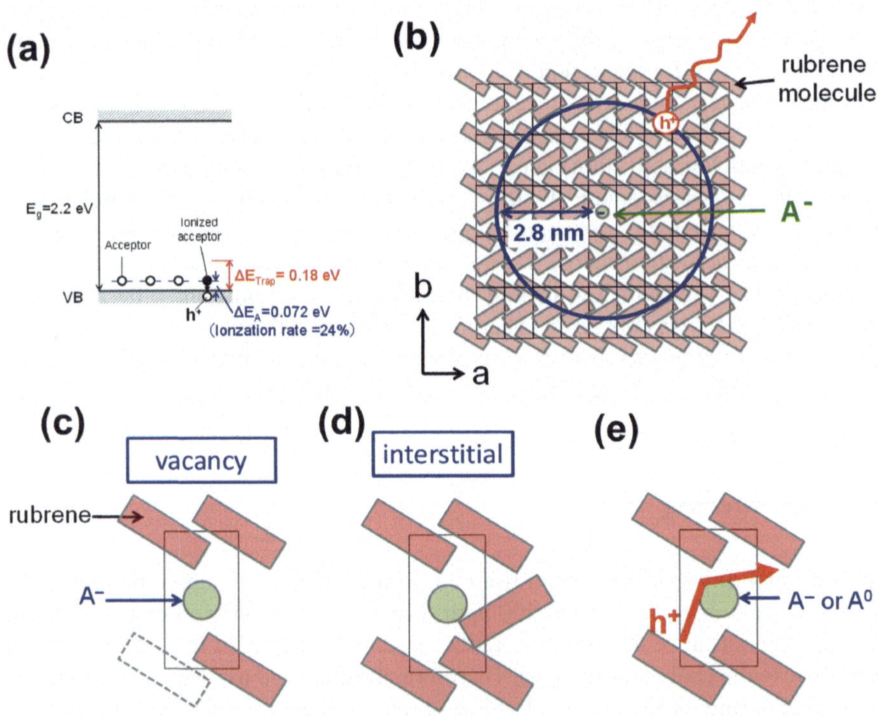

Fig. 9.14 **a** Energetic structure of a 100-ppm Fe_2Cl_3-doped rubrene single crystal. The activation energies of the acceptor (ΔE_A) and trap (ΔE_{trap}) are 72 mV and 180 mV, respectively. **b** The Wannier excitonic nature of doping is quite similar to that in Si (Fig. 9.1c). A hole loosely bound around the negatively ionized acceptor ion ($Fe_2Cl_6^-$) in the rubrene single crystal shows 24% doping efficiency. The size of the rubrene molecules is accurately scaled. **c** Molecular vacancy. **d** Interstitial molecule. **e** Scattering of a hole by the negatively ionized or neutral acceptor molecule. Reproduced with permission from M. Hiramoto et al., *Adv. Mater.*, Copyright 2018 John Wiley and Sons

Holes created by doping were compensated by the doping-induced traps. Methods for suppressing trap formation, such as post-annealing and organic-crystal-friendly dopants, could be introduced.

9.9.6 Future Prospects

9.9.6.1 Inherent High Doping Efficiency

An ionization rate of 24% ($\Delta E_A = 72$ meV) (Fig. 9.14a) corresponds to a Bohr's radius of 2.8 nm (Fig. 9.14b). This resembles the acceptor doping (B) of Si, which has an activation energy of 45 mV [31] and a Wannier excitonic nature (Fig. 9.1c). This model predicts that the observed high ionization rate can be considered the inherent nature of the rubrene single crystal itself and essentially independent of the dopant characteristics. However, our preliminary results suggest that the ionization rates of inorganic and organic dopants are not very different, despite their relatively different dielectric constants (ε). We presume that due to the nature of band conduction, the orbital of a hole is delocalized over the crystal lattice of the rubrene molecules, especially for an ab-crystal plane with strong π–π stacking (Fig. 9.14b). Such anisotropy may affect the dopant ionization. This model also predicts that various band-conductive organic single crystals [81–85] would show a doping efficiency close to 100%, similar to that of Si, which would be beneficial for creating new types of organic electronic devices with a variety of junctions, e.g., *pn*-, *pin*-, and *pnp*-junctions, in the organic bulk single crystal.

9.9.6.2 Defect Science of Organic Single Crystals

The decrease in the ionization rate above 100 ppm can be explained by doping-induced defects, such as vacancies (Fig. 9.14c) and interstitials (Fig. 9.14d), that may be able to act as carrier traps. Carriers created by doping are consumed by these traps. A vacancy-dopant pair ($V_{Molecule}{}^{+}$-A^{-}), i.e., a hole trapped at a molecular vacancy and a negatively ionized acceptor dopant (Fig. 9.14c), may compensate for the doping. In the history of inorganic single-crystal electronics, for example, for oxide semiconductors such as ZnO, many kinds of traps, such as oxygen vacancies (V_o), metal vacancies (V_M), and metal interstitials (M_i), have been identified in terms of their microscopic structures and energetic level in the bandgap [86–94]. In the present stage of organic semiconductors, however, the microscopic spatial structures, energetic levels, and electrical behaviors of defects in organic single crystals are either unknown or poorly understood.

Therefore, we now stand at the entrance of a vast uncultivated field of defect science for organic single crystals, which is comparable to that of inorganic single crystals such as silicon cultivated during the past half century. If we succeed in controlling doping-induced defects, a doping efficiency close to 100% will no longer be a dream.

9.9.6.3 Scattering Science of Organic Single Crystals

The Hall mobility (μ_H) was found to decrease above 10 ppm. The gradual μ_H reduction is explained by carrier scattering due to the ionized dopant molecule, neutral dopant molecule, and crystal lattice deformation (Fig. 9.14e). Carrier scattering at the ionized dopant has been observed previously in hopping-conductive *n*-doped C_{60} thin films [69]. We believe that the physical meaning of carrier scattering during the hopping conduction of carrier should be clarified. Carrier scattering by dopants in band-conductive organic single crystals is a new issue. For organic semiconductors, the carrier scattering causes not only intermolecular vibrations but also intramolecular vibrations, which have many modes, such as C-H stretching and bending [95–97]. The physics of carrier scattering is significantly different from that of inorganic semiconductor crystals. Thus, we also stand at the entrance of the immense undeveloped field of scattering science for organic single crystals, which is comparable to that of inorganic single crystals.

9.9.6.4 Organic Single-Crystal Electronics

The quality of organic single crystals is still low. The observed carrier concentration of an undoped organic single crystal of approximately 10^{15} cm^{-3} (Fig. 9.13d) is significantly higher than that for undoped Si of 10^{10} cm^{-3}. Many carriers of unknown origins exist at the surface and in the bulk. If the science underlying the single-crystal growth of organic semiconductors is developed, defect-minimized organic single crystals could be achieved.

In the future, bulk-doped organic single-crystal wafers, similar to Si wafers, could be introduced. Organic single-crystal electronics could be constructed utilizing not only bulk-doped homoepitaxy but also bulk-doped heteroepitaxy on doped organic single-crystal substrates. Organic single-crystal solar cells with *pn*- and *pin*-homojunctions, organic single-crystal bipolar transistors with *pnp*- and *npn*-homojunctions, and organic single-crystal lasers with double heterojunctions could be developed.

9.9.7 Conclusion

Historically, the true nature of organic semiconductors had been veiled by external impurities, and they had been unresponsive to intentional doping. Progress over the past decade has lifted the veil, and organic semiconductors have become responsive to even 1-ppm-level doping. This progress has led to the creation of the field of bulk-doped organic single crystals. In the future, organic single-crystal electronics including the organic solar cells constructed on the doped organic single-crystal substrates could be constructed.

Acknowledgements The authors thank M. Kubo, Y. Shinmura, C. Ohashi, and N. Ishiyama for their important work on doping and Professors M. Izaki and H. Naito for their helpful discussions. The authors also appreciate A. Adachi and S. Ohashi of Epitech Inc. for their help in the design and construction of the built-in chamber globe box systems with extremely slow deposition apparatuses. Funding from CREST/JST and NEDO is gratefully acknowledged.

References

1. Martin, M., Andre, J.J., Simon, J.: Influence of dioxygen on the junction properties of metallophthalocyanine based devices. J. Appl. Phys. **54**, 2792–2794 (1983)
2. Tada, H., Touda, H., Takada, M., Matsushige, K.: Quasi-intrinsic semiconducting state of titanyl-phthalocyanine films obtained under ultrahigh vacuum conditions. Appl. Phys. Lett. **76**, 873–875 (2000)
3. Hiramoto, M., Kishigami, Y., Yokoyama, M.: Doping effect on the two-layer organic solar cell. Chem. Lett. **1990**, 119–122 (1990)
4. Akamatsu, H., Inokuchi, H., Matsunaga, Y.: Electrical conductivity of the perylene–bromine complex. Nature **173**, 168–169 (1954)
5. Hiramoto, M., Ihara, K., Fukusumi, H., Yokoyama, M.: Conduction type control from n to p type for organic pigment films purified by reactive sublimation. J. Appl. Phys. **78**, 7153–7157 (1995)
6. Hiramoto, M., Ihara, K., Yokoyama, M.: Fermi level shift in photoconductive organic pigment films measured by Kelvin vibrating capacitor method. Jpn. J. Appl. Phys. **34**, 3803–3807 (1995)
7. Hiramoto, M.: Organic solar cells incorporating a p-i-n junction and a p-n homojunction. In: Sun, S. -S., Sariciftci, N. S. (ed.) Organic photovoltaics, Mechanisms, Materials and Devices, p. 268. CRC Press, New York (2005)
8. Huang, S.J., Pfeiffer, M., Werner, A., Blochwitz, J., Leo, K., Liu, Y.S.: Low-voltage organic electroluminescent devices using pin structures. Appl. Phys. Lett. **80**, 139–141 (2002)
9. Blochwitz, J., Pfeiffer, M., Fritz, T., Leo, K.: Low voltage organic light emitting diodes featuring doped phthalocyanine as hole transport material. Appl. Phys. Lett. **73**, 729–731 (1998)
10. Tietze, M.L., Pahner, P., Schmidt, K., Leo, K., Lüssem, B.: Doped organic semiconductors: trap-filling, impurity saturation, and reserve regimes. Adv. Funct. Mater. **25**, 2701–2707 (2015)
11. Koech, P.K., Padmaperuma, A.B., Wang, L., Swensen, J.S., Polikarpov, E., Darsell, J.T., Rainbolt, J.E., Gaspar, D.J.: Synthesis and application of 1,3,4,5,7,8-hexafluorotetracyanonaphthoquinodimethane (F6-TNAP): A conductivity dopant for organic light-emitting devices. Chem. Mater. **22**, 3926–3932 (2010)
12. Cho, S.H., Pyo, S.W., Suh, M.C.: Low voltage top-emitting organic light emitting devices by using 1,4,5,8,9,11-hexaazatriphenylene-hexacarbonitrile. Synth. Met. **162**, 402–405 (2012)
13. Solomeshch, O., Yu, Y.J., Goryunkov, A.A., Sidorov, L.N., Tuktarov, R.F., Choi, D.H., Jin, J. –Il., Tessler, N.: Ground-state interaction and electrical doping of fluorinated C_{60} in conjugated polymers. Adv. Mater. 21, 4456–4460 (2009)
14. Tokito, S., Noda, K., Taga, Y., J. Phys. D: Metal oxides as a hole-injecting layer for an organic electroluminescent device. Appl. Phys. 29, 2750–2753 (1996)
15. Matsushima, T., Kinoshita, Y., Murata, H.: Formation of Ohmic hole injection by inserting an ultrathin layer of molybdenum trioxide between indium tin oxide and organic hole-transporting layers. Appl. Phys. Lett. 91, 253504 (3 pages) (2007)

16. Harada, K., Werner, A.G., Pfeiffer, M., Bloom, C.J., Elliott, C.M., Leo, K.: Organic homojunction diodes with a high built-in potential: interpretation of the current-voltage characteristics by a generalized Einstein relation. Phys, Rev. Lett. 94, 036601 (4 pages) (2005)
17. Chan, C.K., Zhao, W., Barlow, S., Marder, S., Kahn, A.: Decamethylcobaltocene as an efficient n-dopant in organic electronic materials and devices. Org. Electron. **9**, 575–581 (2008)
18. Harada, K., Riede, M., Leo, K.: Pentacene homojunctions: Electron and hole transport properties and related photovoltaic responses. Phys. Rev. B 77, 195212 (9 pages) (2008)
19. Chan, C.K., Amy, F., Zhang, Q., Barlow, S., Marder, S., Kahn, A.: N-type doping of an electron-transport material by controlled gas-phase incorporation of cobaltocene. Chem. Phys. Lett. **431**, 67–71 (2006)
20. Tietze, M.L., Wolzl, F., Menke, T., Fischer, A., Riede, M., Leo, K., Lüssem, B.: Self-passivation of molecular n-type doping during air exposure using a highly efficient air-instable dopant. Phys. Status Solidi A **210**, 2188–2198 (2013)
21. Liao, H. –H., Chen, L. –M., Xu, Z., Li, G., Yang, Y.: Highly efficient inverted polymer solar cell by low temperature annealing of Cs_2CO_3 interlayer. Appl. Phys. Lett. 92, 173303 (3 pages) (2008)
22. Hamwi, S., Riedl, T., Kowalsky, W.: An organic *p-i-n* homojunction as ultra violet light emitting diode and visible-blind photodiode in one. Appl. Phys. Lett. 99, 053301 (3 pages) (2011)
23. Olthof, S., Mehraeen, S., Mohapatra, S.K., Barlow, S., Coropceanu, V., Bre´das, J.-L., Marder, S.R., Kahn, A.: Ultralow doping in organic semiconductors: Evidence of trap filling. Phys, Rev. Lett. 109, 176601 (5 pages) (2012)
24. Ohashi, C., Shinmura, Y., Kubo, M., Hiramoto, M.: Effects of doping at the ppm level in Simple n^+ *p*-homojunction organic photovoltaic cells. Org. Electron. **27**, 151–154 (2015)
25. Ohashi, C., Izawa, S., Shinmura, Y., Kikuchi, M., Watase, S., Izaki, M., Naito, H., Hiramoto, M.: Hall effect in bulk-doped organic single crystals. Adv. Mater. 29, 1605619 (6 pages) (2017)
26. Shinmura, Y., Yamashina, Y., Kaji, T., Hiramoto, M.: Ionization sensitization of doping in codeposited organic semiconductor films. Appl. Phys. Lett. 105, 183306 (5 pages) (2014)
27. Lüssem, B., Riede, M., Leo, K.: Doping of organic semiconductors. Phys. Status Solidi A **210**, 9–43 (2013)
28. Walzer, K., Maennig, B., Pfeiffer, M., Leo, K.: Highly efficient organic devices based on electrically doped transport layers. Chem. Rev. **107**, 1233–1271 (2007a)
29. Hains, A.W., Liang, Z., Woodhouse, M.A., Gregg, B.A.: Molecular semiconductors in organic photovoltaic cells. Chem. Rev. **110**, 6689–6735 (2010)
30. Lüssem, B., Keum, C.M., Kasemann, D., Naab, B., Bao, Z., Leo, K.: Doped organic transistors. Chem. Rev. **116**, 13714–13751 (2016)
31. Ibach, H., Luth, H.: Solid-State Physics. 4th Edition, Chap.12. Springer-Verlag Berlin (2009)
32. Hebard, F.A., Hadon, C.R., Fleming, M.R., Kortan, R.A.: Deposition and characterization of fullerene films. Appl. Phys. Lett. **59**, 2109–2111 (1991)
33. Salzmann, I., Heimel, G., Oehzelt, M., Winkler, S., Koch, N.: Molecular electrical doping of organic semiconductors: Fundamental mechanisms and emerging dopant design rules. Acc. Chem. Res. **49**, 370–378 (2016)
34. Salzmann, I., Heimel, G., Duhm, S., Oehzelt, M., Pingel, P., George, B., Schnegg, A., Lips, K., Blum, R.-P., Vollmer, A., Koch, N.: Intermolecular hybridization governs molecular electrical doping. Phys. Rev. Lett. 108, 035502 (5 pages) (2012)
35. Laudise, R.A., Kloc, C., Simpkins, P.G., Siegrist, T.: Physical vapor growth of organic semiconductors. J. Cryst. Growth **187**, 449–454 (1998)
36. Hiramoto, M., Kubo, M., Shinmura, Y., Ishiyama, N., Kaji, T., Sakai, K., Ohno, T., Izaki, M.: Bandgap science for organic solar cells. Electronics **3**, 351–380 (2014)
37. Saito, S., Soumura, T., Maeda, T.: Improvements of the piezoelectric driven Kelvin probe. J. Vac. Sci. Technol. A, 2, 1389–1391 (1984)
38. Ito, E., Oji, H., Hayashi, N., Ishii, H., Ouchi, Y., Seki, K.: Electronic structures of TPD/metal interfaces studied by photoemission and Kelvin probe method. Appl. Surf. Sci. **175**, 407–411 (2001)

39. Ishiyama, N., Kubo, M., Kaji, T., Hiramoto, M.: Tandem photovoltaic cells formed in single fullerene films by impurity doping. Appl. Phys. Lett. **101**, 233303 (3 pages) (2012)
40. Falkenberg, C., Uhrich, C., Olthof, S., Maennig, B., Riede, M., Leo, K.: Efficient type organic solar cells incorporating 1,4,5,8-naphthalenetetracarboxylic dianhydride as transparent electron transport material. J. Appl. Phys. **104**, 034506 (6 pages) (2008)
41. Yoshida, H.: Low-energy inverse photoemission spectroscopy using a high-resolution grating spectrometer in the near ultraviolet range. Rev. Sci. Instrum. **84**, 103901 (5 pages) (2013)
42. Kubo, M., Iketaki, K., Kaji, T., Hiramoto, M.: Conduction-type control of fullerene films from *n*- to *p*-type by molybdenum oxide doping. Appl. Phys. Lett. **98**, 073311 (3 pages) (2011)
43. Hiramoto, M., Fujiwara, H., Yokoyama, M.: Three-layered organic solar cell with a photoactive interlayer of codeposited pigments. Appl. Phys. Lett. **58**, 1062–1064 (1991)
44. Hiramoto, M., Fujiwara, H., Yokoyama, M.: *p-i-n* like behavior in three-layered organic solar cells having a co-deposited interlayer of pigments. J. Appl. Phys. **72**, 3781–3787 (1992)
45. Kubo, M., Kaji, T., Hiramoto, M.: *pn*-homojunction organic solar cells formed in phase-separated co-deposited films. Appl. Phys. Lett. **103**, 263303 (4 pages) (2013)
46. Kubo, M., Shinmura, Y., Ishiyama, N., Kaji, T., Hiramoto, M.: Junction formation by doping in H_2Pc:C_{60} co-evaporated films for solar cell application. Mol. Cryst. Liq. Cryst. **581**, 13–17 (2013)
47. Ishiyama, N., Kubo, M., Kaji, T., Hiramoto, M.: Tandem organic solar cells formed in co-deposited films by doping. Org. Electron. **14**, 1793–1796 (2013)
48. Hamwi, S., Meyer, J., Winkler, T., Riedl, T., Kowalsky, W.: *p*-type doping efficiency of MoO_3 in organic hole transport materials. Appl. Phys. Lett. **94**, 253307 (3 pages) (2009)
49. Ishiyama, N., Kubo, M., Kaji, T., Hiramoto, M.: Doping-based control of the energetic structure of photovoltaic co-deposited films. Appl. Phys. Lett. **99**, 133301 (3 pages) (2011)
50. Kröger, M., Hamwi, S., Meyer, J., Riedl, T., Kowalsky, W., Kahn, A.: P-type doping of organic wide band gap materials by transition metal oxides: A case-study on Molybdenum trioxide. Org. Electron. **10**, 932–938 (2009)
51. Tietze, M., Burtone, L., Riede, M., Lussem, B., Leo, K.: Fermi level shift and doping efficiency in *p*-doped small molecule organic semiconductors: A photoelectron spectroscopy and theoretical study. Phys. Rev. B, **86**, 035320 (12 pages) (2012)
52. Walzer, K., Maennig, B., Pfeiffer, M., Leo, K.: Highly efficient organic devices based on electrically doped transport layers. Chem. Rev. **107**, 1233–1271 (2007b)
53. deMello, J.C., Tessler, N., Graham, S.C., Friend, R.H.: Ionic space-charge effects in polymer light-emitting diodes. Phys. Rev. B **57**, 12951–12963 (1998)
54. Hiramoto, M., Tomioka, A., Suemori, K.: Yokoyama, M. Formation of ohmic contacts to perylene molecular crystals. Appl. Phys. Lett. **85**, 1852–1854 (2004)
55. Kubo, M., Shinmura, Y., Ishiyama, N., Kaji, T., Hiramoto, M.: Invertible organic photovoltaic cells with heavily-doped organic/metal ohmic contacts. Appl. Phys. Exp. **5**, 092302 (3 pages) (2012)
56. Kubo, M., Kaji, T., Hiramoto M.: *pn*-homojunction formation in single fullerene films. AIP Advances **1**, 032177 (5 pages) (2011)
57. Ishiyama, N., Kubo, M., Kaji, T., Hiramoto, M.: Tandem photovoltaic cells formed in single fullerene films by impurity doping. Appl. Phys. Lett. **101**, 233303 (3pages) (2012)
58. Hiramoto, M., Suezaki, M., Yokoyama, M.: Effect of thin gold interstitial-layer on the photovoltaic properties of tandem organic solar cell. Chem. Lett. **19**, 327–330 (1990)
59. Yuan, Y., Huang, J., Li, G.: Intermediate layers in tandem organic solar cells. Green **1**, 65–80 (2011)
60. Sze, S.M.: Physics of Semiconductor Devices. Wiley, New York (1969)
61. Yamashina, Y., Shinmura, Y., Ishiyama, N., Kaji, T., Hiramoto, M.: Mapping of band-bending in organic *pn*-homojunctions. J. Appl. Phys. **117**, 125501 (5 pages) (2015)
62. Shinmura, Y., Kubo, M., Ishiyama, N., Kaji, T., Hiramoto, M.: *pn*-control and *pn*-homojunction formation of metal-free phthalocyanine by doping. AIP Advances **2**, 032145 (6 pages) (2012)
63. Ishiyama, N., Yoshioka, T., Kaji, T., Hiramoto, M.: Tuning of barrier parameters of n-type Schottky junctions in photovoltaic co-deposited films by doping. Appl. Phys. Exp. **6**, 012301 (3 pages) (2013)

64. Yoshida, H.: Measuring the electron affinity of organic solids: an indispensable new tool for organic electronics. Anal. Bioanal. Chem. **406**, 2231–2237 (2014)
65. Yoshida, H.: Near-ultraviolet inverse photoemission spectroscopy using ultra-low energy electrons. Chem. Phys. Lett. **539**, 180–185 (2012)
66. Pahner, P., Kleemann, H., Burtone, L., Tietze, M.L., Fischer, J., Leo, K., Lussem, B.: Pentacene Schottky diodes studied by impedance spectroscopy: Doping properties and trap response. Phys. Rev. B **88**, 195205 (8 pages) (2013)
67. Tietze, M.L., Benduhn, J., Pahner, P., Nell, B., Schwarze, M., Kleemann, H., Krammer, M., Zojer, K., Vandewal, K., Leo, K.: Elementary steps in electrical doping of organic semiconductors. Nat. Comm. **9**, 1182 (8 pages) (2018)
68. Harada, K., Li, F., Maenning, B., Pfeiffer, M., Leo, K.: Ionized impurity scattering in *n*-doped C_{60} thin films. Appl. Phys. Lett. **91**, 092118 (3 pages) (2007)
69. Harada, K., Sumino, M., Adachi, C., Tanaka, S., Miyazaki, K.: Improved thermoelectric performance of organic thin-film elements utilizing a bilayer structure of pentacene and 2,3,5,6-tetrafluoro-7,7,8,8-tetracyanoquinodimethane (F_4-TCNQ). Appl. Phys. Lett. **96**, 253304 (3 pages) (2010)
70. Lee, B., Chen, Y., Fu, D., Yi, H.T., Czelen, K., Najafov, H., Podzorov, V.: Trap healing and ultralow-noise Hall effect at the surface of organic semiconductors. Nat. Mater. **12**, 1125–1129 (2013)
71. Podzorov, V., Menard, E., Rogers, J.A., Gershenson, M.E.: Hall effect in the accumulation layers on the surface of organic semiconductors. Phys. Rev. Lett. **95**, 226601 (4 pages) (2005)
72. Takeya, J., Kato, J., Hara, K., Yamagishi, M., Hirahara, R., Yamada, K., Nakazawa, Y., Ikehata, S., Tsukagoshi, K., Aoyagi, Y., Takenobu, T., Iwasa, Y.: In-crystal and surface charge transport of electric-field-induced carriers in organic single-crystal semiconductors. Phys. Rev. Lett. **98**, 196804 (4 pages) (2007)
73. Zeng, X., Wang, L., Duan, L., Qiu, Y.: Homoepitaxy growth of well-ordered rubrene thin films. Cryst. Growth Des. **8**, 1617–1622 (2008)
74. Hiramoto, M.: Title. Advances in Chemical Physics. In: Rice S.A., Dinner A. R. John Wiley & sons, New York, vol. 162, p. 160 (2017)
75. Hiramoto, M.: Energetic and nanostructural design of organic solar cells. In the book "Organic Solar Cells, Advances in Research and Applications". Voronov, M. (ed). Nova Science Publishers, New York Chap. 1, p. 7 (2017)
76. Helou, M.E., Medenbach, O., Witte, G.: Rubrene microcrystals: A route to investigate surface morphology and bulk anisotropies of organic semiconductors. Cryst. Growth Des. **10**, 3496–3501 (2010)
77. Kittel, C.: Introduction to Solid State Physics. John Wiley & sons, New York. 5th ed., Chap. 17 (2017)
78. Anderson, B.L., Anderson, R.L.: Fundamentals of Semiconductor Devices. The McGraw-Hill Companies, New York. Chap. 3 (2005)
79. Conwell, E.M., Weisskopf, V.F.: Theory of impurity scattering in semiconductors. Phys. Rev. **77**, 388–390 (1950)
80. Erginsoy, C.: Neutral impurity scattering in semiconductors. Phys. Rev. **79**, 1013–1014 (1950)
81. Takeya, J., Tsukagoshi, K., Aoyagi, Y., Takenobu, T., Iwasa, Y.: Hall effect of 1uasi-hole gas in organic single crystal transistors. Jpn. J. Appl. Phys. **44**, L1393-1396 (2005)
82. Liu, C., Minari, T., Lu, X., Kumatani, A., Takimiya, K., Tsukagoshi, K.: Solution-processable organic single crystals with bandlike transport in field-effect transistors. Adv. Mater. **23**, 523–526 (2011)
83. Xie, H., Alves, H., Morpurgo, A.F.: Quantitative analysis of density-dependent transport in tetramethyltetraselenafulvalene single-crystal transistors: Intrinsic properties and trapping. Phys. Rev. B **80**, 245305 (7 pages) (2009)
84. Sakanoue, T., Sirringhaus, H.: Band-like temperature dependence of mobility in a solution-processed organic semiconductor. Nat. Mater. **9**, 736–740 (2010)
85. Minder, N.A., Ono, S., Chen, Z., Facchetti, A., Morpurgo, A.F.: Band-like electron transport in organic transistors and implication of the molecular structure for performance optimization. Adv. Mater. **24**, 503–508 (2012)

86. Cox, P.A.: The Electronic Structure and Chemistry of Solids. Oxford University Press, New York. Chap.7 (1987)
87. Bylander, E.G.: Surface effects on the low-energy cathodoluminescence of zinc oxide. J. Appl. Phys. **49**, 1188–1195 (1978)
88. Izaki, M., Chizaki, R., Saito, T., Murata, K., Sasano, J., Shinagawa, T.: Hybrid ZnO/phthalocyanine photovoltaic device with highly resistive ZnO intermediate layer. ACS Appl. Mater. Interfaces **5**, 9386–9395 (2013)
89. Shockley, W., Read, W.T., Jr.: Statistics of the recombinations of holes and electrons. Phys. Rev. **87**, 835–842 (1952)
90. Bethke, S., Pan, H., Wessels, B.W.: Luminescence of heteroepitaxial zinc oxide. Appl. Phys. Lett. **52**, 138–140 (1988)
91. Kim, D., Terashita, T., Tanaka, I., Nakayama, M.: Photoluminescence properties of ZnO thin films grown by electrochemical deposition. Jpn. J. Appl. Phys. **42**, L953–L937 (2003)
92. Jang, J.I., Wolfe, J.P.: Auger recombination and biexcitons in Cu_2O: A case for dark excitonic matter. Phys. Rev. B **74**, 045211 (16 pages) (2006)
93. Snoke, D.W., Shields, A.J., Cardona, M.: Phonon-absorption recombination luminescence of room-temperature excitons in Cu_2O. Phys. Rev. B **45**, 11693–11697 (1992)
94. Petroff, Y., Yu, P.Y., Shen, Y.R.: Study of photoluminescence in Cu_2O. Phys. Rev. B **12**, 2488–2495 (1975)
95. Yang, J.P., Bussolotti, F., Kera, S., Ueno, N.: Origin and role of gap states in organic semiconductor studied by UPS: as the nature of organic molecular crystals. J. Phys. D: Appl. Phys. **50**, 423002 (2017)
96. Ishii, H., Kudo, K., Nakayama, T., Ueno, N.: Electronic Processes in Organic Electronics. Springer Japan, Tokyo (2015)
97. Bussolotti, F., Yang, J., Yamaguchi, T., Yonezawa, K., Sato, K., Matsunami, M., Tanaka, K., Nakayama, Y., Ishii, H., Ueno, N., Kera, S.: Hole-phonon coupling effect on the band dispersion of organic molecular semiconductors. Nature Commun. **8**, 173 (7 pages) (2017)

Chapter 10
Proposal for Future Organic Solar Cells

Masahiro Hiramoto

Recent driving forces for increases in efficiency of organic solar cells have relied on the numerous repeats of the circulation around the syntheses of new organic semiconductors, blends, and performance checks. However, neglect of physical principles, which are hidden in the back of the complicated systems, should be prevented for future long-range development of organic solar cells. In Chapter 10, therefore, the proposals for future organic solar cells are discussed based on the unconventional considerations and the fundamental physical aspects.

10.1 Is Blended Junction Necessary?

The primary question, why a blended junction was inevitably introduced, directly offers the alternative methods for constructing organic solar cells. The first answer is the necessity of combining donor and acceptor molecules to dissociate excitons. A logical consequence is exciton dissociation without combining donor and acceptor molecules, i.e., the use of a single organic semiconductor. The second answer is an extremely short exciton diffusion length of approximately 10 nm. A logical consequence is the use of a material with a long exciton diffusion length. The third answer is the very small relative dielectric constant (ε) of about 3 of organic semiconductors, which inhibits exciton dissociation. A logical consequence is the use of the material with large ε.

M. Hiramoto (✉)
National Institutes of Natural Sciences, Institute for Molecular Science, 5-1 Higashiyama, Myodaiji, Okazaki, Aichi 444-8787, Japan
e-mail: hiramoto@ims.ac.jp

M. Hiramoto and S. Izawa (eds.), *Organic Solar Cells*,
https://doi.org/10.1007/978-981-15-9113-6_10

10.1.1 Bipolar Band-Conductive Organic Semiconductor-Exciton Dissociation Using Single Organic Semiconductor

Carrier generation by doping and light excitation is considered as standard since the liberation process between positive and negative charges can be regarded as identical. In the case of doping, a free carrier is generated by the dissociation of a positive charge of a hole from a negatively ionized dopant ion, which is spatially fixed, and vice versa. In the case of photocarrier generation, free electron and hole are generated by the dissociation of an exciton, i.e., bound electron and hole, which are not spatially fixed.

In the case of acceptor doping to the band-conductive rubrene single crystal, a positive charge of a hole loosely bound around the negatively ionized dopant ion can be liberated by the thermal energy of room temperature; i.e., the Wannier excitonic nature is observed in the band-conductive single crystal (Fig. 10.1a) (Chap. 9, Sect. 9.6.1). A negatively ionized dopant is spatially fixed in the crystal lattice. Therefore, for acceptor and donor doping processes, since the liberated carrier is either a hole or an electron, only hole band-conductive nature or only electron band-conductive nature are required, respectively.

On the other hand, in the case of photocarrier generation, both a hole and an electron should be liberated at room temperature by the Wannier excitonic nature. Therefore, the band-conductive nature is required both for the hole and the electron. When two kinds of organic crystals with hole band conduction and electron band conduction are combined (Fig. 10.1b), exciton automatically dissociates to free electron and hole by the thermal energy of room temperature. When single organic crystal possesses both hole band-conductive and electron band-conductive nature, i.e., possesses bipolar band-conductive nature, exciton automatically dissociates to free electron and hole by thermal energy of room temperature in the single organic crystal (Fig. 10.1c). We presume that the bipolar band-conductive organic semiconductor would dissociate exciton and generate photocurrent without the assistance of donor/acceptor (D/A) sensitization.

Concerning holes, a significant number of band-conductive organic semiconductor crystals, such as rubrene [1] and films such as C8-BTBT [2], has been reported in this decade. However, for electrons, only one organic semiconductor (PDIF-CN_2) has been reported [3]. There has been no report on bipolar band-conductive organic semiconductor. Initially, a sign of bipolar band conduction under very low temperature was reported in naphthalene and perylene single crystals in 1985 by Karl [4] (Fig. 10.2). The mobilities for both hole and electron increased with decreasing temperature, suggesting the bipolar band-conductive nature. The bipolar band-conductive organic semiconductor at room temperature would represent a breakthrough in the field of organic solar cells.

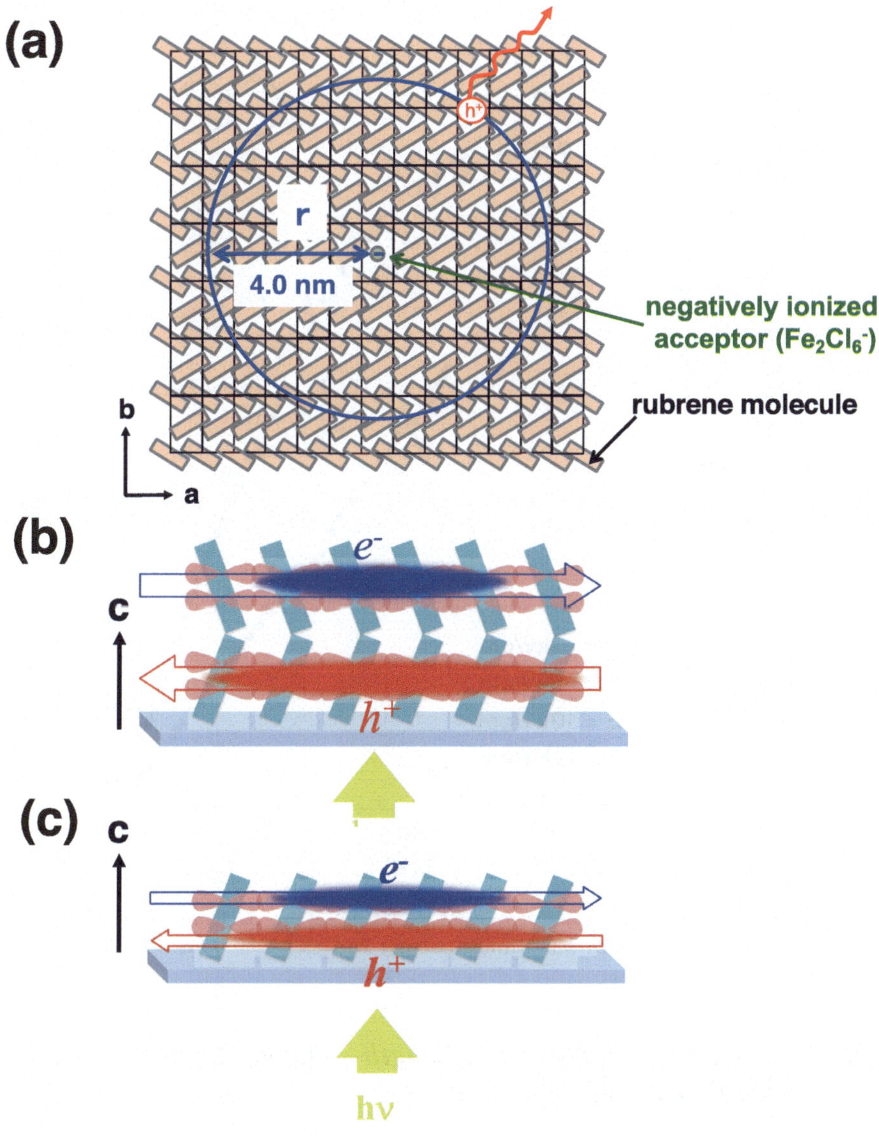

Fig. 10.1 **a** Wannier excitonic nature of doping. A hole loosely bound around the negatively ionized acceptor ion ($Fe_2Cl_6^-$) in the rubrene single crystal showed 24% doping efficiency. Band conduction promotes the carrier generation by doping. **b** Combination of two types of organic crystals exhibiting hole and electron band conduction. Exciton automatically dissociates into free electron and hole at room temperature. **c** Single organic crystal with bipolar band-conductive nature. Exciton automatically dissociates into free electron and hole in the single organic crystals at room temperature. **a** Reproduced with permission from M. Hiramoto et al., *Adv. Mater* Copyright 2018 John Wiley and Sons

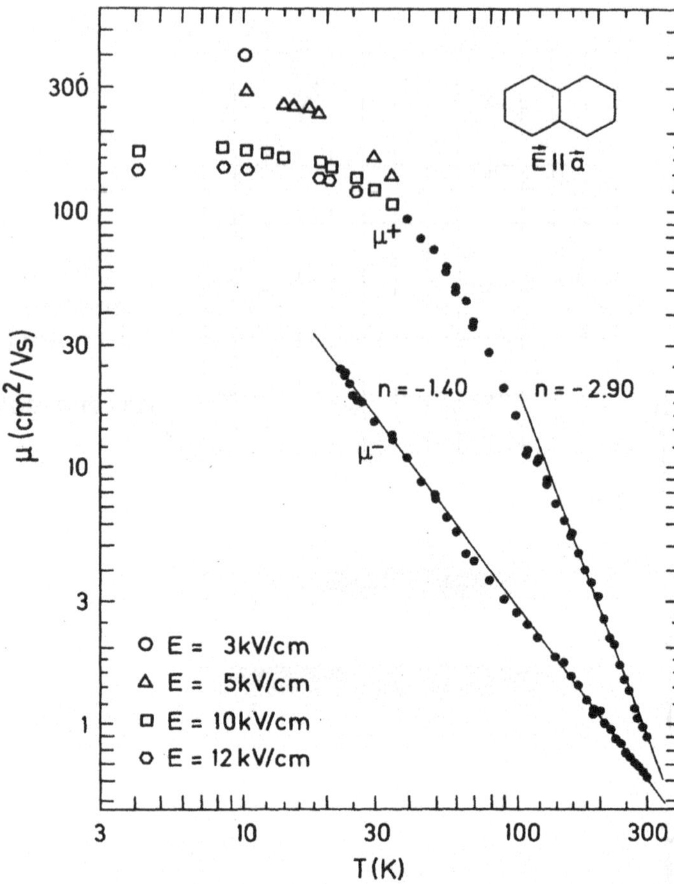

Fig. 10.2 Drift mobility increase of electron and hole by decreasing the temperature observed for naphthalene crystal. Reproduced with permission from W. Warta et al., *Phys. Rev. B*, Copyright 1985 American Physical Society

10.1.2 Long Exciton Diffusion Length—Doped Organic Single-Crystal Solar Cell

If we can use organic semiconductors with significantly long exciton diffusion lengths, exciton diffusion would no longer be a limiting factor for organic solar cells; i.e., the blended junction would not be necessary. A long diffusion length of excitons in organic single crystals, such as anthracene, has been suggested earlier [5–7]. Recently, a long exciton diffusion length of 8 μm was reported for rubrene single crystals [8]. This fact shows the possibility of organic single-crystal solar cells without blended junction (Fig. 10.3).

Organic solar cells can be constructed on the *p*-doped rubrene single-crystal substrate acting as a hole transporting substrate. We have already constructed the

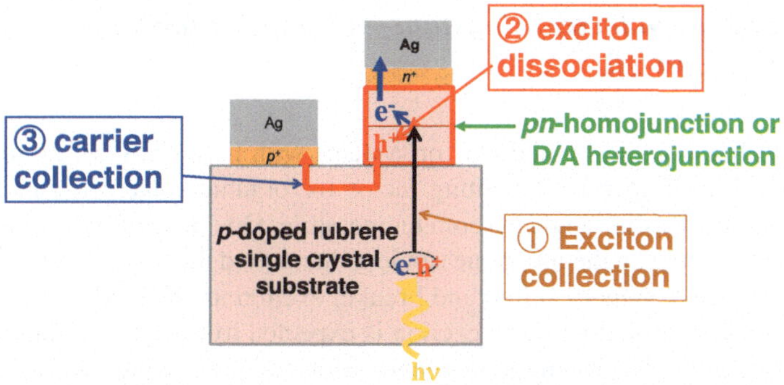

Fig. 10.3 Organic solar cell constructed on the *p*-doped rubrene single-crystal substrate

prototype of the organic single-crystal solar cell [9]. A single-crystal substrate collected excitons to the *pn*-homojunction with the collection efficiency reaching 46%, owing to the long exciton diffusion length of 2.7 μm. Essentially, the blended junction is not necessary when whole excitons are collected to the planar junctions.

The entire photocurrent generated in the *pn*-homojunction with a macroscopic area of 2 mm × 1 mm was confirmed to be collected through the *p*-doped rubrene substrate. The J_{SC} values were revealed to be dominated by the sheet conductivity ($\sigma_\square$) of the *p*-doped substrate. To attain a practical value of $J_{SC} = 20$ mAcm^{-2}, the *p*-doped substrate should have a sheet conductivity ($\sigma_\square$) of 3.1×10^{-5} S. This is equal to the $\sigma_\square$-value of the entire bulk-doped rubrene single crystals (Fe_2Cl_6: 100 ppm) with a thickness of approximately 65 μm [10]. An entire bulk doping technology for organic single crystals should be developed for the fabrication of practical organic single-crystal solar cells. Similar to single-crystal Si substrate, the doped organic single-crystal substrate would be a fundamental element in future organic single-crystal solar cells (Fig. 10.4).

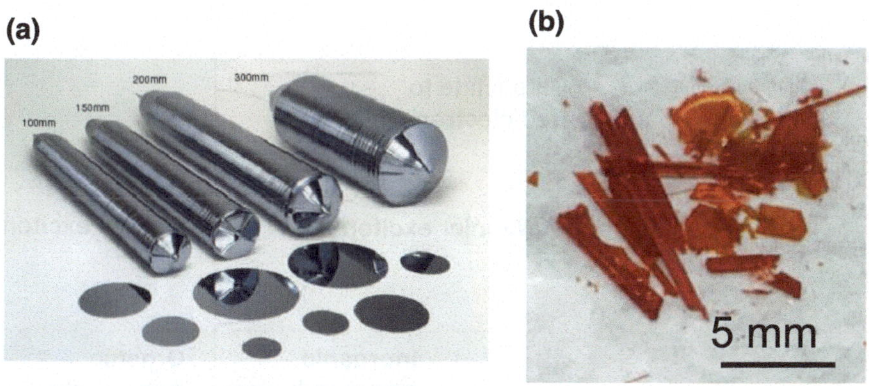

Fig. 10.4 **a** Photograph of doped Si single-crystal ingots and substrates. **b** Photograph of rubrene single crystals

10.1.3 Large Dielectric Constant—Organic/Inorganic Hybrid Cell

If we can utilize organic semiconductor systems with high ε values, exciton dissociation would no longer be a limiting factor for organic solar cells; i.e., exciton would automatically dissociate at room temperature. One possibility is to utilize the heterointerface between the inorganic semiconductor and the organic semiconductor (Fig. 10.5). Since Wannier exciton and Frenkel exciton are formed in the inorganic and organic sides, respectively, an exciton is expected to dissociate efficiently with the assistance of the large ε in the inorganic semiconductor, which allows charge to feel weak Coulombic attraction force in the inorganic semiconductor. For example, *n*-type inorganic semiconductor films such as TiO_2 and ZnO can be incorporated in the lateral cell instead of the electron transporting layer (PTCDI-C8).

10.2 Advanced Lateral Junctions—Beyond Shockley–Queisser Limit

Advanced lateral cells composed of organic semiconductor films with high mobilities for holes (C8-BTBT) and electrons (PTCDI-C8) (Chapter 3, Sect. 3.5.4) are shown in Fig. 10.6a. Between C8-BTBT and PTCDI-C8, visible near-infrared absorbers such as DBP and Pc (phthalocyanines), were inserted. Cascade energetic structure is preferable for the choice of absorbers (Fig. 10.6b).

By utilizing the unlimited choices and sharpness of the absorptions of organic semiconductors, the solar spectrum would be divided finely and absorbed by the appropriately designed organic semiconductors (Fig. 10.6c). For lateral junctions, the unlimited number of layers can be stacked, and different photovoltages depending on the bandgap (E_g) are collected using the parallel circuits. For one bandgap system,

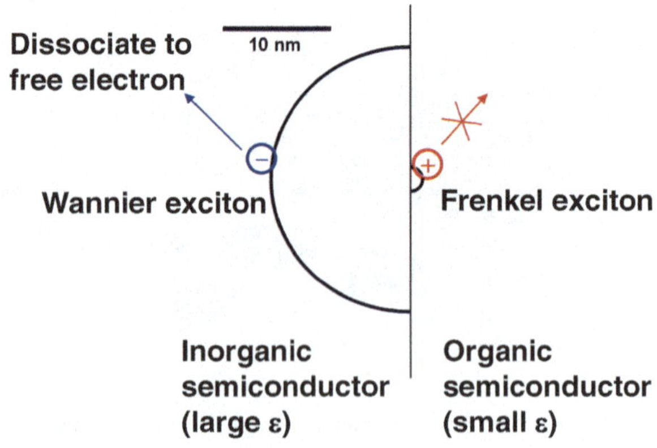

Fig. 10.5 Schematic illustration of heterointerface between the inorganic semiconductor and the organic semiconductor. Exciton is expected to dissociate efficiently with the assistance of the large ε in the inorganic semiconductor

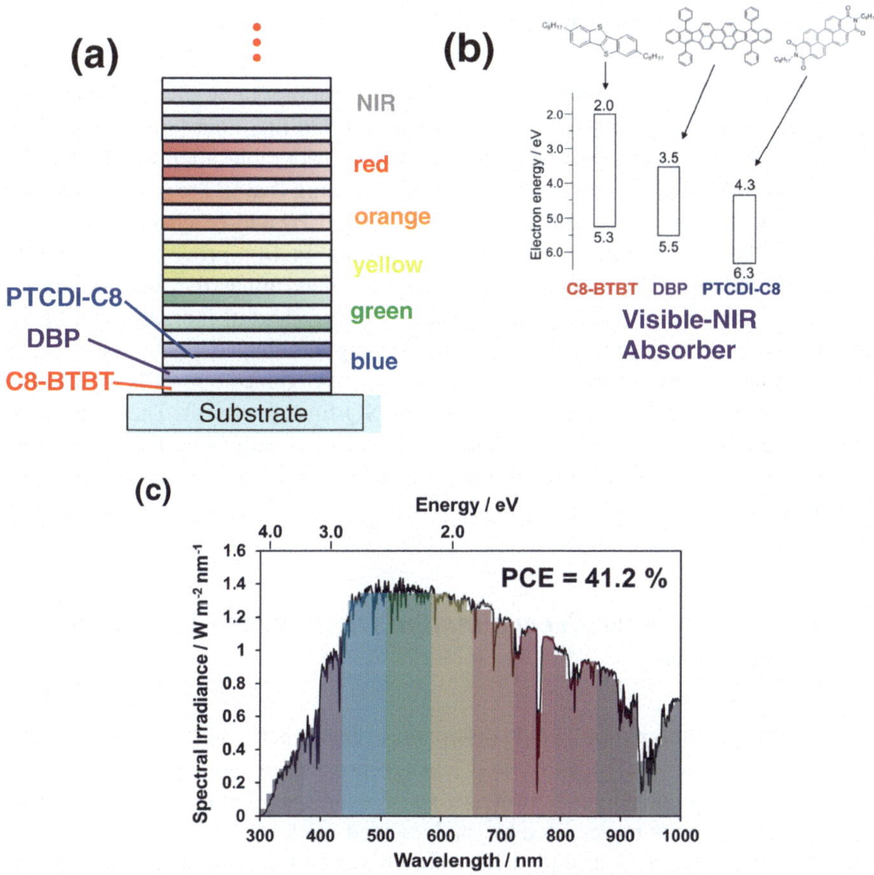

Fig. 10.6 **a** Advanced lateral cell. Between C8-BTBT and PTCDI-C8, visible to near-infrared absorbers such as DBP and Pc (phthalocyanines) are inserted. **b** Cascade energetic structure. **c** Fine division of the solar spectrum into ten regions

the maximum conversion efficiency of 33% was obtained for $E_g = 1.4$ eV (Shockley–Queisser; SQ) limit [11]. When the solar spectrum is divided into ten regions, the conversion efficiency of 41.2% beyond SQ-limit would be obtained under the assumption that the V_{oc}-loss of 0.3 V and the fill factor of 0.8 (Fig. 10.6c).

It should be noted that organic semiconductor is intrinsically suitable for a fine division of the solar spectrum since the sharp absorption can be intentionally tuned to various wavelengths by the molecular design. Moreover, the organic semiconductor is highly suitable for multilayer fabrication at a low price. This would be the utmost advantage of organic solar cells.

10.3 Recombination Suppression

Recombination suppression has been a priority issue for inorganic solar cells since the recombination is the main leakage of photocurrent. On the other hand, for organic solar cells, the recombination has not attracted considerable attention due to the small magnitude of photocurrent. However, considering that in the past decade, the photocurrent density of organic solar cells exceeded 20 and reached 25 $mAcm^{-2}$ [12], researchers have started to focus on the recombination process in relation to the V_{oc}-loss (Chapter 8). Radiative recombination determines the thermodynamic theoretical limitation of V_{oc} (SQ-limit) (Fig. 10.7). V_{oc}-loss of inorganic solar cells such as GaAs is negligible, and V_{oc} value (triangular-shaped point) has already almost reached the SQ-limit. However, V_{oc}-losses of organic solar cells reported so far are more than 0.5 V, while V_{oc} is far below SQ-limit (crosses). Therefore, non-radiative recombination causing V_{oc}-loss for organic solar cells should be suppressed. Presently, non-radiative recombination suppression is a cutting-edge issue for organic solar cells.

10.3.1 Non-radiative Recombination Dissipated to Molecular Vibration

Non-radiative recombination, which dissipates excited energy to the thermal molecular vibrational energy, is an essential leakage process in the organic solar cells. Conventionally, the carrier generation process of organic semiconductors showing the hopping conduction is represented by the localized model (Fig. 10.8a). In this case, since there is a large Stokes shift (Fig. 10.8c) between the ground state and charge transfer (CT) state due to the considerable reorganization energy, strong vibrational coupling induces the non-radiative recombination (geminate recombination) [13].

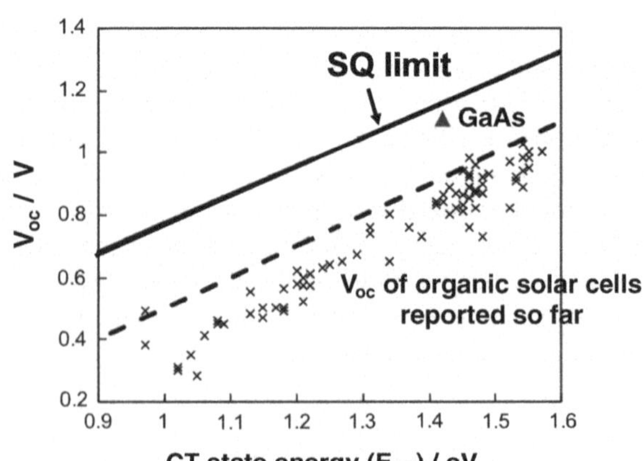

Fig. 10.7 Dependence of open-circuit photovoltage (Voc) on the CT state energy (E_{CT}). The black solid line is the theoretical (SQ) limit of Voc. The dashed black line shows Voc = E_{CT}/e − 0.5 V. The crosses are Voc for all types of organic solar cells reported so far. A closed triangle is Voc for GaAs

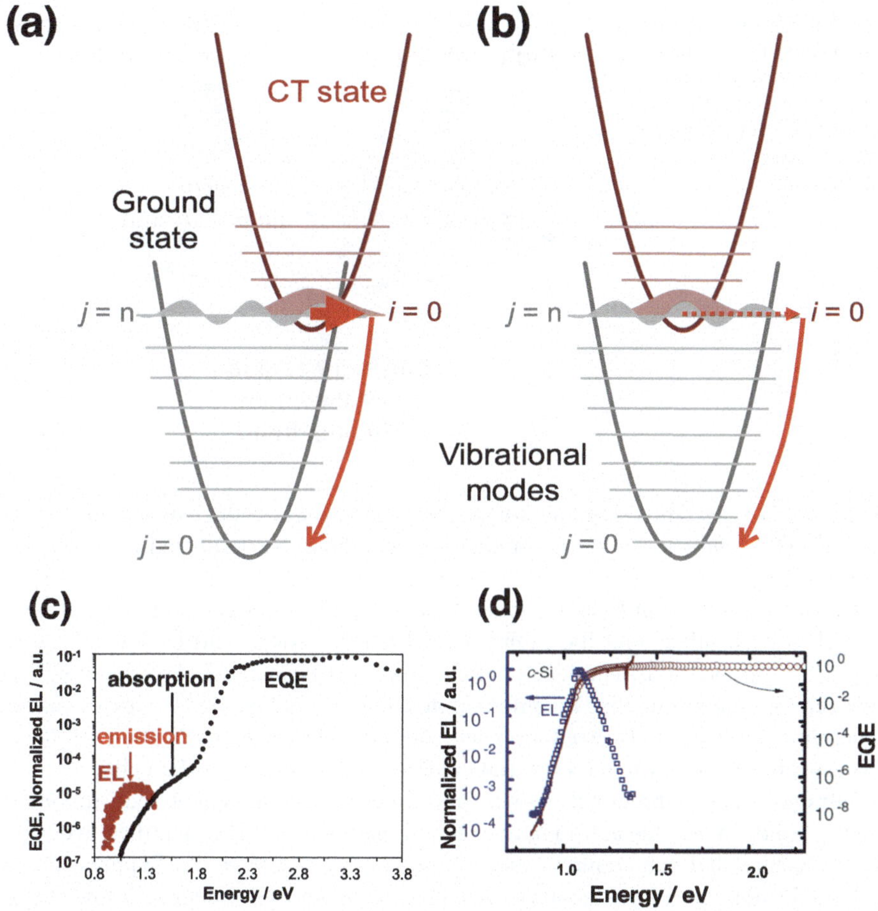

Fig. 10.8 **a** Localized model of the non-radiative recombination from CT state to ground state. **b** Delocalized model of the non-radiative recombination from CT state to ground state. **c** Absorption and EL emission from CT state for 6T/PTCDI-C8 system exhibiting hopping conduction showed a large Stokes shift. Absorption by CT is represented by the edge of the EQE (external quantum efficiency) spectrum. **d** Absorption and EL emission for single-crystal Si exhibiting band conduction showed a little Stokes shift. **d** Reproduced with permission from W. Warta et al., *Phys. Rev. Appl.*, Copyright 2015 American Physical Society

Even if the bimolecular recombination occurs between completely separated electron and hole, non-radiative recombination occurs through the CT state (Fig. 10.8a).

On the other hand, the organic semiconductors showing the band conduction are represented by the delocalized model (Fig. 10.8b). In this case, since there is a small Stokes shift between the ground state and CT state due to the little reorganization energy, little vibrational coupling suppresses the non-radiative recombination. This coincides well with the small Stokes shift observed for the single-crystal Si

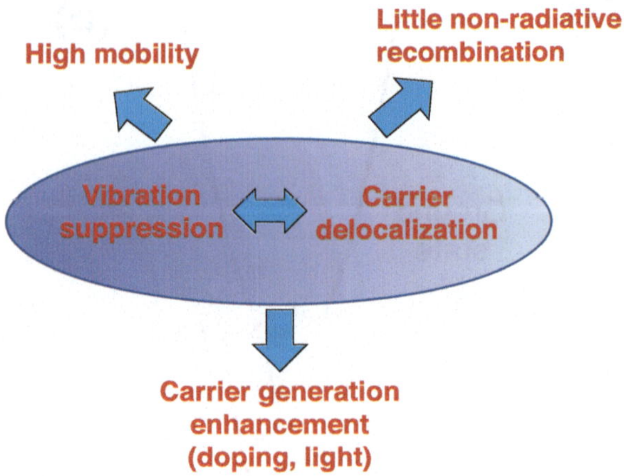

Fig. 10.9 Carrier mobility, non-radiative recombination, and carrier generation are closely related to each other with the molecular vibration through carrier delocalization

(Fig. 10.8d) [14]. Thus, organic solar cells using band-conductive organic semiconductors would show little non-radiative recombination dissipated to molecular vibration.

Carrier mobility, non-radiative recombination, and carrier generation are closely related to each other with the molecular vibration through carrier delocalization (Fig. 10.9). Suppression of molecular vibration causes the increase of carrier mobility, the decrease of non-radiative recombination, and presumably, the increase of carrier generation both by light excitation and doping. Detailed clarification of these relationships is the key to the essential breakthrough of organic solar cells.

Intramolecular vibrational modes and inter-molecular vibrational (phonon) modes, which induce the non-radiative recombination, should be identified to design the molecules that can suppress the non-radiative recombination. Suppression of inter-molecular vibration (phonon) might require an innovative approach since there are no chemical bonds among molecules, contrary to the inorganic semiconductors.

In general, the organic semiconductor showing little non-radiative recombination is highly radiative. Namely, the organic semiconductor suitable for efficient organic solar cells is also suitable for the organic electroluminescent device.

10.3.2 Non-radiative Recombination Via Carrier Traps

Inorganic crystals inevitably have dangling bonds at the surface, the grain boundary, and the heterojunction, which act as the carrier traps. Thus, for inorganic solar cells, the suppression of trap-induced recombination has been a priority issue, because it is the main leakage mechanism of photocurrent. Passivation of the surface states (Fig. 10.10a), the interfacial states at the grain boundary (Fig. 10.10b), and the interfacial states at heterojunction (Fig. 10.10c) has been a key to obtain efficient solar cells.

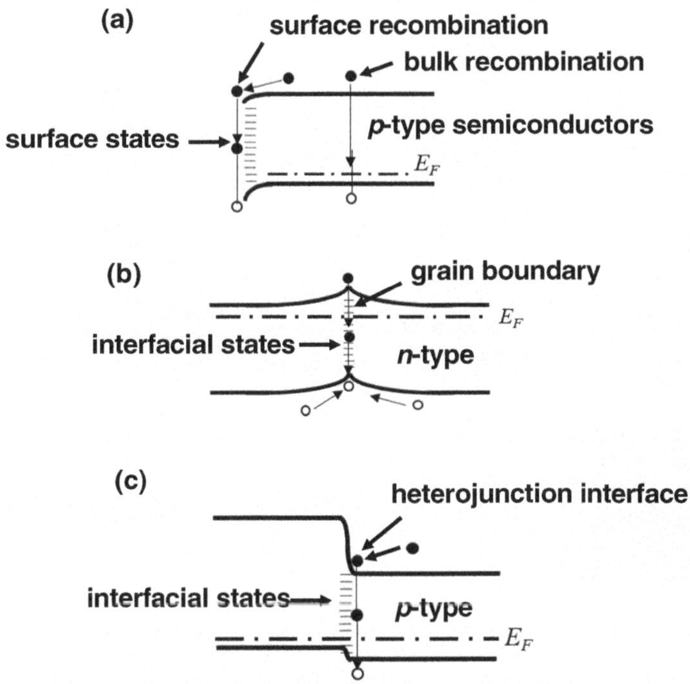

Fig. 10.10 Various recombination processes in the inorganic solar cells. **a** Surface recombination. **b** Recombination at the grain boundary. **c** Recombination at heterojunction interface

On the other hand, organic crystals do not have dangling bonds. However, the various kinds of carrier traps exist in organic crystals. More precisely, they act as carrier traps and accelerate the non-radiative recombination [15] (Fig. 10.11a). The molecular-level spatial structure and energetic structure of these carrier traps can hardly be identified. The concept of traps in organic crystals is not clear. By the analogy of inorganic crystals, they might be a molecular vacancy (Fig. 10.11b) or interstitial molecule (Fig. 10.11c). Steps and kinks at the crystal surface or grain boundary may act as traps (Fig. 10.11d). There should be traps at organic/metal interface and organic/organic heterointerface. Their real nature should close to suppress the trap-induced recombination. This is the next challenge for the field of organic solar cells.

10.4 Conclusion

The following proposal to future organic solar cells based on the fundamental physical aspects is discussed. (i) Exciton dissociation using single bipolar band-conductive organic semiconductors. (ii) The doped organic single-crystal substrates having long

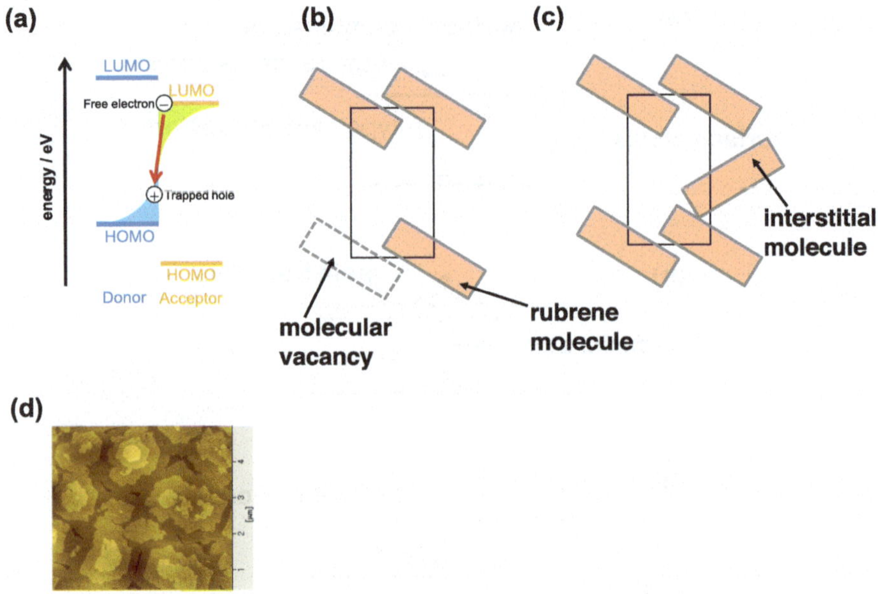

Fig. 10.11 **a** Trap-assisted non-radiative recombination. **b** Molecular vacancy. **c** Interstitial molecule. **d** AFM image of homoepitaxially grown rubrene single crystal. Steps and kinks at the crystal surface or grain boundary may act as traps. **a** Reproduced with permission from [15] Copyright 2018 Elsevier. **d** Reproduced with permission from M. Hiramoto et al., *Adv. Mater.*, Copyright 2018 John Wiley and Sons. Reproduced with permission from C. Ohashi et al., *Adv. Mater.*, Copyright 2017 John Wiley and Sons

exciton diffusion length. (iii) Exciton dissociation using organic/inorganic hybrid cells with a large dielectric constant. (iv) Advanced lateral cells showing the efficiency beyond the SQ-limit. An ultimate advantage of the organic semiconductor is its suitability for a fine division of the solar spectrum.

Suppression of non-radiative recombination is a cutting-edge issue for organic solar cells. V_{OC}-loss by non-radiative recombination would be suppressed close to the SQ-limit. Non-radiative recombination dissipated to molecular vibration would be suppressed by identifying the intramolecular vibrational modes and inter-molecular vibrational phonon modes, related to the non-radiative recombination. Non-radiative recombination via carrier traps has become a key issue for organic solar cells. The spatial and energetic nature of carrier traps would be clarified to suppress the trap-induced recombination.

The research on organic solar cell has advanced gradually, on per with that on inorganic solar cell.

References

1. Takeya, J., Kato, J., Hara, K., Yamagishi, M., Hirahara, R., Yamada, K., Nakazawa, Y., Ikehata, S., Tsukagoshi, K., Aoyagi, Y., Takenobu, T., Iwasa, Y.: In-crystal and surface charge transport of electric-field-induced carriers in organic single-crystal semiconductors. Phys. Rev. Lett. **98**, 196804 (4 pages) (2007)
2. Liu, C., Minari, T., Lu, X., Kumatani, A., Takimiya, K., Tsukagoshi, K.: Solution-processable organic single crystals with bandlike transport in field-effect transistors. Adv. Mater. **23**, 523 (2011)
3. Minder, N.A., Ono, S., Chen, Z., Facchetti, A., Morpurgo, A.F.: Band-like electron transport in organic transistors and implication of the molecular structure for performance optimization. Adv. Mater. **24**, 503–508 (2012)
4. Warta, W., Karl, N.: Hot holes in naphthalene: High, electric-field-dependent mobilities. Phys. Rev. B **32**, 1172–1182 (1985)
5. Avakian, P., Merrield, R.E.: Experimental determination of the diffusion length of triplet excitons in anthracene crystals. Phys. Rev. Lett. **13**, 541–543 (1964)
6. Ern, V., Avakian, P., Merrield, R.E.: Diffusion of triplet excitons in anthracene crystals. Phys. Rev. **148**, 862–867 (1966)
7. Williams, D.F., Adolph, J.: Diffusion length of triplet excitons in anthracene crystals. J. Chem. Phys. **46**, 4252–4254 (1967)
8. Najafov, H., Lee, B., Zhou, Q., Feldman, L.C., Podzorov, V.: Observation of long-range exciton diffusion in highly ordered organic semiconductors. Nat. Mater. **9**, 938–943 (2010)
9. Kikuchi, M., Makmuang, S., Izawa, S., Wongravee, K., Hiramoto, M.: Doped organic single-crystal photovoltaic cells. Org. Electron. **64**, 92–96 (2019)
10. Yabara, Y., Izawa, S., Hiramoto, M.: Donor/acceptor photovoltaic cells fabricated on p-doped organic single-crystal substrates. Materials **13**, 2068 (8 pages) (2020)
11. Shockley, W., Queisser, H.J.: Detailed balance limit of efficiency of *p-n* junction solar cells. J. Appl. Phys. **32**, 510–519 (1961)
12. Liu, L., Kan, Y., Gao, K., Wang, J., Zhao, M., Chen, H., Zhao, C., Jiu, T., Jen, A., Li, Y.: Graphdiyne derivative as multifunctional solid additive in binary organic solar cells with 17.3% efficiency and high reproductivity. Adv. Mater. **32**, 1907604 (7 pages) (2020)
13. Benduhn, J., Tvingstedt, K., Piersimoni, F., Ullbrich, S., Fan, Y., Tropiano, M., McGarry, K.A., Zeika, O., Riede, M.K., Douglas, C.J., Barlow, S., Marder, S.R., Neher, D., Spoltore, D., Vandewal, K.: Intrinsic non-radiative voltage losses in fullerene-based organic solar cells. Nat. Energy **2**, 17053 (2017)
14. Yao, J., Kirchartz, T., Vazie, M. S., Faist, M. A., Gong, W., He, Z., Wu, H., Troughton, J., Watson, T., Bryant, D., Nelson, J.: Quantifying losses in open-circuit voltage in solution-processable solar cells. Phys. Rev. A **4**, 014020 (10 pages) (2015)
15. Shintaku, N., Hiramoto, M., Izawa, S.: Effect of trap-assisted recombination on open-circuit voltage loss in phthalocyanine/fullerene solar cells. Org. Electron. **55**, 69–74 (2018)

Correction to: Organic Solar Cells

Masahiro Hiramoto and Seiichiro Izawa

Correction to:
M. Hiramoto and S. Izawa (eds.),
Organic Solar Cells,
https://doi.org/10.1007/978-981-15-9113-6

In the original version of the book, the following chapters corrections have been incorporated:

1. Chapter 2 Figure 2.1 replaced with new figure.
2. Chapter 4 Figure 4.6 (b) updated with revised figure.
3. Chapter 6 Equation 6.1 Greek symbol corrected for v

The erratum chapters and book have been updated with the changes.

In Chapter 6 Equation 6.1 updated from

$$\tau_{\Delta n} = \tau_{\Delta n_0} \exp\left(-\frac{eV_{\mathrm{OC}}}{v k_{\mathrm{B}} T}\right) \quad (6.1)$$

to

$$\tau_{\Delta n} = \tau_{\Delta n_0} \exp\left(-\frac{eV_{\mathrm{OC}}}{\nu k_{\mathrm{B}} T}\right) \quad (6.1)$$

The updated version of these chapters can be found at
https://doi.org/10.1007/978-981-15-9113-6_2
https://doi.org/10.1007/978-981-15-9113-6_4
https://doi.org/10.1007/978-981-15-9113-6_6

M. Hiramoto and S. Izawa (eds.), *Organic Solar Cells*,
https://doi.org/10.1007/978-981-15-9113-6_11

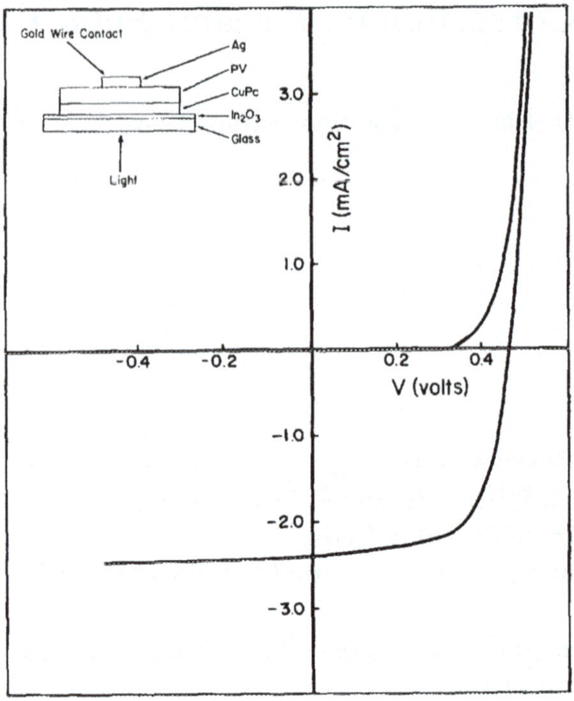

Fig. 2.1 Current–voltage characteristics reported in the paper, "Two-layer organic photovoltaic cell," by Ching. W. Tang [9]. Surprising performance, i.e., J_{sc}: 2.3 mAcm^{-2}, FF: 0.65, V_{oc}: 0.45 V, conversion efficiency: 1%, was reported. Inset shows the cell structure. The PV in this figure corresponds to Im-PTC in this chapter. Reproduced with permission from [9]. Copyright 1986 AIP Publishing

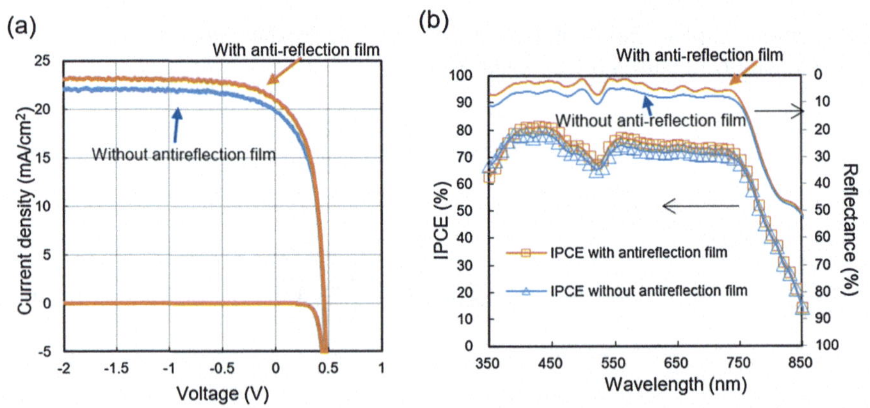

Fig. 4.6 Solar cell characteristics of crystalline ZnPc:C_{60} OPV cell with a thickness of 600 nm with buffer layer optimized for its performance. **a** J–V characteristics, **b** IPCE spectra and the reflectance of the same cell, with the presence or absence of antireflection film on ITO glass substrate

The manufacturer's authorised representative in the EU is Springer
Nature Customer Service Centre GmbH, Europaplatz 3, 69115 Heidelberg,
Germany. If you have any concerns regarding our products, please
contact ProductSafety@springernature.com

Printed and bound by CPI Group (UK) Ltd, Croydon, CR0 4YY
06/10/2026
02225252-0008